D084478ó

GRAPHS

North-Holland Mathematical Library

VOLUME 6
PART 1

NORTH-HOLLAND
AMSTERDAM · NEW YORK · OXFORD

Graphs

CLAUDE BERGE

University of Paris

Second revised edition

NORTH-HOLLAND
AMSTERDAM · NEW YORK · OXFORD

ISBN: 0 444 87603 0

Translation and revised edition of
part one of

GRAPHES ET HYPERGRAPHE
© DUNOD, Paris 1970
First edition: 1973
Revised edition (first printing): 1976
　　　　　　(second printing): 1979
Second revised edition (first printing): 1985
　　　　　　(second printing): 1989

Published by:
ELSEVIER SCIENCE PUBLISHERS B.V.
P.O. Box 103
1000 AC Amsterdam
The Netherlands

Sole distributors for the U.S.A. and Canada:
ELSEVIER SCIENCE PUBLISHING COMPANY, INC.
655 Avenue of the Americas
New York, N.Y. 10010
U.S.A.

Library of Congress Cataloging in Publication Data
Berge, Claude.
Graphs.

(North-Holland mathematical library; v. 6, pt. 1)
Rev. translation of: Graphes et hypergraphes, 1ere
ptie: Graphes.
Bibliography: p.
Includes index.
1. Graph theory. I. Title. II. Series: North-Holland
mathematical library; v. 6. pt. 1.
QA166.B389513 1985 511'.5 84-18669
ISBN 0-444-87603-0

PRINTED IN THE NETHERLANDS

To JEAN-MICHEL

FOREWORD

Graph theory has had an unusual development. Problems involving graphs first appeared in the mathematical folklore as puzzles (e.g. Königsberg bridge problem). Later, graphs appeared in electrical engineering (Kirchhof's Law), chemistry, psychology and economics before becoming a unified field of study. Today, graph theory is one of the most flourishing branches of modern algebra with wide applications to combinatorial problems and to classical algebraic problems (Group Theory, with Cayley, Ore, Frucht, Sabidussi, etc.; Category Theory, with Pultr, Hedrlín, etc.).

Graph theory as a separate entity has had its development shaped largely by operational researchers preoccupied with practical problems. It was with these practical problems in mind that we wrote our first book *Théorie des graphes et ses applications* published by Dunod in January 1958. This text hoped to unify the various results then scattered through the literature. For this purpose, we emphasized two major areas.

The first of these areas was the *network flow theory* of Ford and Fulkerson which was beginning to transcend analytic techniques. This theory gave new proofs for more than a dozen graph theory results including some famous theorems by König and by Menger.

The second area was the *theory of alternating chains* which started with Petersen sixty years earlier, but which appeared in optimization problems only in 1957.

These two areas had many curious similarities; however, the integer linear programs that they solved did not overlap. Now, more than ever, we believe that these two areas should form the foundation of graph theory.

The first mathematicians to work in graph theory (in particular the thriving Hungarian school with D. König, P. Erdös, P. Turán, T. Gallai, G. Hajós, etc.) considered mainly undirected graphs, and this could lead students to believe that there are two theories—one for directed graphs and one for undirected graphs. This book is written with the viewpoint that there is only one kind of graph (directed) and only one theory for graphs. This is reasonable because a result for an undirected graph can be interpreted as a result for a directed graph in which the direction of the arcs does not matter. Conversely, a result for a directed graph can be interpreted for an undirected graph by replacing each edge of the undirected graph with two oppositely directed arcs with the same endpoints.

Since 1957, research in graph theory has assumed astonishing proportions. Results have appeared from all over the world, and some of the conjectures of our first book have been solved notably by our students in Paris from 1959 to 1964 (in particular, the late Alain Ghouila-Houri, whose work frequently appears in this text) and by Soviet mathematicians (in particular, A. A. Zykov, V. G. Vizing, L. M. Vitaver, M. K. Goldberg, L. P. Varvak, etc. following the Russian edition of our first text). Thus, because of this embarrassment of riches, this book is vastly more extensive, but still cannot treat very specific applications.[1]

The concept of a *matroid* due to H. Whitney and developed by W. Tutte has made possible an axiomatic study of cycles and trees. However, we cannot treat this algebraic aspect of graph theory too extensively without straying from our purpose. Similarly, new techniques have appeared for topological graphs, but these would also take us astray. Such strictly topological problems will be the subject of a future work.[2]

On the other hand, this book intends to present a systematic study of the *theory of hypergraphs*. A hypergraph is defined to be a family of *hyperedges* which are sets of vertices of cardinality not necessarily 2 (as for graphs). Given a graph, a hypergraph can be defined by its cliques, or by its spanning trees or by its cycles. Thus, the theory of hypergraphs can generate simultaneously several results for graphs.

The formulation of combinatorial problems in terms of hypergraphs often gives surprisingly simple results that will look very familiar to graph theorists. At the Balatonfüred Conference (1969), P. Erdös and A. Hajnal asked us why we would use hypergraphs for problems that can be also formulated in terms of graphs. The answer is that by using hypergraphs, one deals with

---

[1] There are several graph theory texts that emphasize operational research problems; in particular: C. Berge and A. Ghouila-Houri, *Programmes, jeux, et réseau de transport*, Part II, Dunod, Paris, 1962 (English edition, Methuen, London; Wiley, New York, 1965; German edition, Teubner, Leipzig, 1967; Spanish edition, Compania Edit. Continental, Mexico, 1965); L. R. Ford and D. R. Fulkerson, *Flows in Networks*, Princeton Press, 1962 (French edition, Gauthier-Villars, 1967); R. G. Busacker and T. L. Saaty, *Finite Graphs and Networks*, McGraw-Hill, 1965; A. Kaufman, *Introduction à la combinatorique en vue des applications*, Dunod, 1969 (English edition to appear); B. Roy, *Algèbre moderne et théorie des graphes*, Volume 1, Dunod, 1969; Volume 2, 1970; and, finally, A. A. Zykov, *Graph Theory* (in Russian), NAUKA Publishing House, Siberian Branch, Novosibirsk, 1969.

[2] The topological aspects of graph theory will be treated separately in another book to include such topics as the planar representation of graphs, genus, thickness, crossing number of non-planar graphs, proof of the Heawood conjecture by Ringel and Youngs, Edmonds' methods, etc.

generalizations of familiar concepts. Thus, hypergraphs can be used to simplify as well as to generalize.

This English edition contains some results that appeared too late for the original French edition, especially the Chvátal existence theorem for hamiltonian cycles (Chapter 10) and the Lovász proof for the first perfect graph conjecture (Chapter 20).

An index of all definitions is given at the end of the text so that the reader can pass over chapters without much loss of continuity. Theorems are appended with the name of their first discoverer and the year of discovery. Sometimes, an old or fundamental result is treated as a corollary, and the theorem from which it is derived is attributed to a recent author. This is done purely for didactic purposes and is not intended in any way to diminish the importance of results that have been generalized. A bibliography arranged according to chapters is also found at the end of the book.

We first wish to thank Michel Las Vergnas and Jean-Claude Fournier who have made many notable and original contributions and alterations to this text, also to Pierre Rosenstiehl for the assistance given us during our weekly meetings. We wish to thank all those who have helped with suggestions, in particular, J. C. Bermond, J. A. Bondy, P. Camion, U. S. R. Murty, L. Lovász, J. M. Pla, and W. T. Tutte.

C. Berge

FOREWORD TO THE THIRD EDITION

This book is a revised and updated version of the first part of the widely referenced text *Graphs and Hypergraphs*, published by North-Holland in 1973.

As graph theory continues its explosive growth, conjectures are proved and new theorems formed. The techniques involved, which have applications in a broad spectrum of mathematics ranging from analysis to operations research, have become more sophisticated if not more manageable. It is thus that we present, in this new edition, new theorems (e.g., the Perfect Graph Theorem, due to Lovasz) as well as new proofs of classical results. In particular, sections 9.3, 10.3, 12.2 and 13.3 have been significantly revised.

Topics concerning topological subjects (such as graph planarity, genus and thickness) have been removed from this edition to allow for a more comprehensive treatment in a separate text. Similarly, the second half of the 1983 edition, which dealt with *hypergraphs*, will be published separately, also with substantial, up-to-date revisions.

C. Berge

TABLE OF CONTENTS

PART ONE—GRAPHS

CHAPTER 1

Basic Concepts

1. Graphs

Intuitively speaking, a *graph* is a set of points, and a set of arrows, with each arrow joining one point to another. The points are called the *vertices* of the graph, and the arrows are called the *arcs* of the graph.

The set of vertices of a graph is generally denoted by X, and the set of arcs of a graph is generally denoted by U. For example, in the graph in Fig. 1.1,

$$X = \{ a, b, c, d \}, \quad U = \{ 1, 2, 3, 4, 5, 6, 7, 8, 9, 10 \}.$$

Arc 9, which goes from vertex c to vertex d, is said to be *of the form* (c, d), for short one may write $9 = (c, d)$. Arc 2, which goes from vertex a to vertex b, may similarly be written as $2 = (a, b)$. Note that arcs 3 and 4 have the same form as arc 2, but should not be confused with arc 2. In this book, we shall only consider finite sets X and U.

Note that the position of the vertices in the drawing of the graph is not important: only the way in which the vertices are joined by arcs is important. A graph is completely determined by its vertices and by the family of its arcs.

Formally, a *graph* G is defined to be a pair (X, U), where

(1) X is a *set* $\{ x_1, x_2, ..., x_n \}$ of elements called *vertices*, and

(2) U is a *family* $(u_1, u_2, ..., u_m)$ of elements of the Cartesian product $X \times X$, called *arcs*. This family will often be denoted by the set $U = \{ 1, 2, ..., m \}$ of its indices. An element (x, y) of $X \times X$ can appear more than once in this family. A graph in which no element of $X \times X$ appears more than p times is called a *p-graph*.

The number of vertices in a graph is called the *order* of the graph.

An arc of G of the form (x, x) is called a *loop*. For an arc $u = (x, y)$, vertex x is called its *initial endpoint*, and vertex y is called its *terminal endpoint*.

Vertex y is called a *successor* of vertex x if there is an arc with x as its initial endpoint and y as its terminal endpoint. The set of all successors of x is denoted by

$$\Gamma_G^+(x).$$

Similarly, vertex y is called a *predecessor* of vertex x if there is an arc of the form (y, x). The set of all predecessors of vertex x is denoted by

$$\Gamma_G^-(x).$$

The set of all *neighbours* of x is denoted by

$$\Gamma_G(x) = \Gamma_G^+(x) \cup \Gamma_G^-(x).$$

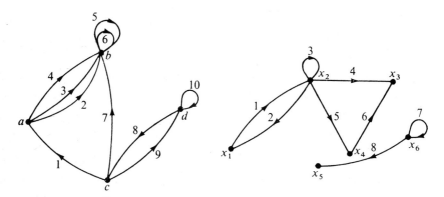

Fig. 1.1. A 3-graph of order 4   Fig. 1.2. A 1-graph of order 6

Note that Γ_G^+ is a *correspondence from X to X* that associates with each $x \in X$ a subset $\Gamma_G^+(x)$ of X. It is possible that $\Gamma_G^+(x) = \varnothing$ (the empty set). If $\Gamma_G(x) = \varnothing$, x is called an *isolated vertex*. For $A \subset X$, let

$$\Gamma_G(A) = \bigcup_{a \in A} \Gamma_G(a),$$

If $x \in \Gamma_G(A)$, $x \notin A$, then x is said to be *adjacent* to set A.

For $p = 1$, a p-graph is called a 1-*graph*. The arcs of a 1-graph are all distinct elements of the cartesian product $X \times X$. In this case,

$$U \subset X \times X, \qquad |U| = m.$$

A 1-graph $G = (X, U)$ is completely defined by X and the correspondence $\Gamma = \Gamma_G^+$. Hence, G can be denoted by (X, Γ).

In a graph $G = (X, U)$, each arc $u_i = (x, y)$ determines a continuous line joining x and y. Such a line, without any specification of its direction, is called an *edge*, and is denoted by $e_i = [x, y]$.

The family $(e_1, e_2, \ldots, e_m)$ of the edges of G is denoted by its set of indices $E = \{1, 2, \ldots, m\}$.

If the directions of the arrows in a graph are not specified, it is convenient, for conceptual reasons, to deal with the pair (X, E), rather than the pair (X, U). Such a pair (X, E) is called a *multigraph* (or *undirected graph*).

A multigraph is called a *simple graph* if:

(1) it has no loops,
(2) no more than one edge joins any two vertices.

In a simple graph, E denotes a subset of $\mathscr{P}_2(X)$, the set of all subsets of X with cardinality 2.

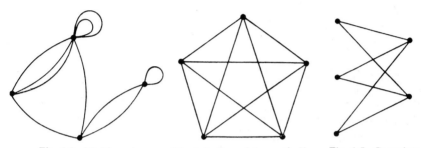

Fig. 1.3. Multigraph   Fig. 1.4. Complete graph K_5 Fig. 1.5. Complete bipartite graph $K_{3,2}$

Graphs and multigraphs often appear under other names: sociograms (psychology), simplexes (topology), electrical networks, organizational charts, communication networks, family trees, etc. It is often surprising to learn that these diverse disciplines use the same theorems. The primary purpose of graph theory was to provide a mathematical tool that can be used in all these disciplines.

It would be convenient to say that there are two theories and two kinds of graphs: directed and undirected. This is not true. *All graphs are directed*, but sometimes the direction need not be specified.

Results for directed graphs can be applied to a multigraph $G = (X, E)$ by replacing G with a directed graph G^* that has two oppositely directed arcs corresponding to each edge in G. Similarly, results for multigraphs can be applied to a directed graph $G = (X, U)$ after removing the direction from each arc in G.

2. Basic definitions

Adjacent arcs, adjacent edges. Two arcs (or two edges) are called *adjacent* if they have at least one endpoint in common.

Multiplicity. The *multiplicity* of a pair x, y is defined to be the number of arcs with initial endpoint x and terminal endpoint y. Denote this number by $m_G^+(x, y)$, and let

$$m_G^-(x, y) = m_G^+(y, x),$$

$$m_G(x, y) = m_G^+(x, y) + m_G^-(x, y).$$

If $x \neq y$, then $m_G(x, y)$ denotes the number of arcs with both x and y as endpoints. If $x = y$, then $m_G(x, y)$ equals twice the number of loops attached to vertex x. If A and B are two disjoint subsets of X, let

$$m_G^+(A, B) = |\{u \,/\, u \in U, u = (x, y), x \in A, y \in B\}|,$$

$$m_G(A, B) = m_G^+(A, B) + m_G^+(B, A).$$

Arc incident to a vertex. If a vertex x is the initial endpoint of an arc u, which is not a loop, the arc u is said to be *incident out of vertex* x. In graph G, the number of arcs that are incident out of x plus the number of loops attached to x is denoted by $d_G^+(x)$ and is called the *outer demi-degree of* x. An arc *incident into vertex* x and the *inner demi-degree* $d_G^-(x)$ are defined similarly.

Degree. The *degree* of vertex x is the number of arcs with x as an endpoint, each loop being counted twice. The degree of x is denoted by $d_G(x) = d_G^+(x) + d_G^-(x)$.

If in a graph each vertex has the same degree, this graph is said to be *regular*.

Arc incident out of a set $A \subset X$. If the initial endpoint of an arc u belongs to A, and if the terminal endpoint of arc u does not belong to A, then u is said to be *incident out of* A, and we write $u \in \omega^+(A)$. Similarly, we define an arc *incident into* A, and the set $\omega^-(A)$. Finally, the set of arcs *incident to* A is denoted by

$$\omega(A) = \omega^+(A) \cup \omega^-(A).$$

Symmetric graph. If $m_G^+(x, y) = m_G^-(x, y)$ for all $x, y \in X$, the graph G is said to be *symmetric*. A 1-graph $G = (X, U)$ is symmetric if, and only if,

$$(x, y) \in U \quad \Rightarrow \quad (y, x) \in U.$$

Anti-symmetric graph. If for each pair $(x, y) \in X \times X$,

$$m_G^+(x, y) + m_G^-(x, y) \leqslant 1,$$

then the graph G is said to be *anti-symmetric*. A 1-graph $G = (X, U)$ is anti-symmetric if, and only if,

$$(x, y) \in U \quad \Rightarrow \quad (y, x) \notin U.$$

An anti-symmetric 1-graph without its direction is a simple graph.

Complete graph. A graph G is said to be *complete* if

$$m_G(x, y) = m_G^+(x, y) + m_G^-(x, y) \geqslant 1$$

for all $x, y \in X$, such that $x \neq y$. A 1-graph is complete if, and only if,

$$(x, y) \notin U \quad \Rightarrow \quad (y, x) \in U.$$

A simple, complete graph on n vertices is called an *n-clique*, and is often denoted by K_n. See Fig. 1.4.

Bipartite graph. A graph is *bipartite* if its vertices can be partitioned into two sets X_1 and X_2 such that no two vertices in the same set are adjacent. This graph may be written as $G = (X_1, X_2, U)$.

Complete bipartite graph. If for all $x_1 \in X_1$ and for all $x_2 \in X_2$, we have $m_G(x_1, x_2) \geqslant 1$, then graph $G = (X_1, X_2, U)$ is said to be a *complete bipartite* graph. A simple, complete bipartite graph with $|X_1| = p$, and $|X_2| = q$ is often denoted by $K_{p,q}$.

Subgraph of G generated by $A \subset X$. The *subgraph of G generated by A* is the graph with A as its vertex set and with all the arcs in G that have both their endpoints in A. If $G = (X, \Gamma)$ is a 1-graph, then the subgraph generated by A is the 1-graph $G_A = (A, \Gamma_A)$ where

$$\Gamma_A(x) = \Gamma(x) \cap A \qquad (x \in A).$$

Partial graph of G generated by $V \subset U$. This is the graph (X, V) whose vertex set is X and whose arc set is V. In other words, it is graph G without the arcs $U - V$.

Partial subgraph of G. A *partial subgraph* of G is the subgraph of a partial graph of G. For example, if G is the graph of all roads in the United States, the set of all 4-lane roads is a partial graph of G; the set of all roads in Illinois is a subgraph, and the set of all 4-lane roads in Illinois is a partial subgraph.

Chain of length $q > 0$. A *chain* is a sequence $\mu = (u_1, u_2, ..., u_q)$ of arcs of G such that each arc in the sequence has one endpoint in common with its predecessor in the sequence and its other endpoint in common with its successor in the sequence. The number of arcs in the sequence is the *length* of chain μ. A chain that does not encounter the same vertex twice is called *elementary*. A chain that does not use the same arc twice is called *simple*.

Path of length $q > 0$. A *path of length q* is a chain $\mu = (u_1, u_2, ..., u_i, ..., u_q)$ in which the terminal endpoint of arc u_i is the initial endpoint of arc u_{i+1} for all $i < q$. For a 1-graph, a path is completely determined by the sequence of vertices $x_1, x_2, ...$ that it encounters. Hence, we often write

$$\mu = ((x_1, x_2), (x_2, x_3), ...) = [x_1, x_2, ..., x_k, x_{k+1}] = \mu[x_1, x_{k+1}].$$

Vertex x_1 is called the *initial endpoint* and vertex x_{k+1} is called the *terminal endpoint* of path μ.

Similarly, for a simple graph, a chain μ with endpoints x and y is determined by the sequence of its vertices, and we may write

$$\mu = \mu[x, y] = [x, x_1, x_2, ..., y].$$

Cycle. A *cycle* is a chain such that

(1) no arc appears twice in the sequence, and
(2) the two endpoints of the chain are the same vertex.

Pseudo-cycle. A *pseudo-cycle* is a chain $\mu = (u_1, u_2, ..., u_q)$ whose two endpoints are the same vertex and whose arcs are not necessarily distinct.

Circuit. A *circuit* is a cycle $\mu = (u_1, u_2, ..., u_q)$ such that for all $i < q$ the terminal endpoint of u_i is the initial endpoint of u_{i+1}.

Connected graph. A *connected graph* is a graph that contains a chain $\mu[x, y]$ for each pair x, y of distinct vertices.

Connected component of a graph. Clearly, the relation [$x = y$, or $x \neq y$ and there exists a chain in G connecting x and y] denoted by $x \equiv y$ is an equivalence relation because

(1) $x \equiv x$ (reflexivity)
(2) $x \equiv y$ $\Rightarrow$ $y \equiv x$ (symmetry)
(3) $x \equiv y, y \equiv z$ $\Rightarrow$ $x \equiv z$ (transitivity)

The classes of this equivalence relation partition X into connected subgraphs of G called the *connected components*. For example, the graph in Fig. 1.2 possesses two connected components.

Articulation set. For a connected graph, a set A of vertices is called an *articulation set* (or a *cutset*) if the subgraph of G generated by $X - A$ is not connected; the term "cutset" will not be used here, in order to avoid confusion with another kind of "cutset" defined in the theory of transportation networks (and used in Chapter 5).

For example, $\{a, c\}$ and $\{c\}$ are two articulation sets of the graph in Fig. 1.1; vertex c is also called an *articulation vertex* (or a *cut-vertex*).

Stable set. A set S of vertices is called a *stable set* if no arc joins two distinct vertices in S; for example, $\{b, d\}$ is a stable set of the graph in Fig. 1.1.

Matrix associated with a graph G. If G has vertices $x_1, x_2, \ldots, x_n$, let

$$a^i_j = m^+_G(x_i, x_j) ;$$

The matrix $((a^i_j))$ is called the *matrix associated* with G. For example, the matrix associated with the graph in Fig. 1.2 is

$$
((a^i_j)) =
\begin{array}{c c}
 & \begin{array}{cccccc} x_1 & x_2 & x_3 & x_4 & x_5 & x_6 \end{array} \\
\begin{array}{c} x_1 \\ x_2 \\ x_3 \\ x_4 \\ x_5 \\ x_6 \end{array} &
\left(\begin{array}{cccccc}
0 & 1 & 0 & 0 & 0 & 0 \\
1 & 1 & 1 & 1 & 0 & 0 \\
0 & 0 & 0 & 0 & 0 & 0 \\
0 & 0 & 1 & 0 & 0 & 0 \\
0 & 0 & 0 & 0 & 0 & 0 \\
0 & 0 & 0 & 0 & 1 & 1
\end{array}\right)
\end{array}
$$

The matrix $((a^i_j)) + ((a^i_j))^*$ is called the *adjacency matrix*.

3. List of symbols

$\mathbb{R}$	Set of all real numbers.
$\mathbb{N}$	Set of all non-negative integers.
$\mathbb{Z}$	Set of all integers.
$\varnothing$	Empty set.
$\lvert A \rvert$	Cardinality of set A. (i.e. number of elements)
$\{x/\ldots\}$	Set of all x such that ...
$a \in A$	a is an element of set A.
$a \notin A$	a is not an element of set A.
$A \cup B$	Union of sets A and B.
$A \cap B$	Intersection of sets A and B.
$A - B$	A less B (the elements of A that are not in B).
$A \subseteq B$	Set A is contained in set B (possibly $A = B$).
$A \nsubseteq B$	Set A is not contained in set B.
$A \times B$	Cartesian product of A and B (the set of all pairs (a, b) where $a \in A$ and $b \in B$).
$\Gamma(a)$	Image of element a in correspondence Γ.
$\Gamma(A)$	Image of set A in correspondence Γ, or $\bigcup_{a \in A} \Gamma(a)$. $(\Gamma(\varnothing) = \varnothing)$.

$\hat{\Gamma}(A)$ Transitive closure of correspondence Γ.

$\Gamma^{-1}(A)$ Inverse correspondence of correspondence Γ, or

$$\Gamma^{-1}(y) = \left\{ x \mid \Gamma(x) \ni y \right\}.$$

$\mathscr{P}(A)$ Set of all subsets of set A.

$\mathscr{P}_k(A)$ Set of all subsets of cardinality k.

$\mathscr{P}_{(k)}(A)$ Set of all non-empty subsets of cardinality $\leqslant k$.

$(1) \Rightarrow (2)$ Property (1) implies property (2).

$(1) \Leftrightarrow (2)$ Property (1) is equivalent to property (2).

$$\binom{p}{q} = \frac{p!}{q!(p-q)!} \qquad \text{Binomial coefficient.}$$

$$\binom{n}{n_1, n_2, \ldots, n_k} = \frac{n!}{n_1! \, n_2! \ldots n_k!} \qquad \text{Multinomial coefficient.}$$

$p \equiv q \,(\text{mod. } k)$ Integer p is equal to q modulo k (the remainder from the division of p by k is equal to the remainder from the division of q by k).

$\log p$ Logarithm of p.

$\left\lfloor \dfrac{p}{q} \right\rfloor$ Integer part of p/q.

$\left\lfloor \dfrac{p}{q} \right\rfloor^{*}$ Smallest integer greater than or equal to p/q.

$((a_j^i))$ Matrix whose entry in the i-th row and j-th column is a_j^i.

$\text{Det } ((a_j^i))$ Determinant.

For a graph G,

$\Gamma_G(x)$ Set of all neighbours of vertex x.

$\Gamma_G^+(x), \Gamma_G^-(x)$ Set of all successors (resp. predecessors) of vertex x.

$d_G(x)$ Degree of vertex x.

$d_G^+(x), d_G^-(x)$ Outer demi-degree, inner demi-degree of vertex x.

$m_G(A, B)$ Number of edges between sets A and B.

$m_G^+(A, B), m_G^-(A, B)$ Number of arcs going from A to B (resp. from B to A).

$\mu[x, y]$ Portion of the chain μ between vertices x and y.

$\omega(A)$ Set of all arcs having exactly one endpoint in A.

$\omega^+(A), \omega^-(A)$ Set of all arcs with only their initial endpoint in A (resp., with only their terminal endpoint in A).

For a family $\mathscr{S}$ of sets, a member $S \in \mathscr{S}$ is defined to be a *minimal set* if it does not contain any other member of $\mathscr{S}$; a member $S \in \mathscr{S}$ is defined to be a *minimum set* if its cardinality has the minimum value. A *maximal set* and a *maximum set* are defined similarly.

EXERCISES

1. Show that if G is a simple graph with n vertices and p connected components, the maximum possible number of edges in G is

$$\frac{1}{2}(n-p)(n-p+1).$$

2. Show that a simple graph with n vertices and more than $\frac{1}{2}(n-1)(n-2)$ edges is connected.

Chapter 2

Cyclomatic Number

1. Cycles and cocycles

In a graph $G = (X, U)$, a *cycle* is a sequence of arcs

$$\mu = (u_1, u_2, \ldots, u_q)$$

such that

(1) each arc u_k, where $1 < k < q$, has one endpoint in common with the preceding arc u_{k-1}, and the other end point in common with the succeeding arc u_{k+1} (i.e., this sequence is a chain),

(2) the sequence does not use the same arc twice,

(3) the initial vertex and terminal vertex of the chain are the same.

An *elementary cycle* is a cycle in which, in addition,

(4) no vertex is encountered more than once (except, of course, the initial vertex which is also the terminal vertex).

Given a cycle μ, we denote by μ^+ the set of all arcs in μ that are in the direction that the cycle is traversed, and we denote by μ^- the set of all the other arcs in μ.

If the arcs in G are numbered 1, 2, ..., m, then cycle μ is defined by a vector

$$\boldsymbol{\mu} = (\mu_1, \mu_2, \ldots, \mu_m),$$

with

$$\mu_i = \begin{cases} 0 & \text{if} \quad i \notin \mu^+ \cup \mu^-, \\ +1 & \text{if} \quad i \in \mu^+, \\ -1 & \text{if} \quad i \in \mu^-. \end{cases}$$

Henceforth, a cycle and its vector μ will be used interchangeably, and when we say that cycle μ is the *sum* of cycles $\boldsymbol{\mu}^1 + \boldsymbol{\mu}^2$, we will mean the vector sum.

Property 1. *A cycle is the sum of elementary cycles that are pairwise arc-disjoint.*

This is evident because as we traverse, $\boldsymbol{\mu}$ an elementary cycle is defined each time we return to a vertex.

10

Property 2. *A cycle is elementary if, and only if, it is a minimal cycle, i.e., no other cycle is properly contained in it.*

The proof is obvious.

If A is a non-empty subset of X, $\omega^+(A)$ denotes the set of arcs that have only their initial endpoint in A and $\omega^-(A)$ denotes the set of arcs that have only their terminal endpoint in A. Let

$$\omega(A) = \omega^+(A) \cup \omega^-(A) \,.$$

A *cocycle* is defined to be a non-empty set of arcs of the form $\omega(A)$, partitioned into two sets $\omega^+(A)$ and $\omega^-(A)$. Corresponding to each cocycle, there is a vector

$$\boldsymbol{\omega} = (\omega_1, \omega_2, \ldots, \omega_m) \,,$$

with

$$\omega_i = \begin{cases} 0 & \text{if} & i \notin \omega(A) \,, \\ +1 & \text{if} & i \in \omega^+(A) \,, \\ -1 & \text{if} & i \in \omega^-(A) \,. \end{cases}$$

A cocycle may be identified by its vector $\boldsymbol{\omega}$.

A cocycle is called *elementary* if it is the set of arcs joining two connected subgraphs A_1 and A_2 such that

$$A_1, \quad A_2 \neq \varnothing$$
$$A_1 \cap A_2 = \varnothing$$
$$A_1 \cup A_2 = C \,,$$

where C is a connected component of the graph.

A *cocircuit* is defined to be a cocycle $\omega(A)$ in which all arcs are directed in the same direction, i.e. into set A, or out of set A.

Property 3. *A cocycle is the sum of elementary cocycles that are pairwise arc-disjoint.*

Let ω be a cocycle of the form $\omega(A)$, and let $A_1, A_2, \ldots, A_k$ be the different connected components of the subgraph generated by A. Then

$$\omega(A) = \omega(A_1) + \omega(A_2) + \cdots + \omega(A_k) \,,$$

and the cocycles $\omega(A_1), \omega(A_2), \ldots, \omega(A_k)$ are pairwise disjoint. It remains to show that $\omega(A_i)$ is the sum of elementary disjoint cocycles.

If C is the connected component that contains A_i, and if the subgraph generated by $C - A_i$ has connected components $C_1, C_2, \ldots,$ then

$$\omega(A_i) = -\omega(C_1) - \omega(C_2) - \cdots$$

where $-\omega(C_1)$ is an elementary cocycle since $\omega(C_1)$ joins the connected subgraphs C_1 and $A_i \cup C_2 \cup C_3$.... Furthermore, $-\omega(C_1)$, $-\omega(C_2)$, ... are pairwise arc-disjoint.

Q.E.D.

Property 4. *A cocycle is elementary if, and only if, it is a minimal cocycle (i.e. no other cocycle is properly contained in the cocycle).*

Let $\omega(A)$ be a minimal cocycle. Therefore, A is contained in a connected component C. Let A_1, A_2, ..., A_k be the connected components of the subgraph generated by $C - A$. If $k \geq 2$, the vector $-\omega(A_1)$ is a cocycle properly contained in $\omega(A)$. But this is impossible, and therefore, $k = 1$, and ω is an elementary cocycle.

Conversely, let ω be an elementary cocycle that joins two connected subgraphs A_1 and A_2. If we remove some, but not all, of the arcs from ω, then we no longer have a cocycle. Therefore ω is a minimal cocycle.

Q.E.D.

Arc Colouring Lemma (Minty [1960]). *Consider a graph with arcs 1, 2, ..., m. Colour arc 1 black, and arbitrarily colour the remaining arcs red, black or green. Exactly one of the following conditions holds :*
  (1) *there is an elementary cycle containing arc 1 and only red and black arcs with the property that all black arcs in the cycle have the same direction,*
  (2) *there is an elementary cocycle containing arc 1 and only green and black arcs, with the property that all black arcs in the cocycle have the same direction.*

Successively label the vertices of the graph using the following iterative procedure:
  (1) Let arc $1 = (b, a)$. Label vertex a,
  (2) If vertex x is labelled, and vertex y is unlabelled, then label y if
      (a) there is a black arc (x, y), or
      (b) there is a red arc (x, y) or (y, x).

When the labelling procedure stops, exactly one of the following two cases occurs:

CASE 1: Vertex b has been labelled. The vertices used by the procedure to label b from a constitute an elementary cycle of red and black arcs with all black arcs having the same direction. (Thus, there cannot exist a cocycle of black and green arcs containing arc 1 with all black arcs in the same direction.) This cycle is the sum of disjoint elementary cycles, one of which contains arc 1.

CASE 2: Vertex b has not been labelled. Let A denote the set of all labelled vertices. Note that $\omega(A)$ contains only black arcs directed into A or green arcs. Thus, there exists a cocycle $\omega(A)$ of green and black arcs containing arc 1 with all black arcs directed into A. (Thus there cannot exist a cycle of red and black arcs with all black arcs having the same direction). This cocycle is the sum of disjoint elementary cocycles, one of which contains arc 1.

Corollary. *Each arc belongs either to an elementary circuit or to an elementary cocircuit, but no arc belongs to both.*

This is shown by applying the lemma with all arcs coloured black.

The cycles μ^1, μ^2, ..., μ^k are said to be *dependent* if there exists a vector equation of the form

$$r_1 \mu^1 + r_2 \mu^2 + \cdots + r_k \mu^k = 0 ,$$

where $r_1, r_2, \ldots, r_k$ are real numbers, not all zero. If the cycles are not dependent, they are said to be *independent*.

A *cycle basis* is defined to be a set $\{ \mu^1, \mu^2, \ldots, \mu^k \}$ of independent elementary cycles such that any cycle μ can be written as

$$\mu = r_1 \mu^1 + r_2 \mu^2 + \cdots + r_k \mu^k ,$$

where $r_1, r_2, \ldots, r_k$ are real numbers. Clearly, k equals the dimension of the subspace of $\mathbb{R}^m$ generated by the cycles and therefore does not depend on the choice of the basis. This constant k is called the *cyclomatic number* of G, and is denoted by $v(G)$.

A *cocycle basis* $\{ \omega^1, \omega^2, \ldots, \omega^l \}$ is defined similarly, and its cardinality l is called the *cocyclomatic number* of G and is denoted by $\lambda(G)$.

EXAMPLE. Consider the graph G in Fig. 2.1; its elementary cycles are;

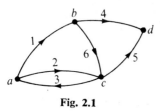

Fig. 2.1

$$\mu^1 = (1, 6, 2) \quad = [abca] ,$$
$$\mu^2 = (1, 6, 3) \quad = [abca] ,$$
$$\mu^3 = (2, 3) \quad = [aca] ,$$
$$\mu^4 = (1, 4, 5, 2) = [abdca] ,$$
$$\mu^5 = (6, 5, 4) \quad = [acdb] ,$$
$$\mu^6 = (1, 4, 5, 3) = [abdca] .$$

These cycles are not independent, since we have, for example:

$$\mu^1 - \mu^2 + \mu^3 = 0 .$$

The cycles μ^2, μ^3, μ^5 form a cycle basis, and therefore, $v(G) = 3$.
A cocycle $\omega(A)$ can be written as a sequence of arcs $(\pm i_1, \pm i_2, \ldots)$ where

each arc i of the sequence is preceded by a $+$ sign if $i \in \omega^+(A)$ or by a $-$ sign if $i \in \omega^-(A)$; it can also be denoted by $\{A\}$. For the graph in Fig. 2.1, the elementary cocycles are:

$$\omega^1 = \{a\} \quad = (+1, +2, -3),$$
$$\omega^2 = \{ab\} \quad = (+6, +2, -3, +4),$$
$$\omega^3 = \{ac\} \quad = (-6, +1, +5),$$
$$\omega^4 = \{abc\} = (+4, +5),$$
$$\omega^5 = \{abd\} = (+6, +2, -3, -5),$$
$$\omega^6 = \{acd\} = (-6, +1, -4).$$

Obviously, these cocycles are not independent. To form a basis, one could take, for example, ω^1, ω^4 and ω^5; hence $\lambda(G) = 3$.

Theorem 1. *Let G be a graph with n vertices, m arcs and p connected components. The cardinality of a cycle basis is $v(G) = m - n + p$. The cardinality of a cocycle basis is $\lambda(G) = n - p$.*

1. *There exist $n - p$ elementary cocycles.*

Suppose first that the graph is connected ($p = 1$). Successively form $n - 1$ independent cocycles $\omega(A_1)$, $\omega(A_2)$, ..., $\omega(A_{n-1})$ in the following way:

(a) Take an arbitrary vertex a_1 and let $A_1 = \{a_1\}$. The cocycle $\omega(A_1)$ contains an elementary cocycle. Let $[a_1, a_2]$ be an edge of this elementary cocycle such that

$$a_1 \in A_1, \qquad a_2 \notin A_1.$$

(b) Let $A_2 = A_1 \cup \{a_2\}$. The cocycle $\omega(A_2)$ contains an elementary cocycle. Let $[x, a_3]$ be an edge in this elementary cocycle such that

$$x \in A_2, \qquad a_3 \notin A_2.$$

(c) Let $A_3 = A_2 \cup \{a_3\}$, and repeat the process until $n - 1$ elementary cocycles have been defined.

These cocycles are independent because each contains an arc not contained in any of the others.

If the graph is not connected ($p > 1$), let $C_1, C_2, ..., C_p$ denote its connected components. Then, there exist

$$(|C_1| - 1) + (|C_2| - 1) + \cdots + (|C_p| - 1) = n - p$$

independent elementary cocycles.

2. *There exist $m - n + p$ independent elementary cycles.*

Let $v(G) = m - n + p$, and construct a sequence $G_0, G_1, ..., G_m = G$ of partial graphs. Graph G_0 consists of the isolated vertices of G. Each G_i is obtained from its predecessor G_{i-1} by the addition of an arc i of $G - G_{i-1}$.

Initially, $v(G_0) = 0$, and there are no cycles. If arc i forms a new cycle $\boldsymbol{\mu}^i$, then

$$v(G_i) = v(G_{i-1}) + 1,$$

since p remains unchanged and m is increased by 1. If arc i does not form a new cycle, then

$$v(G_i) = v(G_{i-1}),$$

since p decreases by 1 and m increases by 1. Upon termination, $v(G) = m - n + p$ cycles $\boldsymbol{\mu}^{i_1}, \boldsymbol{\mu}^{i_2}, ..., \boldsymbol{\mu}^{i_k}$ have been defined. There is no vector of the form

$$r_1 \boldsymbol{\mu}^{i_1} + r_2 \boldsymbol{\mu}^{i_2} + \cdots + r_k \boldsymbol{\mu}^{i_k} = \mathbf{0},$$

with some $r_k \neq 0$ because cycle $\boldsymbol{\mu}^{i_k}$ contains arc i_k which is not contained in any of the other cycles. Thus the $\boldsymbol{\mu}_i$ are $v(G)$ independent cycles.

3. *There cannot exist more than $v(G) = m - n + p$ independent cycles, and there cannot exist more than $\lambda(G) = n - p$ independent cocycles.*

Consider in $\mathbb{R}^m$ the vector space M generated by the cycles and the vector space Ω generated by the cocycles. If $\boldsymbol{\mu}$ is a cycle and if $\boldsymbol{\omega} = \boldsymbol{\omega}(A)$ is a cocycle, their scalar product

$$\langle \boldsymbol{\mu}, \boldsymbol{\omega} \rangle = \sum_{i=1}^{m} \mu_i \omega_i$$

equals zero because

$$\langle \boldsymbol{\mu}, \boldsymbol{\omega}(A) \rangle = \left\langle \boldsymbol{\mu}, \sum_{a \in A} \boldsymbol{\omega}(a) \right\rangle = \sum_{a \in A} \langle \boldsymbol{\mu}, \boldsymbol{\omega}(a) \rangle = 0.$$

Thus M and Ω are orthogonal subspaces of $\mathbb{R}^m$, and their dimensions must satisfy

$$\dim M + \dim \Omega \leqslant m.$$

From Parts 1 and 2,

$$\dim M + \dim \Omega \geqslant v(G) + \lambda(G) = m.$$

Therefore equality holds throughout, and

$$\dim M = v(G),$$
$$\dim \Omega = \lambda(G).$$

Q.E.D.

2. Cycles in planar graphs

Graph G is said to be *planar* if it is possible to represent the graph on a plane in which the vertices are distinct points, the arcs are simple curves and no two arcs cross one another. A representation of G that satisfies the above requirements on a plane is called a *topological planar graph*. Two topological graphs that can be made to coincide by an elastic deformation of the plane are considered to be the same.

EXAMPLE 1. A convex polyhedron in 3-dimensional space defines a simple graph: its "corners" (0-dimensional faces) are the vertices and its "sides" (1-dimensional faces) are the edges of the graph. It has been shown (Steinitz [1922]) that a simple graph G can represent a convex polyhedron in $\mathbb{R}^3$ if, and only if, G is a connected planar graph that cannot be disconnected by the removal of less than three vertices.

EXAMPLE 2.   Problem of three factories and three utilities (Fig. 2.2).

Three factories, a, b, and c rely upon underground supply lines for their water from point d, their gas from point e, and their electricity from point f. Is it possible to arrange the three factories and the three utility stations so that no supply lines cross one another except at their endpoints? It can be shown that 8 supply lines can always be placed but that the 9th supply line must cross at least one other supply line. Thus $K_{3,3}$ is not planar.

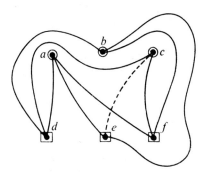

Fig. 2.2

Let G be a topological planar graph. A *face* of G is defined to be a region of the plane bounded by arcs such that any two points in a region can be connected by a continuous curve that meets no arcs or vertices. Let Z denote the set of all faces. The *boundary* of a face z is the set of all arcs that touch face z. Faces z and z' are said to be *adjacent* if their boundaries contain a common arc. (If two faces touch one another only at a vertex, they are not adjacent.)

The *contour* of a face z is defined to be an elementary cycle formed with the edges of the boundary of z that contains in its interior the face z. Note that there is exactly one *unbounded face* and it has no contour. All the other faces are *bounded* and have exactly one contour.

EXAMPLE. A geographic map corresponds to a topological planar multi-graph whose edges are the borders between countries. This graph has no isthmus, and each of its vertices has degree ≥ 3. A given face may be adjacent to another face along several different edges. Note in Fig. 2.3 that faces g and d have a common vertex but are not adjacent.

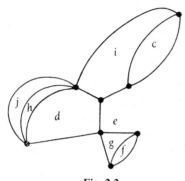

Fig. 2.3

Theorem 2. *In a topological planar graph G, the contours of the different bounded faces constitute a cycle basis.*

Clearly, the theorem is true if G has only 2 bounded faces. If the theorem is true for all graphs with $f - 1$ bounded faces, we shall show that the theorem is also true for a topological planar graph with f bounded faces.

If not all of the contours are edge-disjoint cycles, then the result is evidently true. Suppose that when arc i is removed, a graph G' with $f - 1$ bounded faces is formed. By hypothesis, the contours of this graph G' are a fundamental basis of independent cycles. If arc i is returned to the graph, a new finite face is formed. Its contour is a cycle independent of the cycles of G' because it contains an arc not present in any cycle of G'. Thus the addition of an arc cannot increase the cyclomatic number by more than 1, and the bounded faces of G determine a cycle basis.

Q.E.D.

Corollary 1. *If a connected topological planar graph has n vertices, m arcs and f faces, then*

$$n - m + f = 2 \quad (Euler's \ Formula) .$$

The number of bounded faces equals the cyclomatic number $v(G)$. Thus

$$f = v(G) + 1 = (m - n + 1) + 1 = m - n + 2,$$

and the corollary follows.

Corollary 2. *A simple planar graph G has a vertex x of degree $d_G(x) \leqslant 5$.*

Suppose that G is connected. Otherwise, each connected component will be considered separately. Since G is a simple graph, each face is bounded by at least three distinct edges. Consider the bipartite graph formed by the set A of vertices representing the faces of G and the set B of vertices representing the the edges of G. Place an arc from $a \in A$ to $b \in B$ each time face a is incident to edge b. (This graph is called the *face-edge incidence graph*.) Clearly, the number of arcs is $\leqslant 2m$ and $\geqslant 3f$. Thus

$$f \leqslant \frac{2m}{3}.$$

If each vertex is the endpoint of at least 6 edges, then $n \leqslant \dfrac{2m}{6}$. From Euler's Formula,

$$2 = n - m + f \leqslant \frac{m}{3} - m + \frac{2m}{3} = 0,$$

which yields a contradiction.

$$\text{Q.E.D.}$$

The Euler Formula of Corollary 1 can be used in many proofs.

EXAMPLE 1 (Euler). In 3-dimensional space consider a convex polyhedron with n vertices, m edges and f faces. Obviously, one can represent this polyhedron on the surface of a sphere without having any two edges cross each other. By a stereographic projection whose centre is the middle of one of the faces, we can represent the polyhedron on a plane. Therefore this graph is planar, and we obtain a fundamental relation for convex polyhedra:

$$n - m + f = 2.$$

EXAMPLE 2. Using Euler's Formula, we shall show that the graph for three factories and three utilities cannot be planar. If the graph is planar, then

$$f = 2 - n + m = 2 - 6 + 9 = 5.$$

Each face has at least 4 edges in its contour because if a face s had only 3 edges, then it would be bordered by 3 vertices of which 2 must be in the same class (factories or utilities), but two vertices of the same class cannot be adjacent. For the bipartite face-edge incidence graph, the number of arcs is $\leqslant 2m$ and $\geqslant 4f$. Thus

$$18 = 2m \geqslant 4f = 20,$$

which is a contradiction.

EXAMPLE 3. We shall show that the complete graph K_5 with 5 vertices cannot be planar. If this graph is planar, then

$$f = 2 - n + m = 2 - 5 + 10 = 7 \, .$$

The contour of each face has at least 3 edges. For the bipartite face-edge incidence graph, the number of arcs is $\leqslant 2m$ and $\geqslant 3f$. Thus

$$20 = 2m \geqslant 3f = 21 \, ,$$

which is a contradiction.

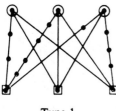
Type 1

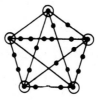
Type 2

Fig. 2.4

Remark. The graph of factories and utilities and the graph K_5 allow us to describe an entire family of non-planar graphs. As shown in Fig. 2.4, unlimited additional vertices may be placed on each edge of these graphs to create new non-planar graphs of type 1 and type 2. Conversely it is now possible to show:

A multigraph G is planar if, and only if, it contains no partial subgraph of type 1 or type 2 (Kuratowski [1930]).

Other characterizations of planar graphs have been made by Whitney [1933], MacLane [1937], Ghouila-Houri [1964], W. T. Tutte [1968], etc. Algorithms to determine if a graph is planar have been described by Demoucron, Malgrange, Pertuiset [1964], Lempel, Even, Cederbaum [1967], etc.

Consider a planar multigraph G that is connected and has no isolated vertices. Construct a planar multigraph G^* that corresponds to G as follows:

Place a vertex x^* of G^* inside each face x of G. For each edge e of G, construct an edge e^* of G^* that joins the vertices corresponding to the faces separated by edge e. Graph G^* is also planar, connected and without isolated vertices. See Fig. 2.5. Graph G^* is called the *topological dual* of G. Note that:

(1) The topological dual of G^* is G, i.e. $(G^*)^* = G$.

(2) A loop in G corresponds to a pendant edge in G^*, and vice versa.

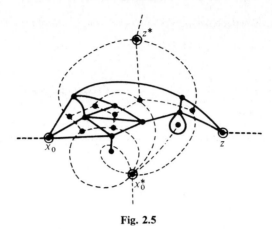

Fig. 2.5

Theorem 3. *Each elementary cycle of G corresponds to an elementary cocycle of the topological dual G*, and vice versa.*

Let $\mu = (u_1, u_2, \ldots)$ be an elementary cycle of G, and let A^* be the set of vertices of G^* that are inside this cycle. Then

$$\omega(A^*) = \{\, u_1^*, u_2^*, \ldots \,\}.$$

Furthermore, $G_{A^*}^*$ is a connected subgraph of G^* because we can always go from one face to another inside cycle μ. Similarly, the complement $G^*_{z^* - A^*}$ is also a connected subgraph of G^*. Thus $\omega(A^*)$ is an elementary cocycle. The other part of the theorem may be proved similarly.

Q.E.D.

Corollary. *For a connected topological planar graph G,*

$$\nu(G^*) = \lambda(G), \quad \lambda(G^*) = \nu(G).$$

The corollary is evident from Theorem 3, but it can also be shown using Euler's Formula, which implies that

$$\nu(G^*) = m - f + 1 = m - (2 + m - n) + 1 = n - 1 = \lambda(G),$$
$$\lambda(G^*) = f - 1 = m - n + 1 = \nu(G).$$

EXERCISES

1. Let ω be a set of edges in G. Show that ω is an elementary cocyle if, and only if, ω meets no elementary cycle of G in only one arc, and if for each pair of arcs $e, e' \in \omega$, there is an elementary cycle μ such that

$$\mu \cap \omega = \{\, e, e' \,\}$$

(P. Rosenstiehl [1970])

2. Consider the following binary operation, denoted by $+$, on the set $K = \{ 0, 1 \}$:.

$$1 + 0 = 0 + 1 = 1 ,$$
$$0 + 0 = 1 + 1 = 0 .$$

A vector $\mu = (\mu_1, \mu_2, \ldots, \mu_n) \in K^m$ is a "topological cycle" if $\mu = 0$ or if μ represents the union of disjoint elementary cycles. A "topological cocycle" is defined similarly. Show that the topological cycles form an abelian group with 0 as the zero element. Do the same for topological cocycles.

Show that a vector $z \in K^m$ is a topological cycle if, and only if, the scalar product $\langle z, \omega \rangle = 0$ for all topological cocycles ω. Do the same for topological cocycles.

CHAPTER 3

Trees and Arborescences

1. Trees and cotrees

A *tree* is defined to be a connected graph without cycles. A tree is a special kind of 1-graph. A *forest* is defined to be a graph whose connected components are trees, i.e., a forest is a graph without cycles.

Theorem 1. *Let $H = (X, U)$ be a graph of order $|X| = n > 2$. The following properties are equivalent (and each characterizes a tree):*

(1) *H is connected and has no cycles*

(2) *H has $n - 1$ arcs and has no cycles,*

(3) *H is connected and contains exactly $n - 1$ arcs,*

(4) *H has no cycles, and if an arc is added to H, exactly one cycle is created,*

(5) *H is connected, and if any arc is removed, the remaining graph is not connected,*

(6) *Every pair of vertices of H is connected by one and only one chain.*

(1) $\Rightarrow$ (2) If p denotes the number of connected components, m denotes the number of arcs and $v(H)$ denotes the cyclomatic number (see Ch. 2), then (1) implies

$$p = 1, \qquad v(H) = m - n + p = 0.$$

Thus,
$$m = n - p = n - 1.$$

(2) $\Rightarrow$ (3) Since $v(H) = 0$, $m = n - 1$, it follows that

$$p = v(H) - m + n = 1.$$

Thus H is connected.

(3) $\Rightarrow$ (4) Since $p = 1$, $m = n - 1$, it follows that

$$v(H) = m - n + p = 0.$$

Thus H contains no cycles. Also, if an arc is added, the cyclomatic number becomes equal to 1, and there is exactly one cycle in the new graph.

(4) $\Rightarrow$ (5) If H were not connected, then two vertices, say a and b, would not

be connected, and an arc (a, b) could be added without creating a cycle, which contradicts (4). Thus $p = 1$, $v(H) = 0$, and therefore $m = n - 1$.

If an arc is removed, we obtain a graph H' with

$$m' = n' - 2, \qquad v(H') = 0 .$$

Hence

$$p' = v(H') - m' + n' = 2 ,$$

and H' is not connected.

(5) $\Rightarrow$ (6) For any two vertices a and b there is a chain connecting them (since H is connected). This chain is unique (otherwise the removal of an arc which belongs only to the second chain would not disconnect the graph).

(6) $\Rightarrow$ (1) Clearly, if H had a cycle, at least one pair of vertices would be joined by two distinct chains, which contradicts (6).

Q.E.D.

Theorem 2. *A vertex is called "pendant" if it is adjacent to exactly one other vertex. A tree of order $n \geqslant 2$ has at least two pendant vertices.*

Let H be a tree that has only 0 or 1 pendant vertices. Consider a traveller who traverses the edges of the graph starting from a pendant vertex (if there is one). If he does not permit himself to use the same edge twice, he cannot go to the same vertex twice (since H has no cycles).

If he arrives at a vertex x, he can always depart using a new edge (since x is not a pendant vertex). Thus the trip lasts indefinitely, and this is impossible since H is finite.

Q.E.D.

Theorem 3. *A graph $G = (X, U)$ has a partial graph that is a tree if, and only if, G is connected.*

If G is not connected, no partial graph of G is connected. Therefore G cannot have a partial graph that is a tree.

If G is connected, look for an arc whose removal does not disconnect the graph. If no such arc exists, G is a tree by virtue of property (5). If such an arc exists, remove it and look for another such arc, etc. ... When no more arcs can be removed, the remaining graph is a tree whose vertex set is X.

Q.E.D.

The tree obtained from G as above is called a *spanning tree*, and Theorem 3 yields a simple algorithm to construct a spanning tree of a connected graph.

A spanning tree can also be constructed in the following way:

Consider any arc u_0. Find an arc u_1 that does not form a cycle with u_0. Then find an arc u_2 that does not form a cycle with $\{u_0, u_1\}$, etc. ... When the procedure cannot continue, a spanning tree has been obtained by property (4)·

Theorem 4. *Let G be a connected graph, let H be a spanning tree of G, and let u_i be an arc of G not in tree H. If arc u_i is added to H, it creates a cycle μ^i by virtue of property* (4). *The different cycles μ^i form a cycle basis of G, called the "basis associated with tree H".*

The cycles μ^i are independent since every one of them contains an arc not contained in any of the others. Moreover, the number of μ^i equals:

$$m(G) - m(H) = m - (n - 1) = m - n + p = v(G).$$

By Theorem (1, Ch. 2), it follows that the μ^i form a cycle basis of G.

Q.E.D.

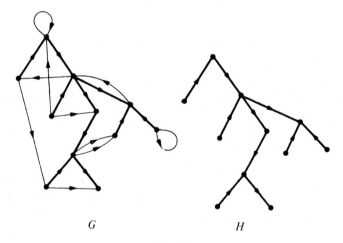

G H

Fig. 3.1

Remark. This theorem yields a simple algorithm to construct a cycle basis of a connected graph G. If G is not connected, each connected component has to be treated separately.

Given a connected graph $G = (X, U)$, Theorem 1 shows that a partial graph $H = (X, V)$ is a spanning tree if it contains no elementary cycles and if upon the addition of any arc in $U - V$, the graph contains an elementary cycle. Similarly, we shall say that a partial graph (X, W) of G is a *cotree* if it contains no elementary cocycles of G and if upon the addition of any arc in $U - W$, it does contain an elementary cocycle of G.

Theorem 5. *Let $G = (X, U)$ be a connected graph, and let (V, W) be a partition of U:*

$$V \cup W = U, \qquad V \cap W = \varnothing.$$

A necessary and sufficient condition for (X, W) to be a cotree is that (X, V) be a tree.

1. *Sufficiency.* If (X, V) is a tree, we shall show that (X, W) is a cotree.

W contains no cycles of G. Suppose that W contains an elementary cocycle $\omega(A)$ of G. Then no chain in the tree (X, V) connects A and $X - A$, which is a contradiction.

If $v \in V$, the set of arcs $W \cup \{v\}$ contains a cocycle of G. Clearly, $(X, V - \{v\})$ has two connected components A and B. Therefore, $\omega(A)$ is an elementary cocycle of G contained in $W \cup \{v\}$.

2. *Necessity.* If (X, W) is a cotree, we shall show that (X, V) is a tree by using the Arc Colouring Lemma (Chap. 2).

V contains no cycles. Let $v \in V$. Colour arc v black. Colour the arcs in $V - \{v\}$ red, and colour the arcs in W green. Since (X, W) is a cotree, G has a cocycle of black and green arcs that contains arc v. Thus there cannot be a cycle of red and black arcs containing arc v. Since v was selected arbitrarily, there cannot be any cycle in V.

If $w \in W$, the set $V \cup \{w\}$ contains a cycle. Colour arc w black. Colour the arcs in V red, and colour the arcs in $W - \{w\}$ green. Since G has no cocycle of black and green arcs containing arc w, there is a cycle of black and red arcs containing arc w. This proves that (X, V) is a spanning tree of G.

Q.E.D.

Theorem 6. *Let $G = (X, U)$ be a connected graph, let $F = (X, W)$ be a cotree, and let u_i be an arc of G not in F. If u_i is added to F, it creates exactly one cocycle ω^i, and the different cocycles ω^i form a cocycle basis of G.*

Clearly, if in the graph $(X, U - W)$ an arc u_i is removed, exactly two connected components A and B are formed, and $\omega(A) = \omega^i$.

The cocycles ω^i are independent since each of them contains an arc not contained in any of the others. The number of cocycles ω^i equals the number of edges in the tree $(X, U - W)$ which equals

$$n - 1 = \lambda(G) \, .$$

From Theorem (1, Ch. 2), it follows that the ω^i form a cocycle basis.

Q.E.D.

This theorem yields a simple algorithm to construct a cocycle basis.

2. Strongly connected graphs and graphs without circuits

Consider a connected graph $G = (X, U)$. A *path of length 0* is defined to be any sequence $[x]$ consisting of a single vertex $x \in X$.

For $x \in X$ and $y \in X$, let the relation $x \equiv y$ signify that there is a path $\mu_1[x, y]$ going from x to y and also a path $\mu_2[y, x]$ going from y to x. This relation is an *equivalence relation*, i.e. it satisfies the following three properties:

$$
\begin{aligned}
x &\equiv x & &\text{for all } x, \\
x &\equiv y & \Rightarrow \quad & y \equiv x, \\
x &\equiv y, \quad y \equiv z & \Rightarrow \quad & x \equiv z \, .
\end{aligned}
$$

The sets of the form

$$A(x_0) = \{ \, x \mid x \in X, \; x \equiv x_0 \, \}$$

partition X and are called the *strongly connected components of G.*

A graph is said to be *strongly connected*, if for all $x, y \in X$, there exists a path $\mu_1[x, y]$ and a path $\mu_2[y, x]$. In other words, *graph G is strongly connected if it has only one strongly connected component.*

Theorem 7. *If G is a connected graph with at least one arc, the following conditions are equivalent:*

(1) *G is strongly connected,*

(2) *Every arc lies on a circuit,*

(3) *G contains no cocircuits.*

(1) $\Rightarrow$ (2) Let (x, y) be an arc of G; since there is a path from y to x, arc (x, y) is contained in a circuit of G.

(2) $\Rightarrow$ (3) If G had a cocircuit that contains arc (x, y), then G cannot have a circuit containing this arc by the Arc Colouring Lemma with all arcs coloured black. This contradicts (2).

(3) $\Rightarrow$ (1) Let G be a connected graph without cocircuits. We shall assume that G is not strongly connected and produce a contradiction.

Since G is not strongly connected, it has more than one strongly connected component. Since G is connected, there exist two distinct strongly connected components that are joined by an arc

(a, b). Arc (a, b) is not contained in any circuit because otherwise a and b would be in the same strongly connected component. By the Arc Colouring Lemma, arc (a, b) is contained in some co-circuit. This contradicts (3).

Q.E.D.

Theorem 8. *If G is a graph with at least one arc, the following conditions are equivalent*:
(1) *G is a graph without circuits,*
(2) *Each arc is contained in a cocircuit.*

The proof is immediate.

Theorem 9. *If G is a strongly connected graph of order n, then G has a cycle basis of $v(G)$ circuits.*

To prove Theorem 9, it is sufficient to show that $v(G)$ independent circuits can be found. This is obviously true if G has order $\leqslant 2$. Assume that this is true for all graphs of order less than $n > 2$. We shall show that this also is true for a graph of order n.

Choose from the circuits of length > 1 a circuit $\mu = (u_1, u_2, ..., u_k)$ of minimum length. No arc joins two non-consecutive vertices of this circuit but there may exist arcs parallel to the arcs of circuit μ.

Replace all the vertices of μ by a single vertex a'. Replace each arc incident to μ but different from $u_1, u_2, ..., u_k$ by an arc with the same index incident to a'. The new graph G' has order $n' = n - k + 1$ with $m' = m - k$ arcs. Graph G' is strongly connected and, by virtue of the induction hypothesis, has a family of $v(G') = q$ independent circuits $\mu'_1, \mu'_2, <, \mu'_q$. These circuits in G' induce in G independent circuits $\mu_1, \mu_2, ..., \mu_q$. We have

$$q = v(G') = m' - n' + 1 = (m - k) - (n - k + 1) + 1$$
$$= v(G) - 1.$$

By adding the circuit μ to the family $\mu_1, \mu_2, ..., \mu_q$ we obtain a family of $q + 1 = v(G)$ independent circuits.

Q.E.D.

Theorem 10. *If graph G contains no circuits, then G has a cocyclic basis of $\lambda(G)$ cocircuits.*

We may assume that G is connected. Otherwise, G has several connected components $C_1, C_2, ..., C_p$; the theorem being true for each connected component, there would be at least

$$\sum_{i=1}^{p} (|C_i| - 1) = n - p = \lambda(G)$$

independent cocircuits, and the theorem would be also true for G.

It suffices to show that *there exists* $\lambda(G) = n - 1$ *independent cocircuits in a connected graph G of order n.* Clearly, this is true for $n \leqslant 2$. Let $n > 2$. If this is true for all graphs with $n - 1$ vertices, we shall show that it is also true for a graph G with n vertices.

Since G contains no circuits, there is at least one vertex b such that the length of the longest path from b equals 1, and there exists a vertex a without successors such that $(b, a) \in U$. Consider the graph G' obtained from G by deleting the arcs from b to a and by replacing a and b by a single vertex a'.

We shall show first that graph G' contains no circuits. If a circuit μ' were present in G', it would necessarily pass through a' and have length > 1. Let d be the vertex that follows a' in this circuit. The cycle μ in G induced by μ' contains arc (a, b) (because G has no circuits). Hence $(b, d) \in U$ and there exists a path of length > 1 from b to a, which is a contradiction. Thus G' is connected, has $n - 1$ vertices, has no circuits. By the induction hypothesis, G' has $n - 2$ independent cocircuits $\omega'(A_1')$, $\omega'(A_2')$, ..., $\omega'(A_{n-2}')$ and we may assume

$$a' \in A_1', A_2', ..., A_{n-2}'.$$

Each of these cocircuits of G' induces a cocircuit $\omega(A_i)$ in G. Since the vector $\omega(A_i)$ has the same coordinates as the vector $\omega'(A_i')$, the $\omega(A_i)$ are linearly independent vectors. The vectors $\omega(a)$, $\omega(A_1)$, $\omega(A_2)$, ..., $\omega(A_{n-2})$ are also linearly independent because $\omega(a)$ contains the arc (b, a) that is not contained in any of the other cocircuits.

Thus $n - 1$ independent cocircuits have been found.

Q.E.D.

Let $G = (X, U)$ be a strongly connected graph without loops and with more than one vertex. For each vertex x, there is a path from it and a path going into it; therefore there exist at least two arcs incident to x. A vertex that has more than two arcs incident to it is called a *node*. Otherwise, it is called an *anti-node*. A path whose only nodes are its endpoints is called a *branch*. A strongly connected graph without loops that has exactly one node is called a *rosace*. A rosace has a very simple structure since each branch leaves from and returns to the only node.

A graph G is said to be *minimally connected* if it is strongly connected and the removal of any arc destroys the strongly connected property. Clearly, a rosace is a minimally connected graph. Furthermore, each minimally connected graph is a 1-graph without loops.

For a graph $G = (X, U)$, the *contraction* of a set A of vertices is the operation defined by replacing A by a single vertex a and by replacing each arc going into A (resp. out of A) by an arc with the same index going into a (resp. out of a).

Lemma 1. *Let G be a minimally connected graph. Let A be a set of vertices that generates a strongly connected subgraph of G. Then the contraction of A yields a minimally connected graph.*

1. We shall show first that the contraction of A yields a 1-graph. If this were not the case, there would exist a vertex $x \notin A$ and two vertices $a, a' \in A$ such that $(x, a), (x, a') \in U$ (or, with $(a, x), (a', x) \in U$ but this would not change the proof). If one of these arcs is removed, the graph remains strongly connected. Thus, G is not minimally connected, which is a contradiction.

2. *We shall show now that the contraction of A yields a graph G' that is minimally connected.* Clearly, graph G' is strongly connected. If an arc u is removed, the remaining graph is not strongly connected, since the graph $(X, U - \{ u \})$ is not strongly connected.

Q.E.D.

Lemma 2. *Let G be a minimally connected graph, and let G' be the minimally connected graph obtained by the contraction of an elementary circuit of G. Then*

$$v(G) = v(G') + 1 .$$

Let μ be the elementary circuit to be contracted. It is of length $k > 1$ and has no chords (otherwise G would not be minimally connected). Let n' and m' respectively denote the number of vertices and arcs in G'. Then

$$v(G') = m' - n' + 1 = (m - k) - (n - k + 1) + 1$$
$$= m - n = v(G) - 1 .$$

Q.E.D.

Theorem 11. *If G is a minimally connected graph of order $n \geq 2$, then G has at least two anti-nodes.*

Since $| X | > 1$, G contains at least one circuit, and $v(G) \geq 1$. If $v(G) = 1$, the result is true because G is an elementary circuit. We shall assume that the result is true for graphs with a cyclomatic number $< k$ and show that it is also true for a graph G with $v(G) = k > 1$.

CASE 1. *G has no circuits of length ≥ 3.* Then any two adjacent vertices of G are necessarily joined in both directions. The simple graph H that has the same vertices as G with two vertices joined by an edge if, and only if, they are adjacent in G, is connected (because G is connected) and has no cycles (because G is minimally connected). Thus, H is a tree of order ≥ 2, and from Theorem 2, H has two pendant vertices x and y. The vertices x and y are both anti-nodes in graph G and this proves the result.

CASE 2. *G has a circuit μ of length* ≥ 3. This circuit has no *chord* (a chord is an arc joining two non-consecutive vertices of a circuit *i* if μ had a chord, *G* would not be minimally connected). Since $v(G) \geq 2$, there exists a vertex of *G* that is not contained in μ.

The graph *G'* obtained from *G* by the contraction of μ has order > 1, and from Lemma 2, $v(G') = v(G) - 1$. Because of the induction hypothesis, *G'* possesses two anti-nodes *x* and *y*. If one of these anti-nodes, say *x*, is the contracted image of μ, then μ contains an anti-node *z* of *G* (because its length is ≥ 3). Thus *G* has at least two anti-nodes, *y* and *z*.

If neither of the vertices *x* and *y* is the contracted image of μ, then *G* has at least two anti-nodes, *x* and *y*.

Q.E.D.

Corollary 1. *Let G be a minimally connected graph that is not an elementary circuit. Then there exists a branch whose anti-nodes form a non-empty set A such that the subgraph G_{X-A} is strongly connected.*

Construct from *G* a graph $\overline{G}$ whose vertices are the nodes of *G* and whose arcs are the branches of *G*. Graph $\overline{G}$ is strongly connected, but it is not minimally connected (because it has no anti-nodes). Thus, an arc can be removed from $\overline{G}$ without destroying the strong connectivity, This arc of $\overline{G}$ is necessarily a branch of *G* of length > 1, because *G* is minimally connected.

Q.E.D.

Corollary 2. *If G is a strongly connected graph without loops having at least one node, then there exists a branch whose arcs and anti-nodes can be removed without destroying the strong connectivity.*

The proof is similar to the proof of Corollary 1.

Theorem 12. *If $G = (X, U)$ is a graph, the graph C' obtained from G by contracting each strongly connected component contains no circuits.*

The proof is immediate.

3. Arborescences

In a graph $G = (X, U)$, a vertex *a* is called a *root* if all the vertices of *G* can be reached by paths starting from *a*. A graph does not always have a root.

A graph *G* is said to be *quasi-strongly connected* if for each pair of vertices *x*, *y*, there exists a vertex $z(x, y)$ from which there is a path to *x* and a path to *y*. A strongly connected graph is quasi-strongly connected because we can let $z(x, y) = x$; the converse is not true. A quasi-strongly connected graph is connected.

Finally, an *arborescence* is defined as a tree that has a root. For example, the family tree of the male descendants of King Henry IV is an arborescence whose root is King Henry IV.

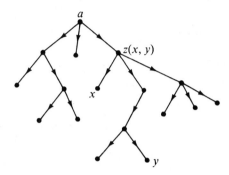

Fig. 3.2: Arborescence

Lemma. *A necessary and sufficient condition that a graph $G = (X, U)$ have a root is that G be quasi-strongly connected.*

Clearly, if G has a root, then G is quasi-strongly connected. Conversely, suppose G is quasi-strongly connected, and consider its vertices $x_1, x_2,...,x_n$. There exists a vertex z_2 from which there is a path to x_1 and a path to x_2. There exists a vertex z_3 from which there is a path to z_2 and a path to x_3, etc. Also, there exists a vertex z_n from which there is a path to z_{n-1} and a path to x_n. Clearly, vertex z_n is a root of G.

Q.E.D.

Theorem 13. *Let H be a graph of order $n > 1$. The following properties are equivalent (and each characterizes an arborescence):*
  (1) *H is a quasi-strongly connected graph without cycles,*
  (2) *H is a quasi-strongly connected graph and has $n - 1$ arcs,*
  (3) *H is a tree having a root a,*
  (4) *There exists a vertex a such that each other vertex is connected with it by one path from a, and only one,*
  (5) *H is quasi-strongly connected and this property is destroyed if any arc is removed from H,*
  (6) *H is quasi-strongly connected and has a vertex a such that*

$$d_H^-(a) = 0,$$

$$d_H^-(x) = 1 \qquad (x \neq a),$$

(7) *H has no cycles and contains a vertex a such that*

$$d_H^-(a) = 0$$

$$d_H^-(x) = 1 \qquad (x \neq a).$$

(1) $\Rightarrow$ (2) From property (1), H is connected and without cycles. Thus H is a tree. Therefore, H has $n - 1$ arcs.

(2) $\Rightarrow$ (3) From property (2), H is connected and has $n - 1$ arcs. Thus H is a tree. From the lemma, H has a root a.

(3) $\Rightarrow$ (4) The root a of tree H has the desired property.

(4) $\Rightarrow$ (5) Suppose that the quasi-strongly connected property is not destroyed when an arc (x, y) is removed. Then, there exist two elementary paths

$$[z, c_1, c_2, \ldots, x] \quad \text{and} \quad [z, d_1, d_2, \ldots, y]$$

that do not use arc (x, y). Thus there are two paths in graph G from z to y, and there are two paths from a to y. This contradicts property (4).

(5) $\Rightarrow$ (6) From the lemma, graph H has a root a because it is quasi-strongly connected. Thus

$$d_H^-(x) \geq 1 \qquad (x \neq a).$$

If a vertex x satisfies $d_H^-(x) > 1$, there exist two distinct arcs $u, v \in \omega^-(x)$ and, therefore, there are two distinct paths from a to x. If arc u is removed, the graph still has a root at a and therefore remains quasi-strongly connected, which contradicts (5). Thus

$$d_H^-(x) = 1 \qquad (x \neq a).$$

Finally, there cannot exist an arc incident into a because the graph obtained from H by removing this arc has a as a root and is quasi-strongly connected, which contradicts (5).

(6) $\Rightarrow$ (7) The number of arcs in H equals

$$\sum_{j=1}^{n} d_H^-(x_j) = n - 1.$$

Since H is connected and has $n - 1$ arcs, it is a tree, and contains no cycles.

(7) $\Rightarrow$ (1) Starting from a vertex $b \neq a$, travel through the graph traversing the arcs against their direction. No vertex is encountered twice because H has no cycles. If a vertex $x \neq a$ is encountered, the trip will

continue because $d_H^-(x) = 1$. Therefore, the trip can only end at vertex a. Thus a is a root, and H is quasi-strongly connected.

Q.E.D.

Corollary. *A graph G has a partial graph that is an arborescence if, and only if, G is quasi-strongly connected.*

If G is not quasi-strongly connected, no partial graph is an arborescence.

Conversely, if G is quasi-strongly connected, we can successively delete all the arcs whose removal does not destroy the quasi-strongly connected property. When no such arcs exist, the graph is an arborescence by virtue of Theorem 13, property (5).

Q.E.D.

The following theorem deals with simple graphs and is a constructive reformulation of a result of P. Camion [1968].

Theorem 14 (Crestin [1969]). *Let $G = (X, E)$ be a simple connected graph, and let $x_1 \in X$. It is possible to direct all the edges of E so that the resulting graph $G_0 = (X, U)$ has a spanning tree H such that:*

1. *H is an arborescence with root x_1,*
2. *The cycles associated with tree H are circuits,*
3. *The only elementary circuits of G_0 are the cycles associated with tree H.*

Construct a sequence $x_1, x_2, \ldots$ of distinct vertices as follows: Given the partial sequence $x_1, x_2, \ldots, x_i$, find the vertex x_j whose index j is as large as possible such that

(1) $$1 \leqslant j \leqslant i ,$$

(2) $$\Gamma_G(x_j) \not\subset \{ x_1, x_2, \ldots, x_i \} .$$

Then take $x_{i+1} \in \Gamma_G(x_j) - \{ x_1, \ldots, x_i \}$, and direct edge $[x_j, x_{i+1}]$ from x_j to x_{i+1}. Stop when all the vertices are in the sequence. Let H be the 1-graph formed by the edges directed by this procedure.

1. From Theorem 13, H is an arborescence with root x_1 because H is connected, and

$$d_H^-(x_1) = 0 ,$$

$$d_H^-(x_i) = 1 \quad \text{for} \quad i \neq 1 .$$

2. Let $[x_j, x_k]$, where $j < k$, be an edge of E that is not in H. This edge determines a cycle of the basis associated with H. We shall show that if this edge is directed from x_k to x_j, then the cycle becomes a circuit.

Since x_k is adjacent to x_j, all vertices x_i with $j \leqslant i \leqslant k$, are in the subarborescence of H rooted at x_j. Therefore, there exists in H a path $\mu[x_j, x_k]$, and by adding arc (x_k, x_j), a circuit is obtained.

3. If in arborescence H, there is a path from x to y, we write $x < y$. Clearly, the relation $<$ is transitive. Let

$$\mu = [a_1, a_2, ..., a_q = a_1]$$

be an elementary circuit of G_0. See Fig. 3.3. Since H is an arborescence,

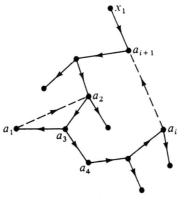

Fig. 3.3

there exists at least one arc in μ, say (a_1, a_2), that belongs to $G_0 - H$. If the circuit μ is not one of the circuits associated with the tree H, then it contains another arc of $G_0 - H$. Let (a_i, a_{i+1}) be the first such arc to occur in μ. Since μ is an elementary circuit, $a_{i+1} < a_2$. Moreover, the vertices a_{i+2}, $a_{i+3}, ..., a_q$ cannot be $> a_2$ (because μ would pass through a_2 twice and would not be elementary). This contradicts that:

$$a_q = a_1 > a_2 .$$

Q.E.D.

Remark. The last part of this proof yields easily a result of Chaty [1966]: *A strongly connected graph G_0 has exactly $v(G_0)$ elementary circuits, if, and only if, there exists in G_0 a spanning tree H such that the elementary cycles associated with H are circuits.*

4. Injective, functional and semi-functional graphs

The concept of an arborescence can be generalized in the following way: A graph G is said to be *injective* if $d_G^-(x) \leqslant 1$ for all vertices x. If G is injective, then G is a 1-graph and may be written as $G = (X, \Gamma)$. Then the correspondence Γ is an *injective correspondence*, i.e.

$$x \neq y \quad \Rightarrow \quad \Gamma(x) \cap \Gamma(y) = \varnothing .$$

A graph G is said to be a *functional* if $d_G^+(x) \leqslant 1$ for each vertex x. If G is a functional, then $G = (X, \Gamma)$ is a 1-graph, and the correspondence Γ is a *function* φ defined on X. For example, if X is a set of states, and if $\varphi(x)$ denotes the unique state that follows x in a deterministic process, then the pair (X, φ) is a functional graph.

Finally, a 1-graph $G = (X, \Gamma)$ is said to be a *semi-functional* if

$$\Gamma(x) \cap \Gamma(y) \neq \varnothing \quad \Rightarrow \quad \Gamma(x) = \Gamma(y) .$$

A functional graph is semi-functional. An injective graph $G = (X, \Gamma)$ is semi-functional, because

$$\Gamma(x) \cap \Gamma(y) \neq \varnothing \quad \Rightarrow \quad x = y \quad \Rightarrow \quad \Gamma(x) = \Gamma(y) .$$

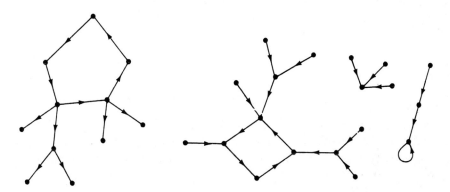

Fig. 3.4. Injective graph   Fig. 3.5. Functional graph

Property 1. *A 1-graph $G = (X, \Gamma)$ is functional if, and only if, its inverse $H = (X, \Gamma^{-1})$ is injective.*

The proof follows since $d_G^+(x) = d_H^-(x)$.

Property 2. *A 1-graph $G = (X, \Gamma)$ is semi-functional if, and only if, its inverse $H = (X, \Gamma^{-1})$ is semi-functional.*

Suppose that G is semi-functional; let y and y' be two vertices of G such that $\Gamma^{-1}(y) \cap \Gamma^{-1}(y') \neq \varnothing$ and let $x_0 \in \Gamma^{-1}(y) \cap \Gamma^{-1}(y')$. We have

$$x \in \Gamma^{-1}(y) \quad \Rightarrow \quad y \in \Gamma(x) \quad \Rightarrow \quad \Gamma(x) \cap \Gamma(x_0) \neq \varnothing \quad \Rightarrow$$
$$\Rightarrow \quad \Gamma(x) = \Gamma(x_0) \quad \Rightarrow \quad y' \in \Gamma(x) \quad \Rightarrow \quad x \in \Gamma^{-1}(y') .$$

Therefore, $\Gamma^{-1}(y) \subset \Gamma^{-1}(y')$, and equality holds. Thus,

$$\Gamma^{-1}(y) \cap \Gamma^{-1}(y') \neq \varnothing \quad \Rightarrow \quad \Gamma^{-1}(y) = \Gamma^{-1}(y').$$

Hence, graph H is semi-functional.

Q.E.D.

If $G = (X, U)$ is a graph, its *adjoint* G^* is defined to be a 1-graph whose vertices $u_1, u_2, ..., u_m$ represent the arcs of G and which has an arc from u_i to u_j if the terminal endpoint of the arc in G corresponding to u_i is the initial endpoint of the arc corresponding to u_j. A path in G that uses all the arcs corresponds to a path in G^* that uses all the vertices. Thus one can relate properties on the arcs of G to properties on the vertices of G^*.

There are many characterizations of adjoints. (See Heuchenne [1964].) The simplest is given by the following theorem:

Theorem 15. *A 1-graph H is the adjoint of a graph if, and only if, H is semi-functional.*

1. If $H = (U, \Gamma)$ is the adjoint of a graph $G = (X, U)$, then H is semi-functional, since

$$\Gamma(u) \cap \Gamma(u') \neq \varnothing$$

implies that in G the terminal endpoints of arcs u and u' coincide, and consequently,

$$\Gamma(u) = \Gamma(u').$$

2. Let $H = (U, \Gamma)$ be a semi-functional graph, and consider the family $\mathscr{C}$ of subsets of U of the form

$$C(u) = \{ v/\Gamma(v) = \Gamma(u) \}, \text{ or } C_0 = \{ v/\Gamma(v) = \varnothing \}$$

The sets $C_0, C_1, C_2, ..., C_q$ of $\mathscr{C}$ form a partition of U.

Consider a partition $\mathscr{D}$ of U formed by the sets

$$D_i = \Gamma(C_i) \quad (i = 1, 2, ..., q)$$

$$D_{q+1} = \{ u / \Gamma_H^-(u) = \varnothing \}.$$

Construct a graph G with vertices $x_0, x_1, x_2, ..., x_{q+1}$ and k arcs from x_i to x_j denoted by $u_{i_1}, u_{i_2}, ..., u_{i_k}$ if

$$D_i \cap C_j = \{ u_{i_1}, u_{i_2}, ..., u_{i_k} \}.$$

Since $\mathscr{C}$ and $\mathscr{D}$ are two partitions of U, each vertex of H is represented by exactly one arc of G. Clearly, in G, the terminal endpoint of arc u coincides with the initial endpoint of arc v if, and only if, $v \in \Gamma(u)$.

Therefore, $H = G^*$.

Q.E.D.

Theorem 16. *An injective* 1-*graph* $G = (X, U)$ *is connected if, and only if, G is quasi-strongly connected.*

1. Recall that a graph G is quasi-strongly connected if for each pair $x, y \in X$, there is a vertex z whose set of descendants

$$\{ z \} \cup \Gamma(z) \cup \Gamma^2(z) \cup \cdots$$

contains both x and y. If G is quasi-strongly connected, it is clearly connected.

2. Let G be injective and connected; then each pair $x, y \in X$, where $x \neq y$, is joined by an elementary chain

$$\mu[x, y] = [x, a_1, a_2, ..., a_k, y] .$$

First, suppose that $(x, a_1) \in U$; then

$$(x, a_1) \in U \quad \Rightarrow \quad (a_k, y) \in U$$

because, otherwise, there exists an a_i with $d_G^-(a_i) > 1$. Thus the vertex $z(x, y) = x$ has both x and y for descendants.

Now suppose that $(y, a_k) \in U$; then

$$(y, a_k) \in U \quad \Rightarrow \quad (a_1, x) \in U .$$

Thus vertex $z(x, y) = y$ has both x and y for descendants.

Finally, suppose $(a_1, x) \in U$ and $(a_k, y) \in U$; then there exists at least one vertex a_i in the chain $\mu[x, y]$ such that

$$(a_i, a_{i-1}) \in U, \qquad (a_i, a_{i+1}) \in U .$$

Since G is injective, $\mu[a_i, x]$ and $\mu[a_i, y]$ are paths, and $z(x, y) = a_i$ has both x and y for descendants.

This shows that G is quasi-strongly connected.

Q.E.D.

Corollary. *A functional* 1-*graph* $G = (X, \Gamma)$ *is connected if, and only if, the inverse* 1-*graph* (X, Γ^{-1}) *is quasi-strongly connected.*

The proof is obvious.

Theorem 17. *A necessary and sufficient condition that the edges of a simple graph* $G = (X, E)$ *can be directed to form an injective (resp. functional) graph is that each connected component of G contains at most one cycle.*

1. *Necessity.* Let H be an injective graph; each cycle is a circuit (otherwise, there would be two arcs leaving the same vertex). No two distinct circuits can have a common vertex because then two arcs would enter the same vertex.

On the other hand no two circuits in the same connected component can be without a common vertex, because each arc incident to the circuit is directed out of the circuit (and there is no vertex z that has descendants in both circuits).

Thus there is at most one cycle in each connected component.

2. *Sufficiency.* Let $G = (X, E)$ be a simple graph with at most one cycle in each connected component. Direct its edges in the following way: If a connected component of G contains no cycles, it is a tree, and its arcs can be directed to form an arborescence. If a connected component of G has only one cycle, direct first the edges of this cycle so that it becomes a circuit. Then, by contracting into a single vertex x_0 all the vertices in the cycle, the connected component becomes a tree: direct the edges of this tree to form an arborescence with root x_0.

Clearly, the directed edges induce in G an injective graph.

Q.E.D.

Theorem 18. *Let φ be a mapping from a subset of X into X. Each connected component C_x of the functional graph $G = (X, \varphi)$ is the union of two connected components D_x and $D_{\varphi(x)}$ of the functional graph $H = (X, \varphi^2)$.*

Moreover, if $D_x = D_{\varphi(x)}$, then D_x is a connected component of H with a circuit of odd length.

Let C_x denote the connected component of G that contains x.

1. If $y \in C_x$, there exist from Theorem 16 integers p and q with $\varphi^{2q}(y) = \varphi^{2p}(x)$ or $\varphi^{2q}(y) = \varphi^{2p+1}(x)$. Let $g(x) = \varphi^2(x)$. We may write

$$\text{either} \quad g^q(y) = g^p(x) \quad \text{or} \quad g^q(y) = g^p[\varphi(x)] .$$

Hence

$$y \in D_x \cup D_{\varphi(x)} .$$

Conversely, $y \in D_x \cup D_{\varphi(x)}$ implies that $\varphi^p(x) = \varphi^q(y)$, and therefore $y \in C_x$.

2. If $D_x = D_{\varphi(x)}$, there exist integers p and q with

$$g^p(x) = g^q[\varphi(x)]$$

or, equivalently,

$$\varphi^{2p}(x) = \varphi^{2q+1}(x).$$

This proves the existence in G of two paths $\mu[x, z]$ and $v[x, z]$, one of which is odd and the other even. Hence, there exists a cycle of odd length. Since G is functional, this cycle is a circuit, and its vertices are the vertices of a circuit in graph $H = (X, \varphi^2)$.

Q.E.D.

APPLICATION (Rufus Isaacs). Find a real valued function $\varphi(x)$ on $\mathbb{R}$ such that $\varphi[\varphi(x)] = ax + b$, where $a, b \in \mathbb{R}$.

If $a > 0$, Menger found:

$$\varphi(x) = \frac{b}{1 + \sqrt{a}} + x\sqrt{a}.$$

This function φ is the required solution, since

$$\varphi[\varphi(x)] = \frac{b}{1 + \sqrt{a}} + \sqrt{a}\left(\frac{b}{1 + \sqrt{a}} + x\sqrt{a}\right) = ax + b.$$

If $a < 0$, the problem is more difficult. For example, take $a = -1$ and $b = 0$. From the preceding theorem, we know that a connected component of the graph $G = (X, \varphi)$ is the union of two distinct connected components of H, for example, D_x and D_{x+1} if $x \in [2k, 2k + 1]$ or D_x and D_{x-1} if $x \in [2k + 1, 2k + 2]$. This immediately gives the graph of the function φ (shown by the dotted lines in Fig. 3.6).

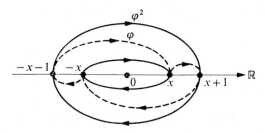

Fig. 3.6. Two connected components of the graph (X, φ^2)

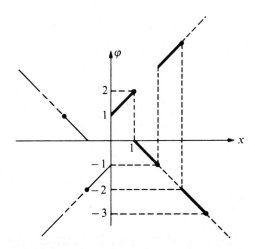

Fig. 3.7. Function φ such that $\varphi^2(x) = -x$

5. Counting trees

Before presenting results about the number of different trees in a graph, we shall state some properties of multinomial coefficients:

$$\binom{n}{n_1, n_2, \ldots, n_p}.$$

By convention, these coefficients will be equal to 0 if we do not have $n_1, n_2, \ldots, n_p \geq 0$ and $n_1 + n_2 + \cdots + n_p = n$.

Proposition 1. *Let X be a set of n distinct objects. Let $n_1, n_2, \ldots, n_p$ be non-negative integers such that $n_1 + n_2 + \cdots + n_p = n$. The number of ways to place the n objects into p boxes $X_1, X_2, \ldots, X_p$, containing $n_1, n_2, \ldots, n_p$ objects respectively, is*

$$\binom{n}{n_1, n_2, \ldots, n_p} = \frac{n!}{n_1! \, n_2! \ldots n_p!}.$$

The set X_1 can be chosen in $\binom{n}{n_1}$ different ways. Suppose the set X_1 is chosen, then the set X_2 can be chosen in $\binom{n - n_1}{n_2}$ different ways, etc. Hence, the required number is

$$\binom{n}{n_1}\binom{n - n_1}{n_2}\binom{n - n_1 - n_2}{n_3} \cdots \binom{n_p}{n_p} =$$

$$= \frac{n!}{n_1! \, (n - n_1)!} \frac{(n - n_1)!}{n_2! \, (n - n_1 - n_2)!} \frac{(n - n_1 - n_2)!}{n_3! \, (n - n_1 - n_2 - n_3!)} \cdots \frac{n_p!}{n_p!}$$

$$= \frac{n!}{n_1! \, n_2! \, n_3! \ldots n_p!}.$$

Proposition 2 (multinomial formula). *Given p real numbers*

$$a_1, a_2, \ldots, a_p \in \mathbb{R},$$

we have

$$(a_1 + a_2 + \cdots + a_p)^n = \sum_{n_1, n_2, \ldots, n_p \geq 0} \binom{n}{n_1, n_2, \ldots, n_p} (a_1)^{n_1} (a_2)^{n_2} \ldots (a_p)^{n_p}.$$

Consider real variables a_j^i, where $1 \leq i \leq n$, $1 \leq j \leq p$, and form the product

$$(a_1^1 + a_2^1 + \cdots + a_p^1)(a_1^2 + a_2^2 + \cdots + a_p^2) \ldots (a_1^n + a_2^n + \cdots + a_p^n).$$

Given the integers $n_1, n_2, \ldots, n_p$ whose sum is n, consider in the above polynomial a monomial of the form

$$(a_1^{i_1} a_1^{i_2} \ldots a_1^{i_{n_1}}) (a_2^{j_1} a_2^{j_2} \ldots a_2^{j_{n_2}}) \ldots (a_p^{k_1} a_p^{k_2} \ldots a_p^{k_{n_p}}).$$

This monomial corresponds uniquely to an arrangement of the set $N = \{1, 2, \ldots, n\}$ into the boxes $N_1, N_2, \ldots, N_p$, where

$$| N_1 | = n_1, | N_2 | = n_2, \ldots, | N_p | = n_p.$$

By proposition 1, the total number of such monomials is therefore

$$\binom{n}{n_1, n_2, \ldots, n_p} = \frac{n!}{n_1! \, n_2! \ldots n_p!}.$$

If we put

$$a_i^1 = a_i^2 = \cdots = a_i^n = a_i$$

for all i, we obtain the desired formula.

Q.E.D.

Proposition 3.

$$\binom{n}{n_1, n_2, \ldots, n_p} = \sum_{i/n_i \geqslant 1} \binom{n-1}{n_1, n_2, \ldots, n_{i-1}, n_i - 1, n_{i+1}, \ldots, n_p}.$$

Clearly,

$$(a_1 + a_2 + \cdots + a_p)^n = (a_1 + a_2 + \cdots + a_p)(a_1 + a_2 + \cdots + a_p)^{n-1},$$

and the general term of this polynomial is

$$\binom{n}{n_1, n_2, \ldots, n_p} a_1^{n_1} a_2^{n_2} \ldots a_p^{n_p} =$$

$$= \sum_{i/n_i \neq 0} a_i \binom{n-1}{n_1, \ldots, n_i - 1, \ldots, n_p} a_1^{n_1} \ldots a_i^{n_i - 1} \ldots a_p^{n_p}.$$

Q.E.D.

We are now ready to consider the problem of counting the number of ways to choose a set $E \subset \mathscr{P}_2(X)$ such that the simple graph (X, E) is a tree.

Theorem 19. *Let $T(n; d_1, d_2, \ldots, d_n)$ denote the number of distinct trees H with vertices $x_1, x_2, \ldots, x_n$ and with degrees $d_H(x_1) = d_1$, $d_H(x_2) = d_2, \ldots,$ $d_H(x_n) = d_n$. Then*

$$T(n; d_1, d_2, \ldots, d_n) = \binom{n-2}{d_1 - 1, d_2 - 1, \ldots, d_n - 1}.$$

1. Clearly, the sum of the degrees is twice the number of edges. From Theorem 1, the sum of the degrees for a tree is $2(n-1)$. Thus, $T \neq 0$ only if

$$\sum_{i=1}^{n} (d_i - 1) = 2(n - 1) - n = n - 2.$$

Without loss of generality, we may suppose that $d_1 \geq d_2 \geq \cdots \geq d_n$; since the above equality implies that $d_n = 1$, vertex x_n is pendant in the tree.

2. We shall show that

$$T(n \, ; d_1, d_2, ..., d_n) = \sum_{i/d_i \geq 2} T(n - 1 \, ; d_1, d_2, ..., d_i - 1, ..., d_{n-1}).$$

Let C_i be the set of the trees H with vertices $x_1, x_2, ..., x_n$ and degrees $d_H(x_k) = d_k$, such that the pendant vertex x_n is joined to x_i. If $d_i \geq 2$, then

$$|\mathscr{C}_i| = T(n - 1 \, ; d_1, d_2, ..., d_i - 1, ..., d_{n-1}).$$

Since the set of all trees is the union of the sets $\mathscr{C}_i$ for $d_i \geq 2$, the above equality follows.

3. The theorem is true for $n = 2$. Assume that $n \geq 3$ and that the theorem is true for $n - 1$. Then

$$T(n \, ; d_1, d_2, ..., d_n) = \sum_{i/d_i \geq 2} T(n - 1 \, ; d_1, d_2, ..., d_i - 1, ..., d_{n-1}) =$$

$$= \sum_{i/d_i \geq 2} \binom{n - 3}{d_1 - 1, d_2 - 1, ..., d_i - 2, ..., d_{n-1} - 1} =$$

$$= \binom{n - 2}{d_1 - 1, d_2 - 1, ..., d_{n-1} - 1} =$$

$$= \binom{n - 2}{d_1 - 1, d_2 - 1, ..., d_n - 1}.$$

Q.E.D.

Corollary 1 (Cayley [1897]). *The number of different trees with vertices* $x_1, x_2, ..., x_n$ *is* n^{n-2}.

Using Proposition 2, the number of trees equals

$$\sum_{d_1, ..., d_n \geq 1} \binom{n - 2}{d_1 - 1, d_2 - 1, ..., d_n - 1} = (1 + 1 + \cdots + 1)^{n-2} = n^{n-2}.$$

Corollary 2 (Clarke [1958]). *The number of different trees* H *with vertices* $x_1, x_2, ..., x_n$ *and with* $d_H(x_1) = k$ *is*

$$\binom{n - 2}{k - 1} (n - 1)^{n-k-1}.$$

The desired number equals

$$\sum_{d_2,d_3,\dots,d_n} \binom{n-2}{k-1,\, d_2-1,\, d_3-1,\, \dots,\, d_n-1} =$$

$$= \frac{(n-2)!}{(k-1)!\,(n-k-1)!} \sum_{d_2,d_3,\dots,d_n \geq 1} \binom{n-k-1}{d_2-1,\, d_3-1,\, \dots,\, d_n-1} =$$

$$= \binom{n-2}{k-1}(n-1)^{n-k-1}$$

(by setting all variables equal to 1 in the multinomial formula).

Corollary 3 (Moon [1967]). *Let $G = (X, E)$ be a simple complete graph of order n. Let $(X_1, X_2, \dots, X_p)$ be a partition of X, and let*

$$H_1 = (X_1, E_1),\ H_2 = (X_2, E_2),\ \dots,\ H_p = (X_p, |E_p|),$$

be pairwise disjoint trees of orders $|X_i| = n_i$. The number of spanning trees of G that have $H_1, H_2, \dots, H_p$ as subgraphs is

$$T(H_1, H_2, \dots, H_p) = n_1 n_2 \dots n_p\, n^{p-2}.$$

If each set X_i were contracted to a unique vertex a_i, then, the number of trees $\bar{H}$ with

$$d_{\bar{H}}(a_i) = d_i \qquad (i = 1, 2, \dots, p)$$

is

$$\binom{p-2}{d_1-1,\, d_2-1,\, \dots,\, d_p-1}.$$

To each tree $\bar{H}$ correspond exactly $(n_1)^{d_1} (n_2)^{d_2} \dots (n_p)^{d_p}$ different spanning trees H of graph G. Hence

$$T(H_1, H_2, \dots, H_p) =$$

$$= \sum_{d_1,d_2\dots,d_p \geq 1} \binom{p-2}{d_1-1,\, d_2-1,\, \dots,\, d_p-1}(n_1)^{d_1}(n_2)^{d_2}\dots(n_p)^{d_p} =$$

$$= n_1 n_2 \dots n_p(n_1 + n_2 + \dots + n_p)^{p-2}.$$

Q.E.D.

Corollary 4 (Cayley [1889]). *The number of forests with vertices $x_1, x_2, \dots, x_n$ and with p connected components such that $x_1, x_2, \dots, x_p$ belong to p different trees is*

$$T'(n\,;\,p) = p n^{n-p-1}.$$

Let C be the set of trees H on the vertices $x_0, x_1, x_2, ..., x_n$ such that $d_H(x_0) = p$. From Corollary 2,

$$| \mathscr{C} | = \binom{n-1}{p-1} n^{n-p}.$$

If $P \subset \{1, 2, ..., n\}$ and $|P| = p$, let $\mathscr{C}_p$ denote the set of trees in $\mathscr{C}$ such that, for all $i \in P$, the vertex x_i is joined to x_0. Then

$$| \mathscr{C} | = \sum_P | \mathscr{C}_P | = \binom{n}{p} T'(n\,;\,p).$$

Hence

$$\binom{n-1}{p-1} n^{n-p} = \binom{n}{p} T'(n\,;\,p).$$

Therefore,

$$T'(n\,;\,p) = \frac{(n-1)!}{(p-1)!\,(n-p)!} \cdot \frac{p!\,(n-p)!}{n!}\, n^{n-p} = p n^{n-p-1}.$$

Q.E.D

Let $X = \{x_1, x_2, ..., x_n\}$ be a set of n vertices, and let $E \subset \mathscr{P}_2(x)$ be a set of q edges that join pairs of vertices in X. We now propose to calculate the number $T(X, E)$ of different trees on X that do not contain any edge of E. Let (X, F) be a graph with n vertices, q edges and p connected components with, respectively, $n_1, n_2, ..., n_p$ vertices. Let

$$v(F) = \begin{cases} 0 & \text{if graph } (X, F) \text{ contains a cycle,} \\ n_1\, n_2\, ...\, n_p & \text{otherwise.} \end{cases}$$

Theorem 20 (Temperley [1964]). *The number of different trees on set X that do not contain any edge in E is*

$$T(X, E) = n^{n-2} \sum_{F \subset E} v(F) \left(\frac{-1}{n}\right)^{|F|}.$$

If $e \in E$, let A_e denote the set of all trees that contain edge e. Let $F \subset E$. If (X, F) has no cycles and has p connected components, Corollary 3 to Theorem 19 states that the number of different trees that contain all the edges of F is

$$\left| \bigcap_{e \in F} A_e \right| = v(F)\, n^{p-2} = v(F)\, n^{n-|F|-2}.$$

If (X, F) contains a cycle, the above formula is still valid, since both sides of the equality are 0. From Sylvester's Formula, we have

$$T(X, E) = n^{n-2} + \sum_{\substack{F \subset E \\ F \neq \emptyset}} (-1)^{|F|} v(F) n^{n-2-|F|} = n^{n-2} \sum_{F \subset E} v(F) \left(\frac{-1}{n}\right)^{|F|} .$$

Q.E.D.

Corollary 1 (Weinberg [1958]). *If E is a set of q pairwise disjoint edges, then*

$$T(X, E) = n^{n-2}\left(1 - \frac{2}{n}\right)^q .$$

In this case, if $F \subseteq E$, then

$$v(F) = 2^{|F|} ,$$

and

$$T(X, E) = n^{n-2} \sum_{k=0}^{q} 2^k \binom{q}{k} \left(\frac{-1}{n}\right)^k = n^{n-2}\left(1 - \frac{2}{n}\right)^q .$$

Q.E.D.

Corollary 2 (O'Neil [1963]). *If E is a set of q edges, all with a common endpoint x_1, then*

$$T(X, E) = n^{n-2}\left(1 - \frac{1}{n}\right)^{q-1}\left(1 - \frac{q+1}{n}\right) .$$

The proof follows, since we have

$$\sum_{F \subset E} v(F) \left(\frac{-1}{n}\right)^{|F|} = \sum_{k=0}^{q} (k+1)\binom{q}{k}\left(\frac{-1}{n}\right)^k =$$

$$= \sum_{k=0}^{q} \binom{q}{k}\left(\frac{-1}{n}\right)^k + \sum_{k-1=0}^{q-1} \left(\frac{-q}{n}\right)\binom{q-1}{k-1}\left(\frac{-1}{n}\right)^{k-1} =$$

$$= \left(1 - \frac{1}{n}\right)^q - \frac{q}{n}\left(1 - \frac{1}{n}\right)^{q-1} =$$

$$= \left(1 - \frac{1}{n}\right)^{q-1}\left(1 - \frac{1}{n} - \frac{q}{n}\right) .$$

Q.E.D.

Corollary 3 *Let $S \subset X$ with $|S| = s$. If E is the set of edges that join all the possible pairs of vertices in S (i.e. (S, E) is a complete graph), then*

$$T(X, E) = n^{n-2}\left(1 - \frac{s}{n}\right)^{s-1} .$$

Let $\mathscr{T}_p$ denote the family of subsets $F \subset E$ such that (S, F) is acyclic and has p connected components. Then

$$\sum_{F \subset E} v(F) \left(\frac{-1}{n} \right)^{|F|} = \sum_{p=1}^{s} \left(\frac{-1}{n} \right)^{s-p} \sum_{F \in \mathscr{F}_p} v(F) .$$

For $P \subset S$, $|P| = p$, and $F \in \mathscr{T}_p$, consider the triples (S, F, P) such that the graph (S, F) has a vertex of P in each connected component. From Theorem 19, Corollary 4,

$$\left| \{ (S, F, P) / F \in \mathscr{F}_p \} \right| = p s^{s-p-1} .$$

Therefore

$$\sum_{F \in \mathscr{F}_p} v(F) = \sum_{V \in \mathscr{F}_p} \left| \{ (S, F, P) / P \subset S, |P| = p \} \right| =$$

$$= \left| \{ (S, F, P) / P \subset S, |P| = p , F \in \mathscr{F}_p \} \right| =$$

$$= \sum_{\substack{P \subset S \\ |P| = p}} p s^{s-p-1} = \binom{s}{p} p s^{s-p-1} ,$$

Consequently,

$$\sum v(F) \left(\frac{-1}{n} \right)^{|F|} = \sum_{p=1}^{s} \left(\frac{-1}{n} \right)^{s-p} \binom{s}{p} p s^{s-p-1}$$

$$= \sum_{p-1=0}^{s-1} \left(\frac{-s}{n} \right)^{s-p} \binom{s-1}{p-1} = \left(1 - \frac{s}{n} \right)^{s-1} .$$

Q.E.D.

Corollary 4 (Scoin [1962]). *If graph (X, E) is the union of two disjoint complete graphs (S, V) and (T, W) with $|S| = s$ and $|T| = t$, then*

$$T(X, E) = s^{t-1} t^{s-1} .$$

From Theorem 19, it follows that

$$\frac{T(X, V \cup W)}{n^{n-2}} = \frac{T(X, V)}{n^{n-2}} \cdot \frac{T(X, W)}{n^{n-2}} .$$

Therefore, by Corollary 3,

$$T(X, E) = n^{n-2} \left(1 - \frac{s}{n} \right)^{s-1} \left(1 - \frac{t}{n} \right)^{t-1} =$$

$$= (s + t)^{s+t-2} \left(\frac{s+t-s}{s+t} \right)^{s-1} \left(\frac{s+t-t}{s+t} \right)^{t-1} = s^{t-1} t^{s-1} .$$

Q.E.D.

Corollary 5 (Moon [1967]). *If E is a set of $m - 1$ edges that forms an elementary chain on a set Y of m vertices, then*

$$T(X, E) = n^{n-2} \sum_{p=1}^{m} \binom{m + p - 1}{m - p} \left(\frac{-1}{n}\right)^{m-p}.$$

If $F \subset E$ determines a graph (Y, F) with p connected components, then

$$|F| = |Y| - p = m - p.$$

If $m_1, m_2, \ldots, m_p$ are respectively the numbers of vertices of these connected components, then $m_1 + m_2 + \cdots + m_p = m$. For $|F| = m - p$, there are as many graphs (Y, F) as there are ways to choose positive integers $m_1, m_2, \ldots, m_p$ that sum to m. Thus

$$\sum_{F \subset E} v(F) \left(\frac{-1}{n}\right)^{|F|} = \sum_{p=1}^{m} \left(\frac{-1}{n}\right)^{m-p} \sum_{\substack{|F| = m - p \\ F \subset E}} v(F) =$$

$$= \sum_{p=1}^{m} \left(\frac{-1}{n}\right)^{m-p} \sum_{\substack{m_1, m_2, \ldots > 0 \\ m_1 + m_2 + \ldots + m_p = m}} m_1\, m_2 \ldots m_p.$$

The last summation equals the coefficient of x^m in the expansion of

$$(x + 2\, x^2 + 3\, x^3 + \cdots)^p = x^p (1 - x)^{-2p}.$$

From the binomial formula, this coefficient equals

$$(-1)^{m-p} \frac{-2\, p(-2\, p - 1)(-2\, p - 2) \ldots (-2\, p - (m - p - 1))}{(m - p)!} =$$

$$= \frac{(m + p - 1)(m + p - 2) \ldots (2\, p + 1)\, 2\, p}{(m - p)!} = \binom{m + p - 1}{m - p}.$$

$$\text{Q.E.D.}$$

We now turn to the problem of counting the partial subgraphs of a given graph G that are arborescences.

Let $A = ((a_j^i))$ be the matrix associated with graph G, where

$$a_j^i = m_G^+(x_i, x_j)$$

denotes the number of arcs in G from x_i to x_j. Let $D = ((d_j^i))$ be the diagonal matrix defined by

$$d_j^i \begin{cases} = 0 & \text{if} \quad i \neq j \\ = m_G^+(X - \{x_i\}, x_i) & \text{if} \quad i = j. \end{cases}$$

The matrix $D - A = ((d_j^i - a_j^i))$ can be written as

$$D - A = \begin{pmatrix} \sum_{i \neq 1} a_1^i & - a_2^1 & \cdots & - a_n^1 \\ - a_1^2 & \sum_{i \neq 2} a_2^i & \cdots & - a_n^2 \\ - a_1^3 & - a_2^3 & \cdots & - a_n^3 \\ \vdots & \vdots & & \vdots \\ - a_1^n & - a_2^n & \cdots & \sum_{i \neq n} a_n^i \end{pmatrix}.$$

The determinant of this matrix equals 0 because the sum in each row is 0. The minor obtained by removing the first row and first column of this matrix is denoted by

$$\Delta_1 = \begin{vmatrix} \sum_{i \neq 2} a_2^i & - a_3^2 & - a_4^2 & \cdots & - a_n^2 \\ - a_2^3 & \sum_{i \neq 3} a_3^i & - a_4^3 & \cdots & - a_n^3 \\ \cdots\cdots & \cdots\cdots & \cdots\cdots & & \cdots\cdots \\ - a_2^n & - a_3^n & - a_4^n & \cdots & \sum_{i \neq n} a_n^i \end{vmatrix}.$$

Lemma. *Let $G = (X, U)$ be a graph with $m = n - 1$ arcs and no loops. Then G is an arborescence with root x_1 if, and only if, $\Delta_1 = 1$. Otherwise $\Delta_1 = 0$.*

1. If G is an arborescence with root x_1, then d_i^i is equal to $d_G^-(x_i) = 1$ for $i = 2, 3, ..., n$. Index the vertices so that the indices increase along any path (this is possible since G is an arborescence). Then

$$\Delta_1 = \begin{vmatrix} 1 & - a_3^2 & - a_4^2 & \cdots & - a_n^2 \\ 0 & 1 & - a_4^3 & \cdots & - a_n^3 \\ 0 & 0 & 1 & \cdots & - a_n^4 \\ \cdots\cdots & \cdots\cdots & \cdots\cdots & & \cdots\cdots \\ 0 & 0 & 0 & \cdots & 1 \end{vmatrix} = 1.$$

2. We shall now show by induction on n that if G is a graph with n vertices, $m = n - 1$ arcs and with $\Delta_1 \neq 0$, then G is an arborescence with root x_1.

Each vertex x_k, where $k \neq 1$, is the terminal endpoint of at least one arc of G (because, otherwise, the k-th column vector of Δ_1 is the zero vector and $\Delta_1 = 0$). Since $m = n - 1$, the inner demi-degrees (see Chapter 1) satisfy the equalities:

(i)
$$\begin{cases} d_G^-(x_1) = 0 \\ d_G^-(x_k) = 1 \quad (k = 2, 3, \ldots, n) . \end{cases}$$

Hence, by Property (6) of Theorem 13, it is sufficient to show that G is connected in order to show that G is an arborescence. Suppose that G is not connected. Then Δ_1 can be decomposed into two square matrices B' and B'' in the following way:

$$\Delta_1 = \begin{array}{c} \\ S\{ \\ T\{ \end{array} \begin{array}{c} \overbrace{\quad}^{S} \quad \overbrace{\quad}^{T} \\ \left| \begin{array}{cc} B' & 0 \\ 0 & B'' \end{array} \right| \end{array} = \mathrm{Det}\,(B')\mathrm{Det}\,(B'') .$$

From equation (i), we see that the vertices x_s with $s \in S$ and the vertex x_1 generate a subgraph G' with $m' = n' - 1$. The vertices x_t with $t \in T$ and the vertex x_1 generate a subgraph G'' with $m'' = n'' - 1$. Since $\mathrm{Det}\,(B') \neq 0$ and $\mathrm{Det}\,(B'') \neq 0$, we know by the induction hypothesis that G' and G'' are arborescences with root x_1. This contradicts the assumption that G is not connected.

Q.E.D.

Theorem 21 (Tutte [1948]). *Let $G = (X, U)$ be a graph, and let $x_1 \in X$. The number of partial subgraphs of G that are arborescences with root x_1 equals a_2^i*

$$\Delta_1 = \left| \begin{array}{cccc} \displaystyle\sum_{i \neq 2} a_2^i & - a_3^2 & \cdots & - a_n^2 \\ - a_2^3 & \displaystyle\sum_{i \neq 3} a_3^i & \cdots & - a_n^3 \\ \cdots\cdots\cdots\cdots\cdots\cdots\cdots\cdots\cdots \\ - a_2^n & - a_3^n & \cdots & \displaystyle\sum_{i \neq n} a_n^i \end{array} \right| .$$

where $a_j^i = m_G^+(x_i, x_j)$.

Without loss of generality, we may assume that G has no loops. Note that $\Delta_1 = \Delta_1(\mathbf{a}_2, \mathbf{a}_3, \ldots, \mathbf{a}_n)$ is a linear function of the last $n - 1$ column vectors of the matrix $((a_j^i))$, i.e.,

$$\Delta_1(\mathbf{a}_2' + \mathbf{a}_2'', \mathbf{a}_3, \ldots, \mathbf{a}_n) = \Delta_1(\mathbf{a}_2', \mathbf{a}_3, \ldots, \mathbf{a}_n) + \Delta_1(\mathbf{a}_2'', \mathbf{a}_3, \ldots, \mathbf{a}_n)$$

$$\Delta_1(\lambda\mathbf{a}_2, \mathbf{a}_3, \ldots, \mathbf{a}_n) = \lambda\Delta_1(\mathbf{a}_2, \mathbf{a}_3, \ldots, \mathbf{a}_n) .$$

Denote by $\mathbf{e}_k = (0, 0, \ldots, 1, 0, \ldots, 0)$ the n-vector with its k-th coordinate equal to 1 and with all other $n - 1$ coordinates equal to 0. Then

$$\Delta_1(\mathbf{a}_2, \mathbf{a}_3, \ldots, \mathbf{a}_n) = \Delta_1\left(\sum_{k_2 \neq 2} a_2^{k_2} \mathbf{e}_{k_2}, \sum_{k_3 \neq 3} a_3^{k_3} \mathbf{e}_{k_3}, \ldots, \sum_{k_n \neq n} a_n^{k_n} \mathbf{e}_{k_n}\right) =$$

$$= \sum_{k_2, k_3, \ldots, k_n} a_2^{k_2} a_3^{k_3} \ldots a_n^{k_n} \Delta_1(\mathbf{e}_{k_2}, \mathbf{e}_{k_3}, \ldots, \mathbf{e}_{k_n}).$$

From the lemma, $\Delta_1(\mathbf{e}_{k_2}, \mathbf{e}_{k_3}, \ldots, \mathbf{e}_{k_n})$ equals 0 or 1. For each term in the sum, the partial graph defined by

$$\mathbf{a}_2 = \mathbf{e}_{k_2}, \quad \mathbf{a}_3 = \mathbf{e}_{k_3}, \ldots, \mathbf{a}_n = \mathbf{e}_{k_n},$$

has no loops (because G has no loops) and is formed from the $n - 1$ arcs

$$(x_{k_2}, x_2), (x_{k_3}, x_3), \ldots, (x_{k_n}, x_n).$$

Finally, from the lemma, we know that $\Delta_1(\mathbf{e}_{k_2}, \mathbf{e}_{k_3}, \ldots, \mathbf{e}_{k_n}) = 1$ if, and only if, the graph is an arborescence with root x_1. This completes the proof.

$$\text{Q.E.D.}$$

Corollary. *If $G = (X, E)$ is a simple graph, the number of spanning trees in G is equal to the minor (which is independent of the coefficients of the principal diagonal) of the square matrix $((b_j^i))$ of order n, where*

$$b_j^i \begin{cases} = d_G(x_i) & \text{if} \quad i = j \\ = -1 & \text{if} \quad i \neq j \quad \text{and} \quad [x_i, x_j] \in E \\ = 0 & \text{if} \quad i \neq j \quad \text{and} \quad [x_i, x_j] \notin E. \end{cases}$$

Let G^* be the graph obtained from G by replacing each edge by two oppositely directed arcs. A spanning tree of G corresponds uniquely to an arborescence of G^* rooted at x_1 (say). Therefore, by Theorem 21, the number of spanning trees in G equals

$$\Delta_1 = \begin{vmatrix} d_G(x_2) & b_3^2 & & b_n^2 \\ b_2^3 & d_G(x_3). & & b_n^3 \\ \vdots & \vdots & \ddots & \vdots \\ b_2^n & b_3^n & \cdots & d_G(x_n) \end{vmatrix}.$$

$$\text{Q.E.D.}$$

EXAMPLE. Consider the graph G in Fig. 3.8. As shown in Fig. 3.9, the number of spanning trees in G equals

$$\Delta_1 = \begin{vmatrix} 3 & -1 & -1 \\ -1 & 3 & -1 \\ -1 & -1 & 3 \end{vmatrix} = 16.$$

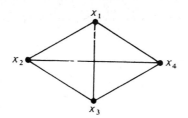

Fig. 3.8

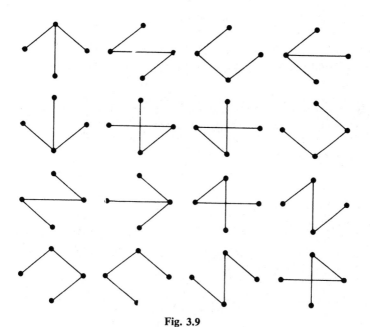

Fig. 3.9

EXERCISES

1. Let m and n be two integers.

(1) State a necessary and sufficient condition for the existence of a strongly connected 1-graph with n vertices and m edges.

(2) Show that, for all strongly connected 1-graphs with n vertices and m edges, the lower bound on the number of edges whose removal can destroy the strong connectivity is $\left[\dfrac{m}{n}\right]$.

2. Show that if $G = (X, E)$ is a simple graph such that each edge is contained in some elementary cycle, then the anti-symmetric graph G' constructed from G as in Theorem 14 is strongly connected; show that G' has only $m - n + 1 = \nu(G)$ elementary cycles, and that a graph without circuits can be constructed from G' by reversing the direction of a subset of its arcs with cardinality

$$\min \{ m - n + 1, n - 1 \}.$$

(Chaty [1968])

3. Show that the number of trees on n vertices that have exactly k pendant vertices is $\dfrac{n!}{k!} S_{n-2}^{n-k}$, where S_p^n denotes the Stirling number of the second kind.

(Rényi [1959])

4. Consider a set $X = \{x_1, x_2, \ldots, x_n\}$ and the group S_n of permutations on X. A set T of transpositions $[x_i, x_j]$ defines a simple graph (X, T). Show that:

(1) A set T of $n - 1$ transpositions generates the symmetric group S_n if and only if the graph (X, T) is a tree.

(2) (Denès) If f is a circular permutation of degree n, then the number of ways of writing f as a product of $n - 1$ transpositions equals n^{n-2}. (Use Corollary 2, Theorem 19; for a detailed proof, see C. Berge, *Principles of Combinatorics*, Academic Press, New York, 1971, p. 143.)

5. The following extension of Theorem 11 is due to Las Vergnas: If G is a minimally connected graph of order $n \geq 2$, then each circuit of G includes a cocycle that contains exactly two arcs. Also, each cycle that is not a circuit contains at least two such cocycles.

6. A graph G is called strongly k-connected if, for every pair of vertices x, y, there exist k disjoint paths from x to y.

Mader showed that Theorem 11 can be generalized as follows: If G is minimally strongly k-connected, then there exist two vertices x and y, such that

$$d^+(x) = d^+(y) = d^-(x) = d^-(y) = k.$$

7. Let G be a regular planar graph of degree 4. Use the corollary to Theorem 21 to count the number of crossing-free closed walks of G.

(Smith – Tutte)

CHAPTER 4

Paths, Centres and Diameters

1. The path problem

The *path problem* is the following: *Find (as quickly as possible) a path from a given vertex a to a given vertex b in a* 1-*graph* $G = (X, U)$. The *chain problem* is similarly defined for a simple graph $G = (X, E)$. Note that the chain problem becomes a path problem in the 1-graph $G^* = (X, U)$ obtained from $G = (X, E)$ by replacing each edge in E by two oppositely directed arcs.

EXAMPLE. Problem of the hunter, wolf and Brussels sprout. A hunter, a wolf and a Brussels sprout arrive simultaneously at a river bank. The ferry boat is too small to take more than one passenger (in addition to the boatman) at the same time. For obvious reasons, the boatman cannot leave the hunter and the wolf alone together, nor can he leave the wolf and the Brussels sprout alone together. How should he arrange their passage across the river?

This well-known problem can be solved mentally by considering only a small number of states. Nonetheless, it is a typical example of the path problem. A graph of the various states can be constructed, and a path must be found from state a (the hunter H, wolf W, Brussels sprout S and boatman B are all on the right bank) to state b (all are on the left bank). One solution to the problem is shown in Fig. 4.1.

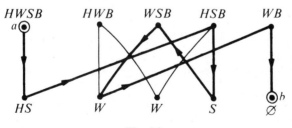

Fig. 4.1

For more complicated cases, several systematic algorithms have been proposed. If the graph is already known, it is always possible to find all the elementary paths starting from vertex a by constructing all the different arbo-

53

rescences rooted at a. Such an arborescence for the above example is shown
in Fig. 4.2.

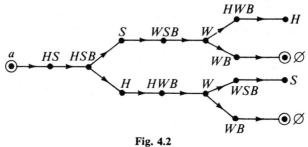

Fig. 4.2

By inspection we see that there are two paths from a to b.

A formal statement of an algorithm for finding all elementary paths from a
was given by B. Roy [1960], and others.

Of more interest are *local algorithms* applicable when the entire graph is not
known. Ideally, a local algorithm will not trace through the entire graph.

Let $G = (X, U)$ be a graph, and let $a \in X$. If $u_1, u_2, \ldots, u_m$ are the arcs of G,
define an alphabet whose letters are $+ u_1, + u_2, \ldots, + u_m, - u_1, - u_2, \ldots,$
$- u_m$ (positive and negative letters). A *word* is a sequence of letters, written
in the form

$$v_1 + v_2 + \cdots + v_i + v_{i+1} + \cdots + v_k.$$

If $v_{i+1} = - v_i$, the word

$$v_1 + v_2 + \cdots + v_{i-1} + v_{i+2} + \cdots + v_k$$

is called a *reduction* of the preceding word.

A word $\mu^k = v_1 + v_2 + \cdots + v_k$ is called the *trajectory* of graph G when:

(1) v_1 is an arc with a as its initial endpoint, and tracing through the
arcs corresponding to the consecutive letters of a word (in the direction of the
arc if the letter is positive, and in the opposite direction if the letter is negative)
defines a chain μ^k.

(2) If $u \in U$, then μ^k contains the letters $+ u$ and $- u$ at most one time
each.

(3) If the letter $- u$ is in the word, then the letter $+ u$ precedes $- u$ in the
sequence.

(4) After word μ^k has been reduced as much as possible, the reduced word
contains no more negative letters (and, therefore, the reduced word $\bar{\mu}^k$ defines
an elementary path starting at a).

The Trémaux Algorithm given below is a rapid, local algorithm to construct a trajectory terminating at vertex b.

Trémaux algorithm

We shall construct successively a sequence of the trajectories $\mu^1, \mu^2, ...$ Each time an arc is used or a vertex is encountered, it is labelled. The algorithm consists of three rules:

RULE 1. Label vertex a. If $u = (a, x)$ is an arc with initial endpoint a, put $\mu^1 = + u$. (In other words, we advance along arc u and label it.)

RULE 2. Let $\mu^i = v_1 + v_2 + \cdots + v_i$ be a trajectory terminating at x with $v_i > 0$. If vertex x has not been previously encountered, and if x is the initial endpoint of an arc $u_{i+1} = (x, y)$, let $\mu^{i+1} = \mu^i + u_{i+1}$ and label arc u_{i+1} and vertex y. Otherwise, let $\mu^{i+1} = \mu^i - v_i$, and label arc v_i again.

RULE 3. Let $\mu^i = v_1 + \cdots + v_i$ be a trajectory terminating at x with $v_i < 0$. If x is the initial endpoint of an arc $u_{i+1} = (x, y) \notin \mu^i$, let $\mu^{i+1} = \mu^i + u_{i+1}$. Otherwise, let $\mu^{i+1} = \mu^i - u_j$ where u_j is the last letter of the reduced word $\bar{\mu}^i$.

Trémaux's Theorem. *When the above algorithm terminates, the terminal trajectory μ^k has the property that for each arc u in a path starting from a, μ^k contains the letters $+ u$ and $- u$ exactly once.*

1. When the procedure ends, the trajectory has returned to a, and all the arcs incident out of a have been labelled (by Rule 3).

2. We shall show that all arcs u labelled by the procedure have been traversed in reverse. In fact, if, each time the procedure reaches a labelled vertex via an unlabelled arc, we detach the arc from its terminal endpoint before traversing it in reverse (Rule 2), the graph of labelled arcs would become an arborescence rooted at a (by Theorem 13, Ch. 3, Property (6)). Hence, if this arborescence is explored through the same trajectories, all its arcs will have been traversed in reverse.

3. We shall show that if there exists an elementary path from a, say

$$v = (u_1, u_2, ..., u_p) = [a, a_1, a_2, ..., a_p],$$

then the arc u_p has been labelled by the procedure.

If not, then vertex a_{p-1} has not been labelled (because, otherwise the arc preceding vertex a_{p-1} in the trajectory has not been traversed in reverse by either Rule 2 or Rule 3, which contradicts Part 2 of this proof). Ther' .ore, arc u_{p-1} has not been labelled. For the same reason, arc u_{p-2} has not been labelled, and arc u_1 has not been labelled. But this contradicts Part 1 of this proof.

Q.E.D.

Remark 1. With Trémaux's Algorithm, a path from a to b will be found by purely local methods, if such a path exists. In fact, the procedure never labels the same arc more than twice—once in each direction. This gives a bound on the number of steps in the algorithm.

Remark 2. If a *chain* between two vertices in graph $G = (X, U)$ is sought, the algorithm can be applied to the symmetric graph G^* constructed from G by replacing each edge in G by two oppositely directed arcs. To avoid the possibility of using an edge of G four times, we can, after labelling an arc in G for the first time, remove the oppositely directed arc from the graph. This can also simplify the procedure.

Remark 3. For a graph $G = (X, U)$, it is easy to see that the Trémaux Algorithm described above can be used to construct a maximal arborescence rooted at a given vertex. This arborescence is defined by the set of all labelled arcs which have been traversed in reverse by using only Rule 2.

Algorithm (P. Rosenstiehl, J. C. Bermond)

P. Rosenstiehl [1966] noted that the Trémaux Algorithm is a special case of a class of algorithms described by Tarry [1895]. He also devised another labelling procedure, called the *algorithm for new arcs*, which is described below.

We shall successively construct the trajectories μ^1, μ^2, ... by the following rules:

RULE 1. If $u = (a, x)$ is an arc whose initial endpoint is a, let $\mu^1 = + u$. In other words, advance along arc u and label it.

RULE 2. Let $\mu^i = v_1 + \cdots + v_i$ be a trajectory with x as terminal vertex. If x is the initial endpoint of an unlabelled arc $u_{i+1} = (x, y)$, let $\mu^{i+1} = \mu^i + u_{i+1}$. Label the arc u_{i+1}. If there does not exist an unlabelled arc with initial endpoint x, and if the reduced word $\bar{\mu}_i$ is not null, let $\mu^{i+1} = \mu^i - u$, where u is the last letter of the reduced word.

Theorem. *When the above algorithm terminates, the terminal trajectory μ^k has the property that for each arc u in a path from a, μ^k contains the letters $+ u$ and $- u$ exactly once.*

(The proof is identical to the proof of the Trémaux theorem.)

In other words, a path from a to b, if it exists, will be found by purely local methods and without labelling the same arc more than twice. Note that if G is an arborescence rooted at a, the algorithm for new arcs and Trémaux's Algorithm are equivalent.

Algorithm for a planar graph.

If a chain is sought between two vertices a and b in a planar graph G, and if both a and b are on the unbounded face, we have the *maze problem*. A traveller, who tries to get out of the maze by following the passageways, can find a chain between a and b by the following rule:

At a junction, always take the passageways on the extreme right.

This algorithm always locates the exit to the labyrinth without traversing any passageway more than twice.

2. The shortest path problem

The *shortest path problem* is the following:

Consider a graph G, and for each arc u, a number $l(u) \geqslant 0$, called the length of u. Find an elementary path from a to b that minimizes

$$l(\mu) = \sum_{u \in \mu} l(u)$$

EXAMPLE. Find the shortest route on a map from city a to city b. To do this, construct a graph G by representing each road between two localities on the map by oppositely directed arcs in the graph; let the road's mileage correspond to the arc's length. Thus, this example reduces to a shortest path problem. (Similarly, we might search for the fastest or most economical journey.)

Shortest path algorithm (Dantzig [1960])

Let the starting vertex be denoted by a_1; we shall determine, for each vertex x, the length $t(x)$ of the shortest path from a_1 to x by the following rules:

(1) To begin with, let $t(a_1) = 0$. Function t is therefore defined on the set $A_1 = \{a_1\}$.

(2) Suppose that on the k-th step, the function t has been defined on the set $A_k = \{a_1, a_2, ..., a_k\}$. For each vertex $a_j \in A_k$, select a vertex $b_j \in X - A_k$ such that $(a_j, b_j) \in U$ and such that the length $l(a_j, b_j)$ is minimum. Find a vertex $a_q \in A_k$ such that

$$t(a_q) + l(a_q, b_q) = \min_j \left\{ t(a_j) + l(a_j, b_j) \right\} .$$

Then, put

$$A_{k+1} = A_k \cup \{ b_q \}, \cdot$$
$$t(b_q) = t(a_q) + l(a_q, b_q) .$$

It remains to show that $t(b_q)$ is the length of the shortest path to b_q, and that this path passes through a_q. Assume that $t(a_j)$ is the length of the shortest path to a_j, for all $a_j \in A_k$. All paths that leave A_k have a length $\geq t(a_q) + l(a_q, b_q) = t(b_q)$. To reach b_q one must surely pass through A_k, since $a_1 \in A_k$. Therefore $t(b_q)$ is the length of the shortest path from a to b.

Q.E.D.

Note that each time a vertex b_q is selected, all the arcs with terminal endpoint b_q may be deleted.

Remark. Suppose that the vertex b_j that is associated with $a_j \in A_k$ can be found without difficulty. Then the selection of the $(k + 1)$-th vertex requires only k comparisons. The maximum number of comparisons needed for a graph with n vertices is

$$1 + 2 + \cdots + (n - 1) = \frac{1}{2} n(n - 1).$$

This bound can be improved if we search simultaneously for the length of the shortest path from b_i to b.

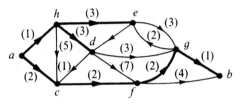

Fig. 4.3

EXAMPLE. Consider the graph in Fig. 4.3. The number next to each arc u is its length $l(u)$.

1. $t(a) = 0$. Compare $ah = (0 + 1)$ and $ac = (0 + 2)$. Choose ah.
2. $t(h) = 1$. Compare $he = (1 + 3)$, $hd = (1 + 3)$ and $ac = (0 + 2)$. Choose ac.
3. $t(c) = 2$. Compare $he = (1 + 3)$, $hd = (1 + 3)$ and $cf = (2 + 2)$. Choose he, hd and cf.
4. $t(e) = t(d) = t(f) = 4$. Compare $eg = (4 + 3)$, $dg = (4 + 3)$ and $fg = (4 + 2)$. Choose fg.
5. $t(g) = 6$. Compare $gb = (6 + 1)$ and $fb = (4 + 4)$. Choose gb.
6. $t(b) = 7$.

Thus a shortest path from a to b is $acfgb$.

The next theorem follows immediately from the above algorithm:

Theorem 1. *If $G = (X, U)$ is a graph with root a, and with a length $l(u) \geq 0$ for all $u \in U$, then there exists in G a spanning tree $H = (X, U)$ which is an* *arborescence with root a, and such that each path in H is a shortest path from a in G.*

Clearly, the different arcs chosen by the algorithm form an arborescence H (since H has no circuits and is connected).

Analogous algorithm for symmetric graphs

For a symmetric graph, a simple algorithm to find a shortest path is to represent the arcs of the graph by strings of the appropriate length and to let the vertices be represented by knots that tie the arcs together. To find a shortest path between knots a and b, simply pull knots a and b apart. The taut strings between a and b will represent the shortest path.

3. Centres and radii of quasi-strongly connected graphs

Consider a 1-graph $G = (X, U)$ and two vertices x and y of G. The *directed distance* $d(x, y)$ is defined to be the length of the shortest path from x to y. (If no such path exists, let $d(x, y) = \infty$.) The *associated number* $e(x)$ of a vertex x is defined to be

$$e(x) = \max_{\substack{y \in X \\ y \neq x}} d(x, y).$$

A traveller at vertex x can reach any other vertex in $e(x)$ or less steps.

A *centre* of G is defined to be a vertex x_0 with the smallest associated number. The associated number $e(x_0)$ of vertex x_0 is called the *radius* of G, and is denoted by $\rho(G)$.

These concepts are important in telecommunications. A communication network may be represented by a graph (not necessarily symmetric), and a centre of the graph represents an optimal site for a transmitting station.

Proposition 1. *The directed distances $d(x, y)$ satisfy*

(1) $$d(x, x) = 0$$

(2) $$d(x, y) + d(y, z) \geqslant d(x, z).$$

If the graph is symmetric, we have also

(3) $$d(x, y) = d(y, x).$$

The proof is immediate. If the graph is symmetric, the function $d(x, y)$ satisfies (1), (2), and (3) and is a *distance* in the topological sense.

Recall (Ch. 3, § 3) that a "root" of a graph is a vertex x_0 such that for each vertex y, there is a path from x_0 to y. Also, recall that a graph has a root if, and only if, it is quasi-strongly connected.

Clearly, a centre has a finite associated number if, and only if the graph has a root, and therefore, if, and only if, the graph is quasi-strongly connected. Henceforth, we shall assume that graph G is quasi-strongly connected.

Proposition 2. *If x_0 and y_0 are two centres of graph G, then they both belong to the same strongly connected component.*

The proof is obvious.

Theorem 2. *If G is a 1-graph of order n, without loops, and such that*

$$\max d_G^+(x) = p > 1,$$

then its radius $\rho(G)$ satisfies

$$\rho(G) \geqslant \frac{\log(np - n + 1)}{\log p} - 1.$$

If $\rho(G) = +\infty$, the theorem is obvious. Suppose that $\rho(G) < \infty$. Let x_0 be a centre of G. The number of vertices with a directed distance of 1 from x_0 is $\leqslant p$. The number of vertices with a directed distance of 2 from x_0 is $\leqslant p^2$. Thus

$$n \leqslant 1 + p + p^2 + \cdots + p^\rho = \frac{p^{\rho+1} - 1}{p - 1},$$

or

$$n(p - 1) + 1 \leqslant p^{\rho+1},$$

and hence,

$$\log(np - n + 1) \leqslant (\rho + 1) \log p.$$

This gives the formula.

$$\text{Q.E.D.}$$

Lemma. *Let $G = (X, U)$ be a strongly connected graph of order n, and let a be a root of G. Consider a spanning arborescence $H = (X, V)$ of G with root a such that each path of H is a shortest path of G. Let B denote the set of all terminal vertices in arborescence H. Then*

$$|B| \, e(a) \geqslant n - 1.$$

Equality holds if, and only if, H consists of $|B|$ paths of length $e(a)$ starting from a without common vertices (except a).

The vertices of H (except a) can be placed on $e(a)$ horizontal lines such that a vertex x with $d(a, x) = i$ is placed on the i-th horizontal line.

If $b \in B$, let $X(a, b)$ denote the set of all vertices (except vertex a) in the path of H that goes from a to b. Then,

$$n - 1 = \left| \bigcup_{b \in B} X(a, b) \right| \leqslant \sum_{b \in B} \left| X(a, b) \right| \leqslant |B| \, e(a) \, .$$

Q.E.D.

Theorem 3 (Goldberg [1965]). *If G is a strongly connected* 1-*graph with n vertices and m arcs, then*

$$\rho(G) \geqslant \left[\frac{n - 1}{m - n + 1} \right]^*$$

where $[r]^*$ *denotes the smallest integer* $\geqslant r$. *For all m and all n, there exists a strongly connected* 1-*graph with n vertices and m arcs such that the above equation holds with equality.*

1. Let a be a centre of G, and let H be a spanning arborescence with root a as defined in the lemma.

Since G is strongly connected, each pendant vertex $b \in B$ in arborescence H is the initial endpoint of an arc in $G - H$. Since the number of arcs in $G - H$ is $m(G - H) = m(G) - m(H) = m - n + 1$, by Theorem (1, Ch. 3), we have

$$|B| \leqslant m - n + 1 \, .$$

Then, from the lemma,

$$\rho(G) = e(a) \geqslant \frac{n - 1}{|B|} \geqslant \frac{n - 1}{m - n + 1} \, .$$

This yields the required inequality.

2. We shall construct a strongly connected graph G with n vertices and m arcs, with radius

$$\rho(G) = \left[\frac{n - 1}{m - n + 1} \right]^* .$$

This graph G will, in fact, be a rosace with centre a (see Fig. 4.4). It consists of $m - n + 1$ circuits (the "branches" of the rosace) having a common vertex a. Let

$$\left[\frac{n - 1}{m - n + 1} \right]^* = q \, .$$

Thus, we can write:

$$n - 1 = (m - n + 1)(q - 1) + r \, ; \qquad 0 < r \leqslant m - n + 1 \, .$$

Distribute the $n - 1$ vertices (other than a) among the branches of the rosace by placing either q or $q - 1$ vertices on each branch. One branch will have at least q different vertices other than a, since $r > 0$. Hence, G is strongly connected with n vertices, and $\rho(G) = q$.

Q.E.D.

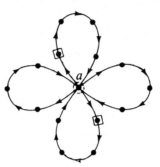

Fig. 4.4. Rosace with centre a,
$n = 15$, $m = 18$, $\rho = 4$.

We shall now study the properties of the centres of a complete 1-graph.

Theorem 4. *If* $G = (X, \Gamma)$ *is a complete* 1-*graph, each vertex* x_0 *such that*

$$\left| \Gamma(x_0) - \{ x_0 \} \right| = \max_{x \in X} \left| \Gamma(x) - \{ x \} \right|$$

is a centre, and $\rho(G) \leqslant 2$.

If

$$\max_x \left| \Gamma(x) - \{ x \} \right| = | X | - 1 \, ,$$

the theorem is true, and $\rho(G) = 1$.

Otherwise, consider a vertex x_0 for which $\left| \Gamma(x_0) - \{ x_0 \} \right|$ is maximum. Since $e(x) \geqslant 2$ for all x, we have only to show that $e(x_0) = 2$.

Suppose that $e(x_0) > 2$. Then there exists a vertex $y \neq x_0$ that cannot be

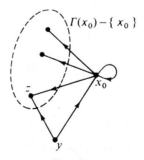

Fig. 4.5

reached from x_0 by any path of length 1 or length 2. Since $y \notin \Gamma(x_0)$, then $x_0 \in \Gamma(y)$. Moreover, if z is a vertex of $\Gamma(x_0) - \{x_0\}$, then $y \notin \Gamma(z)$, since, otherwise, there exists a path of length 2 from x_0 to y. Then $z \in \Gamma(y)$, and $z \in \Gamma(y) - \{y\}$. Hence,

$$\Gamma(x_0) - \{x_0\} \subset \Gamma(y) - \{y\}.$$

Since $x_0 \in \Gamma(y) - \{y\}$, the above inclusion is strict, and consequently,

$$\left| \Gamma(y) - \{y\} \right| > \left| \Gamma(x_0) - \{x_0\} \right| = \max_{x \in X} \left| \Gamma(x) - \{x\} \right|,$$

which is clearly a contradiction. Q.E.D.

Theorem 5 (Maghout [1962]). *If $G = (X, U)$ is a complete 1-graph with radius 2, then for each $y \in X$ there exists a centre x_0 such that $(x_0, y) \in U$.*

Let $y \in X$. Since the radius of the graph is greater than 1, there exists a vertex $x_1 \neq y$ such that

$$(y, x_1) \notin U.$$

Hence, $(x_1, y) \in U$.

If x_1 is a centre, the theorem follows. Otherwise, there exists a vertex $x_2 \neq x_1, y$, with $(x_1, x_2) \notin U$ and $(y, x_2) \notin U$. Hence,

$$(x_2, x_1) \in U, \quad (x_2, y) \in U.$$

If x_2 is a centre, the theorem follows; otherwise, there exists a vertex $x_3 \neq x_1, x_2, y$ such that

$$(x_2, x_3) \notin U, \quad (x_1, x_3) \notin U, \quad (y, x_3) \notin U.$$

Hence

$$(x_3, x_2) \in U, \quad (x_3, x_1) \in U, \quad (x_3, y) \in U.$$

If x_3 is a centre, the theorem follows; otherwise, there exists a vertex x_4, etc.

At least one vertex x_k located by this procedure is a centre. Otherwise,

$$\Gamma_G^-(y) = X - \{y\}.$$

Therefore a centre of the subgraph generated by $X - \{y\}$ is also a centre of graph G, and by Theorem 4, a centre would have been located during the procedure.

Thus, this vertex x_k is the required centre. Q.E.D.

Corollary. *A complete 1-graph G with radius 2 has at least 3 centres.*

Let G' be a complete, anti-symmetric graph obtained from G by removing some of its arcs. This new graph G' has a centre, y_1 and, by Theorem 5, there exists another centre y_2 with $(y_2, y_1) \in U$. Also, there exists a centre y_3 with $(y_3, y_2) \in U$. Since the graph G' is anti-symmetric,

$$y_3 \neq y_1 .$$

Thus, G has at least three distinct centres y_1, y_2 and y_3.

Q.E.D.

4. Diameter of a strongly connected graph

The *diameter* $\delta(G)$ of a graph G is the maximum of the directed distances, i.e.:

$$\delta(G) = \max_{\substack{x, y \in X \\ x \neq y}} d(x, y) ,$$

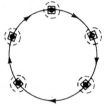

$m = 5, \rho = 4, \delta = 4.$

$m = 9, \rho = 2, \delta = 4.$

$m = 6, \rho = 2, \delta = 4.$

Rosace
$m = 8, \rho = 1, \delta = 2.$

Rosace
$m = 6, \rho = 3, \delta = 4.$

Rosace
$m = 7, \rho = 2, \delta = 3.$

$m = 6, \rho = 3, \delta = 4.$

$m = 7, \rho = 2, \delta = 4.$

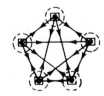

$m = 20, \rho = 1, \delta = 1.$

Fig. 4.6

The diameter is finite if, and only if, G is strongly connected. In this section, we shall assume that G is always strongly connected. Also, without loss of generality, we may assume in this section that G is a 1-graph without loops.

EXAMPLE. In Fig. 4.6, each graph has 5 vertices and is strongly connected. The number of arcs is denoted by m, the radius by ρ, and the diameter by δ. A circle is drawn around each centre. A square is drawn around each vertex x such that $e(x) = \delta(G)$.

Note that for $n = 5$, the equation $\delta + m \geqslant 9$ is always satisfied. Only the complete symmetric graph has a diameter equal to 1.

Since the graph is finite, then clearly, $\delta < \infty$, and if the graph has a centre, $\delta \geqslant \rho$. In a graph representing the avenues of communication between the various members of an organization, the diameter δ represents the maximum number of times that a message must be relayed before it reaches its destination.

The problem of constructing strongly connected graphs with n vertices and m arcs whose diameter is as large as possible (or as small as possible) was considered by Bratton [1955]. The problem of maximum diameter graphs has been solved by Ghouila-Houri [1960], and the problem of minimum diameter graphs has been solved by Goldberg [1966].

Before proceeding, note that if G is a strongly connected graph without loops and with n vertices and m arcs, the numbers n and m cannot be chosen arbitrarily: If $n > 1$, G has at least one cycle, hence, the cyclomatic number $v(G) = m - n + 1 \geqslant 1$, and consequently, we have

$$m \geqslant n .$$

From Theorem (9, Ch. 3), we know that equality holds only if G is an elementary circuit.

Furthermore, since the number of arcs in G cannot be greater than the number of arcs in a symmetric complete graph with n vertices, we have

$$m \leqslant n(n - 1) .$$

Theorem 6 (Goldberg [1966]). *If G is a strongly connected 1-graph without loops and with n vertices and m arcs, and if G is not an elementary circuit, then*

$$\delta(G) \geqslant \left[\frac{2(n - 1)}{m - n + 1} \right]^{*} .$$

Furthermore, this is the best possible result.

If G is not an elementary circuit, then, from Theorem (9, Ch. 3),

$$m - n + 1 = v(G) \geqslant 2 .$$

Consider a strongly connected graph G with cyclomatic number $v(G) \geqslant 2$; we shall show first that

$$v(G)\,\delta(G) \geqslant 2\,(n - 1)\,.$$

A *trail* of G is defined to be any elementary path $\mu = [x_0, x_1, ..., x_s]$ such that for $i = 0, 1, ..., s - 1$, only one arc of G is incident out of x_i. Let x_0 denote the initial vertex of a longest trail of G, say:

$$\mu_0 = [x_0, ..., x_s]\,.$$

Let s be the length of μ_0. Construct an arborescence H rooted at x_0 as defined in the lemma to Theorem 3. Clearly, H begins with path μ_0, and its first node after x_0 is x_s. Let B denote the set of terminal vertices of arborescence H.

1. We shall show that $|\,B\,| + 1 \leqslant v(G)$.

The number of arcs in $G - H$ is $v(G)$. Since each vertex $b \in B$ is the initial endpoint of at least one arc of G (because G is strongly connected), we have $v(G) \geqslant |\,B\,|$. If $v(G) = |\,B\,|$, then each vertex of B is the initial endpoint of only one arc of G, and there exists a vertex $b_1 \in B$ such that $(b_1, x_0) \in U$. Clearly, since $|\,B\,| = v(G) \geqslant 2$, the vertex b_1 is not on $\mu_0 = [x_0, x_1, ..., x_s]$, and $[b_1, x_0, x_1, ..., x_s]$ would be a trail of length $s + 1$. This contradicts the maximality of μ_0. Hence, $v(G) > |\,B\,|$.

2. We shall show that $s \geqslant \frac{1}{2}\,\delta(G)$ implies $n - 1 \leqslant \frac{1}{2}\,\delta(G)\,v(G)$.

Let $v(G) = v$, and let $\delta(G) = \delta$. If we denote the associated number of a vertex x in the arborescence H by $e_H(x)$, then, from the lemma to Theorem 3,

$$n - 1 \leqslant s + e_H(x_s)\,|\,B\,| \leqslant s + (\delta - s)\,|\,B\,| =$$

$$= \delta\,|\,B\,| - s(|\,B\,| - 1) \leqslant \delta|\,B\,| - \frac{\delta}{2}(|\,B\,| - 1) =$$

$$= \frac{\delta}{2}(|\,B\,| + 1) \leqslant \frac{\delta}{2}\,v\,.$$

3. We shall show that $s < \dfrac{\delta}{2}$ implies

$$n - 1 \leqslant \frac{\delta v}{2}\,.$$

Let

$$X_1 = \{\,x_0, x_1, ..., x_s\,\}\,,$$
$$X_2 = \{\,x\,/\,x \in X,\quad e_H(x) \leqslant s - 1\,\}\,,$$
$$X_3 = X - (X_1 \cup X_2)\,.$$

Note that X_2 can be empty (if $s = 0$), and that $x_0 \notin X_2$. If $X_3 \neq \varnothing$, note that H_{X_3} is the union of arborescences; denote by B_3 the set of all their terminal vertices. The number of maximal paths in H_{X_3} is $|B_3|$, and the number of vertices encountered by such a path μ_3 is $n(\mu_3) \leqslant \delta - 2s$ (because $\delta \geqslant l(\mu[x_0, b]) \geqslant s + n(\mu_3) + s$). Hence,

$$n \leqslant |X_1| + |X_2| + |X_3| \leqslant (s+1) + s|B| + (\delta - 2s)|B_3|.$$

Let $\hat{\Gamma}_H(z)$ be the set of all the descendants of z in the arborescence H, and let $t(z)$ denote the number of arcs in $G - H$ with initial endpoint in $\hat{\Gamma}_H(z)$.

We shall show that $z_0 \in B_3$ implies $t(z) \geqslant 2$. If, for a vertex $z_0 \in B_3$, $t(z_0) = 1$, then z_0 is the initial endpoint of only one maximal path $\mu[z_0, b]$ in H; furthermore only one arc of $G - H$ is incident out of the set of vertices encountered by $\mu[z_0, b]$. This arc is of the form (b, c) with $c \in [z_0, b]$ (because, otherwise, C would contain a trail of length $s + 1$).

Hence, there is no path from z_0 to x_0, and the graph is not strongly connected, as there is no path from z_0 to x_0, which contradicts the hypothesis. Thus, $t(z) \geqslant 2$ for $z \in B_3$. Therefore

$$v \geqslant \sum_{z \in B_3} t(z) \geqslant 2|B_3|.$$

Finally, we have

$$n \leqslant s + 1 + s|B| + (\delta - 2s)|B_3| \leqslant s + 1 + s|B| + (\delta - 2s)\frac{v}{2}.$$

Hence,

$$n - 1 \leqslant s(1 + |B|) + (\delta - 2s)\frac{v}{2} \leqslant sv + (\delta - 2s)\frac{v}{2} = \frac{\delta v}{2}.$$

$m = 7$
$\delta = 5$

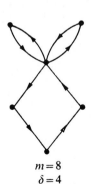

$m = 8$
$\delta = 4$

$m = 9$
$\delta = 3$

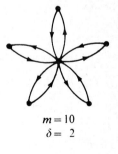

$m = 10$
$\delta = 2$

Fig. 4.7. Strongly connected graphs with

$$n = 6 \quad \text{and} \quad \delta = \left[\frac{2(n-1)}{m-n+1}\right]^* = \left[\frac{10}{m-5}\right]^*$$

4. Parts 2 and 3 of this proof have shown that if $v(G) \geqslant 2$, then

$$\delta(G) \, v(G) \geqslant 2 \, (n - 1),$$

and thus,

$$\delta(G) \geqslant \left[\frac{2(n - 1)}{m - n + 1} \right]^* .$$

Since it is not difficult to construct a rosace G for which equality holds (see Fig. 4.7), the proof is complete

Q.E.D.

We shall now discuss the problem of maximum possible diameter.

Lemma 1. *A necessary and sufficient condition that a strongly connected 1-graph $G = (X, U)$ have diameter $\delta(G) \geqslant p$ is that X can be partitioned into $p + 1$ classes $X_0, X_1, ..., X_p$ with:*

$$x \in X_i, \quad y \in X_j, \quad (x, y) \in U \quad \Rightarrow \quad j \leqslant i + 1 .$$

Sufficiency. If such a partition exists, a path from $x_0 \in X_0$ to $x_p \in X_p$ is of length $\geqslant p$, and, thus, $\delta(G) \geqslant p$.

Necessity. If $\delta(G) \geqslant p$, consider two vertices a and b with $d(a, b) \geqslant p$. The sets:

$$X_i = \{ x \mid x \in X, d(a, x) = i \} \quad (i = 0, 1, ..., p - 1),$$
$$X_p = \{ x \mid x \in X, d(a, x) \geqslant p \}$$

are non-empty and form the required partition.

Q.E.D.

Lemma 2. *Let $p \geqslant 2$. There exists a strongly connected 1-graph with n vertices and m arcs whose vertex set can be partitioned into $p + 1$ classes $X_0, X_1, ..., X_p$ as in Lemma 1, if, and only if,*

(1) $$m \leqslant \frac{n^2}{2} - n + \psi(n, p) ,$$

(2) $$m \geqslant n ,$$

(3) $$p \leqslant n - 1 ,$$

where $$\psi(n, p) = \frac{n^2}{2} - n(p - 2) + \frac{p^2}{2} - \frac{p}{2} - 2 .$$

Let X be a set of cardinality n, and let $(X_0, X_1, ..., X_p)$ be a partition of X in $p + 1 \geqslant 3$ classes. Let $\mid X_i \mid = n_i$. Clearly, $p \leqslant n - 1$. We shall show that for each m that satisfies (1) and (2), there exists a strongly connected 1-graph on X with m arcs that has $(X_0, X_1, ..., X_p)$ as a partition. First, construct a strongly connected graph by placing, between these n vertices, n arcs with the property of Lemma 1: for instance, take an elementary circuit that first encounters all the vertices of X_0, then all the vertices of X_1, etc.

Now successively add to this graph $m - n$ new arcs so that the property of Lemma 1 is preserved.

Clearly the maximum possible value for m equals:

$$n_0(n_0 - 1) + n_0 n_1 +$$
$$+ n_1 n_0 + n_1(n_1 - 1) + n_1 n_2 +$$
$$+ \cdots$$
$$+ n_p n_0 + n_p n_1 + \cdots + n_p n_{p-1} + n_p(n_p - 1) =$$
$$= -n + \frac{n^2}{2} + \frac{1}{2} \sum_{k=0}^{p} n_k^2 + n_0 n_1 + \cdots + n_{p-1} n_p.$$

For integers $n \geqslant 3$ and p with $2 \leqslant p \leqslant n - 1$, let

$$f(n_0, n_1, ..., n_p) = \frac{1}{2} \sum_{k=0}^{p} n_k^2 + (n_0 n_1 + \cdots + n_{p-1} n_p),$$

$$\psi(n, p) = \max_{\substack{n_0 + \cdots + n_p = n \\ n_i \geqslant 1}} f(n_0, n_1, ..., n_p).$$

If $p = 2$, then

$$f(n_0, n_1, n_2) = \frac{1}{2} (n_0 + n_1 + n_2)^2 - n_0 n_2,$$

and

$$\psi(n, p) = \frac{n^2}{2} - 1.$$

Let $p > 2$; let $(n_0, n_1, ..., n_p)$ be a $(p + 1)$-tuple that maximizes f. If $n_i > 1$ and $n_j > 1$ with $j - i > 1$, then the value of f can be increased either by taking $n_i' = n_i - 1$ and $n_j' = n_j + 1$, or by taking $n_i' = n_i + 1$ and $n_j' = n_j - 1$, since

$$f(n_0, ..., n_i \pm 1, ..., n_j \mp 1, ..., n_p) =$$
$$= f(n_0, ..., n_p) + 1 \pm (n_{i-1} + n_i + n_{i+1}) \mp (n_{j-1} + n_j + n_{j+1}).$$

It follows that there exists an index k, with $1 \leqslant k \leqslant p - 2$ such that

$$\left. \begin{array}{c} i \neq k \\ i \neq k + 1 \end{array} \right\} \Rightarrow n_i = 1.$$

Therefore,

$$\psi(n, p) = \frac{p-1}{2} + \frac{n_k^2 + n_{k+1}^2}{2} + p - 3 + n_k + n_k n_{k+1} + n_{k+1}$$

$$= \frac{p-1}{2} + p - 3 + \frac{1}{2}(n_k + n_{k+1} + 1)^2 - \frac{1}{2}.$$

By replacing $n_k + n_{k+1} + 1$ by $n - (p - 2)$, we obtain

$$\psi(n, p) = \frac{n^2}{2} - n(p - 2) + \frac{p^2}{2} - \frac{p}{2} - 2.$$

Since this equality is also true for $p = 2$, the theorem follows.

Q.E.D.

Theorem 7 (Ghouila-Houri [1960]). *The maximum value for the diameter of a strongly connected loopless 1-graph with n vertices and m arcs is*

$$\varphi\,(m, n) \begin{cases} = n - 1 & \text{if} \quad n \leqslant m \leqslant \dfrac{n^2 + n - 2}{2} \\[2ex] = \left[n + \dfrac{1}{2} - \sqrt{2m - n^2 - n + \dfrac{17}{4}} \right] \\[2ex] & \text{if} \quad \dfrac{n^2 + n - 2}{2} < m \leqslant n(n - 1). \end{cases}$$

Let G be a strongly connected loopless graph with n vertices and m arcs; then $m \geqslant n$ because each vertex is the initial endpoint of at least one arc. Furthermore, $m \leqslant n(n - 1)$, since the graph cannot have more arcs than a complete symmetric 1-graph. If $m = n(n - 1)$, the proof is immediate; therefore, we may assume $m < n(n - 1)$ and $m \geqslant n$.

If conditions (1) and (3) of Lemma 2 are fulfilled, then from Lemmas 1 and 2 there exists a strongly connected 1-graph with $\delta(G) \geqslant p$. Conversely, if conditions (1) and (3) are not satisfied, then no such graph exists. Thus, the greatest possible value for the diameter is the greatest value $\varphi(m, n)$ for an integer p satisfying (1) and (3).

Inequality (1) is equivalent to

$$\frac{p^2}{2} - \left(n + \frac{1}{2} \right) p + (n^2 + n - 2 - m) \geqslant 0.$$

Then, by elementary calculus, we obtain the above value $\varphi(m, n)$ for p.

Q.E.D.

For a simple graph $G = (X, E)$, we define the *undirected radius* $\rho^*(G)$ as the radius $\rho(G^*)$ of the graph $G^* = (X, U)$ obtained from G by replacing each edge in G by two oppositely directed arcs. Similarly, we define the *undirected diameter* $\delta^*(G)$ of a simple graph G by

$$\delta^*(G) = \delta(G^*) .$$

The following theorem is due to Camille Jordan [1869].

Theorem 8. *If G is a tree, and if $\delta^*(G)$ is even, then G has a unique centre, and all the elementary chains of maximum length pass through it; furthermore,*

$$\rho^*(G) = \frac{1}{2} \delta^*(G).$$

If $\delta^*(G)$ is odd, then G has exactly two centres (which are adjacent vertices) and all the elementary chains of maximum length pass through them; furthermore,

$$\rho^*(G) = \frac{1}{2} \left(\delta^*(G) + 1 \right) .$$

The theorem is obvious for trees of order $\leqslant 3$. Let $n > 3$, and suppose that the theorem is true for trees of order $< n$; we shall show that the theorem is true for a tree $G = (X, E)$ of order n.

Let B be the set of pendant vertices of tree G. From Theorem (2, Ch. 3), we have $| B | \geqslant 2$, and hence, the subgraph G_{X-B} is a tree of order $< n$. Clearly,

$$\rho^*(G_{X-B}) = \rho^*(G) - 1 ,$$
$$\delta^*(G_{X-B}) = \delta^*(G) - 2 .$$

Each centre of G_{X-B} is a centre of G, and vice versa. Each elementary chain of maximum length in G_{X-B} induces in graph G an elementary chain of length $\delta^*(G)$. Since the theorem is true for G_{X-B}, by the induction hypothesis, the result follows.

Q.E.D.

Remark. The two equalities of Theorem 8 can be summarized by:

$$\rho^*(G) = \left[\frac{1}{2} \left(\delta^*(G) + 1 \right) \right] ,$$

where $[r]$ denotes the greatest integer $\leqslant r$.

For simple graphs, several results regarding $\rho^*(G)$ and $\delta^*(G)$ have been obtained: Vizing [1967] has determined the maximum number of edges in a simple graph with $\rho^*(G) = r$. Murty [1968] has determined the minimum number of edges in a simple graph with $\delta^*(G) \leqslant k$ such that the diameter remains $\leqslant l$ after the elimination of any s vertices. The diameter of a planar 3-connected graph has been studied by M. Balinski [1966], B. Grünbaum [1967], Bollobas [1968], Bosák, Kotzig, Znám [1968].

5. Counting paths

This section presents results about the number of different paths between each pair of vertices x, y in a graph $G = (X, U)$. It will be necessary to use matrix products as defined in linear algebra. The principal result for path counting is:

Theorem 9. *Consider two graphs $G = (X, U)$ and $H = (X, V)$ with the same vertex set, and let $A = ((a_j^i))$ and $B = ((b_j^i))$ denote respectively their associated matrices. The matrix product AB corresponds to a graph $G \cdot H$ with vertex set X, and an arc from x to y for each distinct path from x to y composed of an arc of U followed by an arc of V. The graph $G \cdot H$ is called the composition product of G and H.*

The number of distinct paths of the form $[x_i, x_k, x_j]$ with $(x_i, x_k) \in U$, $(x_k, x_j) \in V$, equals $a_k^i b_j^k$. Thus, the total number of different paths from x_i to x_j formed from an arc of U followed by an arc of V is

$$\sum_{k=1}^{n} a_k^i b_j^k = \langle \mathbf{a}^i, \mathbf{b}_j \rangle ,$$

where $\langle \mathbf{a}^i, \mathbf{b}_j \rangle$ denotes the scalar product of the row vector $\mathbf{a}^i$ and the column vector $\mathbf{b}_j$, which is also the general coefficient of the matrix $A \cdot B$.

Q.E.D.

Corollary 1. *If G is a graph and A is its associated matrix, the general coefficient p_j^i of the matrix $P = A^k$ (the product of A with itself k times) equals the number of distinct paths of length k from x_i to x_j in G.*

The theorem is true for $k = 1$. Let $k > 1$, and suppose the theorem is true for the power $k - 1$. Then it is also true for the power k, since $A^k = A(A^{k-1})$ shows the number of paths of length $1 + (k - 1) = k$ from x_i to x_j (by Theorem 9).

Q.E.D.

Corollary 2. *A graph G possesses a path of length k if, and only if, $A^k \neq 0$. G possesses no circuits if, and only if, $A^k = 0$ for k sufficiently large.*

The proof follows immediately from Corollary 1.

APPLICATION. *The power index of a participant in a partial tournament.*

Let $G = (X, U)$ be a 1-graph of a partial tournament with n participants. Let arc $(x, y) \in U$ if participant x has beaten or had a draw with participant y. It is tempting to choose as the winner the most dominating participant, i.e., the vertex x whose outer demi-degree is as large as possible.

However, there is a good chance that this participant x has beaten a large number of very weak participants, and would lose to a participant y who has beaten only a few very strong players. If $p_j^i(k)$ is the general coefficient of matrix A^k, i.e., the number of paths of length k from x to x_i then a better estimation of the power of participant x_i is the sum

$$p^i(k) = p_1^i(k) + p_2^i(k) + \cdots + p_n^i(k) .$$

Consequently, the *power index* of participant x_i can be defined as

$$\pi^i = \lim_{k \to \infty} \frac{p^i(k)}{p^1(k) + p^2(k) + \cdots + p^n(k)} .$$

Note that by virtue of the Perron-Frobenius Theorem, this limit always exists.

EXERCISES

1. Let $\delta^*(G) = \sup d(x, y)$ be the diameter of a simple connected graph G. Show that the following conditions are equivalent:
  (1) Any two vertices of G are connected by at most one elementary chain of length $\leqslant \delta^*(G)$.
  (2) G has no cycles of length $\leqslant 2 \, \delta^*(G)$.
  (3) Either there exist no cycles in G, or the length of the shortest cycle of G is $2 \, \delta^*(G) + 1$.

2. Given two integers h and δ, the number n of vertices of a simple graph regular of degree h with diameter δ satisfies

$$n \leqslant 1 + h \sum_{i=1}^{\delta} (h - 1)^{i-1} .$$

A simple graph with n vertices, regular of degree h, with diameter δ, for which the equality holds is called a *Moore graph* of type (h, δ).

Show that the Petersen graph (Fig. 10.11) is a Moore graph of type (3, 2). Show that the Moore graphs of type $(h, 1)$ are the $(h + 1)$-cliques. Show that the cycles without chords of length $2 \, \delta + 1$ are Moore graphs of type $(2, \delta)$.

Show that there exists a Moore graph of type (7, 2) and that it is unique (A. Hoffman, R. Singleton [1960]). (Hoffman and Singleton have characterized the Moore graphs of types $(h, 2)$ and $(h, 3)$ with the exception of type (57, 2) for which no example is known.)

3. Show that a simple connected graph G satisfies the equivalent properties in Exercise 1 if, and only if, it is either a tree or a Moore graph.

(J. Bosák, A. Kotzig, S. Znám [1968])

4. Show that for a simple connected graph, two elementary chains of maximum length always have common vertices. Also show that if l_0 is the length of a longest elementary chain, then each vertex is the initial endpoint of an elementary chain of length $\geqslant [\frac{1}{2}(l_0 + 1)]$.

5. Show that a simple graph G of order $n > 1$ which is isomorphic to its complementary graph $\bar{G}$ has a diameter equal to 2 or 3. (H. Sachs [1962])

CHAPTER 5

Flow Problems

1. The maximum flow problem

Consider a graph G with arcs denoted by $1, 2, ..., m$, and consider a set of numbers $b_1, b_2, ..., b_m, c_1, c_2, ..., c_m$ in $\mathbb{Z}$ such that

$$- \infty \leqslant b_i \leqslant c_i \leqslant + \infty .$$

A *flow* in G is defined as a vector $\varphi = (\varphi_1, \varphi_2, ..., \varphi_m) \in \mathbb{Z}^m$ such that:

(1) $\varphi_i \in \mathbb{Z}$ for $i = 1, 2, ..., m$. (The integer φ_i is called an *arc flow*, and may be regarded as the number of vehicles travelling through arc i along its direction if $\varphi_i \geqslant 0$ or against its direction if $\varphi_i < 0$.)

(2) For each vertex x, the sum of the arc flows entering x equals the sum of the arc flows leaving x, i.e.,

$$\sum_{i \in \omega^-(x)} \varphi_i = \sum_{j \in \omega^+(x)} \varphi_j \qquad (x \in X) .$$

In other words, there is a conservation of arc flows at each vertex (Kirchhoff's Law).

We shall study the following problems:

The compatible flow problem. *Given a graph G with an interval $[b_i, c_i]$ associated with each arc i, find a flow φ such that*

$$b_i \leqslant \varphi_i \leqslant c_i \qquad (i = 1, 2, ..., m) .$$

c_i is called the *capacity of arc i*, and represents the maximum number of vehicles that can use the arc i along its direction.

The maximum flow problem. *Given a graph G with an interval $[b_i, c_i]$ associated with each arc i, find a flow φ such that*

(1) $b_i \leqslant \varphi_i \leqslant c_i \qquad (i = 1, 2, ..., m)$,

(2) *the arc flow φ_1 is as large as possible.*

Note that, for both of these problems, we may assume without loss of generality that G is a 1-graph.

74

The maximum flow problem occurs frequently with the following additional conditions:

(1) $b_i = 0$ for $i = 1, 2, ..., m$,

(2) $c_i > 0$ for all i, and $c_1 = +\infty$

(3) arc 1 is an arc joining a vertex b, called the *sink*, to a vertex a, called the *source*, where

$$\omega^-(a) = (1, 0, 0, ..., 0),$$
$$\omega^+(b) = (1, 0, 0, ..., 0).$$

(4) G is an anti-symmetric 1-graph.

In other words, only arc $i = 1$ enters vertex a, and only arc $i = 1$ leaves vertex b. This arc (b, a), generally omitted in the illustrations, is called the *return arc*. (This arc has no function other than the maintenance of the conservation of flow at vertices a and b.) The most important flow problem is to find the maximum number of vehicles that can be sent from a to b without violating the arc capacities.

A graph G with a capacity c_i associated with each arc i and which satisfies the conditions (1), (2), (3), (4) is called a *transportation network* and is often denoted by $R = (X, U, c(u))$.

Below are some examples that reduce to maximum flow problems in a transportation network.

EXAMPLE 1. *Maritime traffic.* Let the seaports $a_1, a_2, ..., a_p, b_1, b_2, ..., b_q$, be represented by vertices, and suppose that bananas are ready for shipment at ports $a_1, a_2, ..., a_p$ to ports $b_1, b_2, ..., b_q$. Let s_i denote the quantity available at a_i, and let d_j denote the quantity demanded at b_j.

Shipping routes can be represented by arcs of the form (a_i, b_j) with a capacity equal to the shipping capacity between the two seaports. Is it possible to satisfy all the demands? How should the bananas be shipped? To answer these questions, create a source a, and join a to each a_i by an arc with capacity

$$c(a, a_i) = s_i .$$

Next, create a sink b, and join each vertex b_j to b by an arc with capacity

$$c(b_j, b) = d_j .$$

A maximum flow for this transportation network yields the number of bananas to ship along each route in order to satisfy all demands, if this is possible.

EXAMPLE 2. *The battle of the Marne.* The towns $a_1, a_2, ..., a_n$ each have motor cars to be sent to town b. If there is a relay route from town a_i to town a_j, let c_{ij} denote the number of motor cars that can leave a_i for a_j each time period. Let t_{ij} denote the traverse time from a_i to a_j, let s_i denote the number

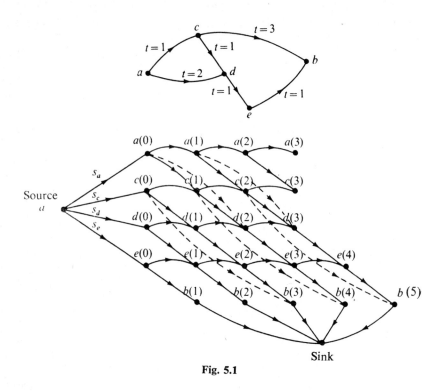

Fig. 5.1

of motor cars initially available at a_i, and let c_i denote the number of motor cars that can park at a_i' simultaneously. How can we direct the transport of these motor cars so that as many as possible arrive at b within T time periods? This type of problem has been studied by R. Fulkerson under the name of "dynamic networks". The problem can be reduced to a maximum flow problem in a transportation network with vertices $a_i(t)$, where $i = 1$, $2, \ldots, n$, and $t = 0, 1, \ldots, T$.

Vertex $a_i(t)$ and vertex $a_i(t + 1)$ are joined by an arc of capacity c_i. If there is a route from a_i to a_j, it is represented by an arc from $a_i(t)$ to $a_j(t + t_{ij})$ with a capacity of c_{ij}. Add a source $\bar{a}$, sink $\bar{b}$, arcs $(a, a_i(0))$ with capacity s_i and arcs $(b(t), b)$ with capacity ∞. The maximum flow in this transportation network determines the optimal routes.

EXAMPLE 3. *The selection of representatives.* A set X of residents belong to various clubs $C_1, C_2, ..., C_q$ (which are not necessarily disjoint subsets of X) and to various political parties $P_1, P_2, ..., P_r$ (which are disjoint subsets of X). Each club must choose one of its members to represent it, and no person can represent more than one club, no matter how many clubs he belongs to. How should one choose a system of distinct representatives $A = \{ a_1, a_2, ..., a_q \}$, such that the numbers of representatives belonging to each party P_j satisfies

$$b_j \leqslant | A \cap P_j | \leqslant c_j .$$

A solution to this problem is given by a maximum flow in the transportation network shown in Fig. 5.2 (for $q = 4$ and $r = 3$).

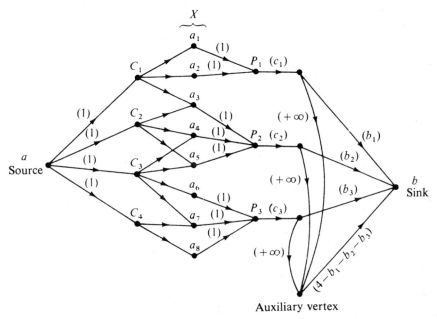

Fig. 5.2

The arc capacities are marked in parenthesis. It is left to the reader to verify that a maximum flow that saturates all source arcs determines a set of distinct representatives as required.

General maximum flow algorithm (when a compatible flow is known) (Ford and Fulkerson [1956]).

Consider an anti-symmetric 1-graph G with arc $1 = (b, a)$. Suppose that a compatible flow φ satisfying $b_i \leqslant \varphi_i \leqslant c_i$ is known. We shall augment successively the value of the arc flow φ_1 by the following labelling procedure:

RULE 1. Label vertex a, the terminal endpoint of the arc 1, with the index $+1$.

RULE 2. If x is labelled and y is unlabelled, label y with the index $+k$ if (x, y) is arc k, and if

$$\varphi(x, y) < c(x, y) \,.$$

RULE 3. If x is labelled and y is unlabelled, label y with the index $-k$ if (y, x) is arc k, and if

$$\varphi(y, x) > b(y, x) \,.$$

If sink b is labelled by this procedure, we shall show that the arc flow φ_1 can be augmented, i.e., we shall construct a new flow $\boldsymbol{\varphi}'$ such that $\varphi_1' > \varphi_1$.

Let

$$\mu = [a, a_1, a_2, \ldots, a_k, b]$$

be a chain from a to b in which each vertex a_{i+1} has been labelled from its predecessor a_i.

(1) If the edge $[a_i, a_{i+1}]$ is directed from a_i to a_{i+1}, then we have

$$\varphi(a_i, a_{i+1}) < c(a_i, a_{i+1}) \,,$$

Then, put

$$\varphi'(a_i, a_{i+1}) = \varphi(a_i, a_{i+1}) + 1 \,.$$

(2) If the edge $[a_j, a_{j+1}]$ is directed from a_{j+1} to a_j, then we have

$$\varphi(a_{j+1}, a_j) > b(a_{j+1}, a_j) \,.$$

Then, put

$$\varphi'(a_{j+1}, a_j) = \varphi(a_{j+1}, a_j) - 1 \,;$$

(3) For arc $1 = (b, a)$, put

$$\varphi'(b, a) = \varphi(b, a) + 1 = \varphi_1 + 1 \,.$$

(4) For all other arcs, put

$$\varphi'(x, y) = \varphi(x, y) \,.$$

In this way a new flow φ' is constructed. Flow φ' is compatible because only the flow around a cycle $\mu' = \mu + [b, a]$ has been changed, and at each vertex of this cycle, the conservation of flow has been maintained. In fact, the new flow can be expressed as the vectorial sum

$$\varphi' = \varphi + \mu' .$$

It remains to show that *if the algorithm cannot label sink b, then the arc flow φ_1 is maximum.*

Lemma 1. *In graph $G = (X, U)$, let $A \subset X$ be a set with $a \in A$ and $b \notin A$. Then, for each flow φ, such that $b_i \leqslant \varphi_i \leqslant c_i$,*

$$\varphi_1 \leqslant \sum_{i \in \omega^+(A)} c_i - \sum_{\substack{i \neq 1 \\ i \in \omega^-(A)}} b_i .$$

Since the algebraic sum of the flow entering set A equals the algebraic sum of the flow leaving set A, it follows that

$$\varphi_1 + \sum_{\substack{i \neq 1 \\ i \in \omega^-(A)}} \varphi_i = \sum_{i \in \omega^+(A)} \varphi_i$$

Hence,

$$\varphi_1 = \sum_{i \in \omega^+(A)} \varphi_i - \sum_{\substack{i \neq 1 \\ i \in \omega^-(A)}} \varphi_i \leqslant \sum_{i \in \omega^+(A)} c_i - \sum_{\substack{i \neq 1 \\ i \in \omega^-(A)}} b_i .$$

Q.E.D.

Lemma 2. *If the Ford and Fulkerson algorithm cannot label sink b, then φ_1 is maximum.*

Let A denote the set of vertices labelled by the algorithm. Clearly, $a \in A$ and $b \notin A$. Since no more labelling is possible, an arc (x, y) with $x \in A$ and $y \notin A$ satisfies $\varphi(x, y) = c(x, y)$, and an arc (y, x) with $y \notin A$ and $x \in A$ satisfies $\varphi(y, x) = b(y, x)$. Thus,

$$\varphi_1 = \sum_{i \in \omega^+(A)} \varphi_i - \sum_{\substack{i \neq 1 \\ i \in \omega^-(A)}} \varphi_i = \sum_{i \in \omega^+(A)} c_i - \sum_{\substack{i \neq 1 \\ i \in \omega^-(A)}} b_i .$$

From Lemma 1, it follows that φ_1 is maximum.

Q.E.D.

Theorem 1. (Ford, Fulkerson [1957]). *In a graph G with numbers b_i, c_i where $-\infty \leqslant b_i \leqslant c_i \leqslant +\infty$, the maximum value of a compatible flow in arc 1 is*

$$\max_{\varphi} \varphi_1 = \min_{\substack{A \ni a \\ A \not\ni b}} \left(\sum_{i \in \omega^+(A)} c_i - \sum_{\substack{i \neq 1 \\ i \in \omega^-(A)}} b_i \right) .$$

The proof follows from Lemma 1 and 2.

In a transportation network $R = (X, U, c(u))$, a *cut* between a and b is defined to be a set of arcs of the form $\omega^+(A)$ with $a \in A$ and $b \notin A$. The *capacity of a cut* is defined to be the sum of the capacities of its arcs, i.e.,

$$c(\omega^+(A)) = \sum_{i \in \omega^+(A)} c_i .$$

Maximum Flow Theorem (Ford, Fulkerson [1957]). *In a transportation network, the maximum value of the arc flow φ_1 equals the minimum capacity of a cut.*

EXAMPLE. Consider the transport network in Fig. 5.3.

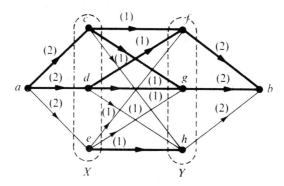

Fig. 5.3

The return arc (b, a) has been deleted to simplify the figure. The capacity of each source arc and sink arc is 2. The capacity of each intermediate arc (arcs that are neither source arcs nor sink arcs) is 1. A compatible flow φ is easily found, and the saturated arcs are indicated by heavy print. Its arc flows are

$\varphi(a, c) = \varphi(a, d) = 2$,

$\varphi(a, e) = \varphi(e, h) = \varphi(d, f) = \varphi(d, g) =$

$\qquad\qquad\qquad\qquad = \varphi(c, f) = \varphi(c, g) = \varphi(h, b) = 1$,

$\varphi(f, b) = \varphi(g, b) = 2$.

This flow is *maximal* in the sense that there are no paths from a to b composed entirely of unsaturated arcs. However, the flow is not *maximum* because the labelling algorithm can be used to construct a flow φ' with $\varphi_1 = \varphi'_1 + 1$. For example, the vertices can be labelled as follows:

$a(+1)$; $e(+ae)$; $f(+ef)$; $g(+eg)$; $c(-cf)$; $d(-dg)$;

$\qquad\qquad\qquad\qquad\qquad h(+ch)$; $b(+hb)$.

This determines the chain:

$$\mu[a, b] = + ae + ef - cf + ch + hb.$$

Let

$$\varphi'(a, e) = \varphi(a, e) + 1$$
$$\varphi'(e, f) = \varphi(e, f) + 1$$
$$\varphi'(c, f) = \varphi(c, f) - 1, \text{ etc.}$$

The reader can easily verify that the new flow φ' is maximum.

Note that for a network of this type, a maximum flow can be obtained directly by using the following rule: during the sequential flow augmentation, send an additional unit of flow toward the vertex that has the greatest capacity to receive flow. When this flow has been constructed, it will be maximum. However, this simplification is possible only for very special networks.

In fact, the Maximum Flow Theorem simplifies even further when the network R is *bipartite*, i.e., the vertices form four disjoint sets, X, Y, $\{a\}$, $\{b\}$ and the arcs are of the following types:

type 1. (x, y) with $x \in X$, $y \in Y$,

type 2. (a, x) with $x \in X$,

type 3. (y, b) with $y \in Y$,

type 4. (b, a), the return arc, denoted by 1.

To simplify, let

$$d(y) = \begin{cases} c(y, b) & \text{if } y \in Y \\ 0 & \text{otherwise,} \end{cases}$$

$d(y)$ is called the *demand* at vertex y. If $B \subset Y$, then the *demand of set B* is defined to be

$$d(B) = \sum_{y \in B} d(y).$$

If $B \subset Y$, let $F(B)$ denote the *maximum quantity of flow that can be sent into B*, i.e., the maximum flow for a network R' obtained from R by changing the capacities in the following way:

$$c'(y, b) = + \infty \qquad \text{if } y \in B$$
$$c'(y, b) = 0 \qquad \text{if } y \in Y - B$$
$$c'(x, y) = c(x, y) \qquad \text{for all other arcs } (x, y).$$

We now have the notation needed to state:

Theorem 2. *In a bipartite transportation network $R = (X, Y, U)$, the maximum value of a compatible flow in arc 1 is*

$$\varphi_1 = d(Y) + \min_{B \subset Y} \left[F(B) - d(B) \right].$$

1. Consider a set $B \subset Y$, and construct from R the network R' as described above. The Ford and Fulkerson Theorem states that

$$F(B) = \min_{\substack{S \ni b \\ S \not\ni a}} c'\left[\omega^-(S) \right].$$

We may restrict our attention to sets S containing B because, otherwise, $c'[\omega^-(S)] = +\infty$. Without changing in the right side of the above equation, S can be replaced by $S - \{b\}$. Since we are only interested in the minimum, we may also restrict our attention to sets S of the form $S = A \cup B$, where $A \subset X$. Thus

$$F(B) = \min_{A \subset X} c'\left[\omega^-(A \cup B) \right] = \min_{A \subset X} c\left[\omega^-(A \cup B) \right].$$

2. If P and Q are two disjoint sets of vertices, in order to simplify, let

$$c(P, Q) = \sum_{\substack{p \in P \\ q \in Q}} c(p, q).$$

Consider a set S with $b \in S$, $a \notin S$. Let

$$S \cap X = A,$$
$$S \cap Y = B.$$

Then,

$$c\left[\omega^-(S) \right] = c(a, A) + c(X - A, B) + c(Y - B, b)$$
$$= c(\omega^-(A \cup B)) + d(Y - B) = c(\omega^-(A \cup B)) - d(B) + d(Y).$$

From the Ford and Fulkerson theorem, the maximum value of a flow in arc 1 equals

$$\varphi_1 = \min_{\substack{S \ni b \\ S \not\ni a}} c\left[\omega^-(S) \right] = \min_{B \subset Y} \min_{A \subset X} \left[d(Y) + c(\omega^-(A \cup B)) - d(B) \right]$$

$$= d(Y) + \min_{B \subset Y} \left(F(B) - d(B) \right).$$

This proves the theorem.

$$\text{Q.E.D.}$$

Corollary (Gale [1958]). *A bipartite transportation network $R = (X, Y, U)$ has a compatible flow that saturates all the sink arcs if, and only if,*

$$F(B) \geqslant d(B) \qquad (B \subset Y).$$

In fact, this condition is equivalent to

$$\min_{B \subset Y} \big(F(B) - d(B)\big) = 0 \ .$$

Q.E.D.

Algorithm for planar networks

If the transportation network R is planar, the following algorithm may be used:

1. Place arc (b, a) horizontally in the plane, with vertex a at the left with vertex b at the right. Draw the graph above arc (b, a) such that the arcs do not cross one another. In this topological planar graph, find a *superior path* μ' from a to b by always choosing the left-most arc (and back tracing when an impasse occurs).

2. Reduce the capacity of each arc in the superior path μ^1 from a to b by the amount

$$\varepsilon_1 = \min_{i \in \mu^1} c_i \ .$$

Eliminate from the graph all arcs with zero capacity.

3. Similarly determine the superior path μ^2 in the new network. Reduce the capacity of each of its arcs by the amount

$$\varepsilon_2 = \min_{i \in \mu^2} c_i \ .$$

4. Repeat this process until there are no more paths from a to b. The required flow is obtained by sending ε_k units along each path μ^k.

To validate this algorithm, it is first necessary to establish the following lemma.

Lemma. *Let A be a set of vertices such that $a \in A$, $b \notin A$ and such that $\omega^+(A)$ is a minimal cut. The superior path μ^1 encounters $\omega^+(A)$ exactly once.*

Clearly, path μ^1 encounters the cut $\omega^+(A)$ at least once. Suppose that the path encounters the cut at two arcs, say i and j, in that order. We shall show that there exists a path μ from a to b that meets $\omega^+(A)$ only at arc i. Colour arc i red, the other arcs of $\omega^+(A)$ green, and colour all other arcs black. Since $\omega^+(A)$ is a minimal cut, arc 1 is not contained in any green and black cocycle in which all black arcs are similarly directed. By the Arc Colouring Lemma, (Ch. 2), it follows that arc 1 belongs to a red and black cycle with all black arcs having the same direction.

Thus, this cycle is a circuit, and it induces the required path μ.
Similarly, there exists a path v from a to b that meets $\omega^+(A)$ only at arc j.

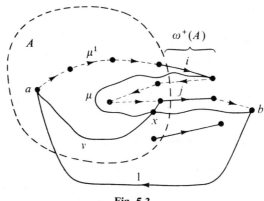

Fig. 5.3

Since μ^1 is the superior path, the portion of path μ following arc i and the
portion of v preceding arc j have a common vertex x. This implies that
$x \in A$ and $x \notin A$, which is a contradiction.

Q.E.D.

Proposition. *The flow* φ *obtained by the Algorithm for planar networks is*
maximum.

Let A_0 denote the set of vertices that can be reached by a path from a.
Then

$$a \in A_0, \qquad b \notin A_0 .$$

Consequently, $\omega^+(A_0)$ is a cut. Let $\omega^+(A)$ be a minimal cut contained in
$\omega^+(A_0)$. From the lemma, the k-th step reduces the capacity of the cut $\omega^+(A)$
by exactly ε_k, and the procedure terminates with a zero cut capacity. Thus,

$$\varphi_1 = \sum_k \varepsilon_k = \sum_{i \in \omega^+(A)} c_i = c(\omega^+(A)) .$$

By Lemma 1 to Theorem 1, flow φ is maximum.

Q.E.D.

2. The compatible flow problem

Clearly, there always exists a compatible flow in a transportation network R
since the flow $\varphi = 0$ is compatible. However, there may not always exist a

flow compatible with the constraints $b_i \leqslant \varphi_i \leqslant c_i$. First, we have:

Lemma. *A necessary condition for the existence of a compatible flow is that for each cocycle* $\omega(A)$,

$$\sum_{i \in \omega^+(A)} c_i - \sum_{i \in \omega^-(A)} b_i \geqslant 0 .$$

Clearly, if a compatible flow φ exists, then

$$0 = \sum_{i \in \omega^+(A)} \varphi_i - \sum_{i \in \omega^-(A)} \varphi_i \leqslant \sum_{i \in \omega^+(A)} c_i - \sum_{i \in \omega^-(A)} b_i .$$

Q.E.D.

The sufficiency of this condition will be demonstrated later.

Compatible flow algorithm (J. C. Herz [1967]). Let $G = (X, U)$ be a graph with numbers b_i and c_i associated with each arc i, By using an iterative procedure similar to the Ford and Fulkerson algorithm, we shall construct a compatible flow starting with any non-compatible flow φ.

Let the *distance of* φ_i *from the interval* $[b_i, c_i]$ be defined to equal

$$d_i(\varphi_i) \begin{cases} = 0 & \text{if} & \varphi_i \in [b_i, c_i] , \\ = b_i - \varphi_i & \text{if} & \varphi_i < b_i , \\ = \varphi_i - c_i & \text{if} & \varphi_i > c_i . \end{cases}$$

The algorithm successively reduces the value of

$$d(\varphi) = \sum_{i=1}^{n} d_i(\varphi) .$$

If $d(\varphi) = 0$, flow φ is compatible and the procedure stops.

If $d(\varphi) > 0$, there exists an arc i with $d_i(\varphi_i) > 0$. Suppose for example that $d_1(\varphi_1) > 0$, and that $\varphi_1 < b_1$. Then, sequentially, label the vertices according to the following rules:

RULE 1. Label vertex a, the terminal endpoint of arc 1 with the index $+1$.

RULE 2. If x is labelled and y is not labelled, label y with the index j if (x, y) is arc j and if $\varphi_j < c_j$.

RULE 3. If x is labelled and y is not labelled, label y with the index $-j$ if (y, x) is arc j and if $\varphi_j > b_j$.

If vertex b, the initial endpoint of arc 1, is labelled, then a new flow φ' such that $d(\varphi') < d(\varphi)$ can be constructed by using the method of the Ford and

Fulkerson algorithm. If the initial endpoint of arc 1 cannot be labelled, then the set A of labelled vertices satisfy $a \in A$ and $b \notin A$. Since $\varphi_1 < b_1$, it follows that

$$0 = \sum_{i \in \omega^+(A)} \varphi_i - \sum_{i \in \omega^-(A)} \varphi_i > \sum_{i \in \omega^+(A)} c_i - \sum_{i \in \omega^-(A)} b_i .$$

From the lemma, no compatible flow exists, and the problem has no solution.

This algorithm establishes the following result:

Compatible Flow Theorem (Hoffman [1958]). *For a graph G with arc numbers b_i and c_i such that $-\infty \leqslant b_i \leqslant c_i \leqslant +\infty$ for all i, a necessary and sufficient condition that there exists a flow φ with $b_i \leqslant \varphi_i \leqslant c_i$ for all i is that, for each set $A \subset X$,*

$$\sum_{i \in \omega^+(A)} c_i - \sum_{i \in \omega^-(A)} b_i \geqslant 0 .$$

The necessity of the condition follows from the lemma.

The condition is sufficient because each flow φ with $d(\varphi) > 0$ can be improved by using the preceding algorithm, until we obtain a flow φ' with $d(\varphi) = 0$.

Q.E.D.

The following algorithm is in general more efficient.

Second compatible flow algorithm

Let G be a graph with vertices $x_1, x_2, ..., x_n$ and with arc numbers b_i, c_i. The following rules construct a flow φ in G such that

$$b_i \leqslant \varphi_i \leqslant c_i \qquad (i = 1, 2, ..., m) .$$

RULE 1. Construct a transportation network R' from G by adding a source a, a sink b, a return arc (b, a), and the various source and sink arcs.

RULE 2. If $b(x, y) \geqslant 0$, let the capacity of arc (x, y) in network R' equal

$$c'(x, y) = c(x, y) - b(x, y) ,$$

and create a sink arc (x, b) with capacity

$$c'(x, b) = b(x, y) .$$

Also, create a source arc (a, y) with capacity

$$c'(a, y) = b(x, y) .$$

RULE 3. If $b(x, y) < 0$, let the capacity of arc (x, y) in network R' equal

$$c'(x, y) = c(x, y) - b(x, y),$$

and create a source arc (a, x) with capacity

$$c'(a, x) = -b(x, y).$$

Also, create a sink arc (y, b) with capacity

$$c'(y, b) = -b(x, y).$$

We shall now show that *a compatible flow exists in G if, and only if, in transportation network R', there exists a maximum flow φ' that saturates all source arcs and sink arcs.*

If we let $\varphi(x, y) = \varphi'(x, y) + b(x, y)$, then clearly φ is a flow in G because for each vertex x in G,

$$\sum_{i \in \omega^+(x)} \varphi_i - \sum_{i \in \omega^-(x)} \varphi_i = \sum_{\substack{i \in \omega^+(x) \\ b_i \geq 0}} (\varphi'_i + b_i) +$$

$$+ \sum_{\substack{j \in \omega^+(x) \\ b_j < 0}} (\varphi'_j + b_j) - \sum_{\substack{i \in \omega^-(x) \\ b_i \geq 0}} (\varphi'_j + b_i) - \sum_{\substack{j \in \omega^-(x) \\ b_j < 0}} (\varphi'_j + b_j) = 0.$$

Flow φ is compatible in G, because

$$0 \leq \varphi'_i \leq c_i - b_i \qquad (i = 1, 2, ..., m),$$

and consequently,

$$b_i \leq \varphi_i \leq c_i \qquad (i = 1, 2, ..., m)$$

Conversely, it is easily seen that each compatible flow φ in G corresponds in R' to a flow φ' that saturates all source and sink arcs. This completes the proof.

Q.E.D.

3. An algebraic study of flows and tensions

Each flow considered above is a vector in $\mathbb{Z}^m$. Flows could also be considered in any ring R, with

(1) $s, t \in R \quad \Rightarrow \quad s + t \in R$,
(2) $0 \in R \quad$ (zero element),
(3) $s \in R \quad \Rightarrow \quad -s \in R$,
(4) $s, t \in R \quad \Rightarrow \quad s.t \in R$.

The space $\mathbb{Z}^m$ is not a "vector space" on $\mathbb{Z}$, because $\mathbb{Z}$ is not a "field", but a "module" on $\mathbb{Z}$, and

$$\mathbf{s}, \mathbf{t} \in \mathbb{Z}^m \qquad \Rightarrow \qquad \mathbf{s} + \mathbf{t} = (s_1 + t_1, ..., s_m + t_m) \in \mathbb{Z}^m ,$$
$$\lambda \in \mathbb{Z}, \mathbf{s} \in \mathbb{Z}^m \qquad \Rightarrow \qquad \lambda \mathbf{s} = (\lambda s_1, ..., \lambda s_m) \in \mathbb{Z}^m .$$

Consequently, the set Φ of all flows in graph G constitutes a *submodule* of $\mathbb{Z}^m$, i.e., we have:

$$\varphi^1, \varphi^2 \in \Phi \qquad \Rightarrow \qquad \varphi^1 + \varphi^2 \in \Phi ,$$
$$s \in \mathbb{Z}, \varphi \in \Phi \qquad \Rightarrow \qquad s\varphi \in \Phi .$$

Theorem 3. *Let $G = (X, U)$ be a connected graph. Let $H = (X, V)$ be a spanning tree of G. Denote the arcs of $U - V$ by $1, 2, ..., k$, and denote the associated cycles of H by $\mu^1, \mu^2, ..., \mu^k$. A flow φ is uniquely defined from the values $\varphi_1, \varphi_2, ..., \varphi_k$ by*

$$\varphi = \varphi_1 \mu^1 + \varphi_2 \mu^2 + \cdots + \varphi_k \mu^k .$$

Consider the vector

$$\varphi' = \varphi - \sum_{i=1}^{k} \varphi_i \mu^i .$$

This vector φ' is a flow, since it is a linear combination of flows. Clearly φ' takes only zero values outside of tree H.

Let $W \subset V$ be the set of arcs i for which $\varphi'_i \neq 0$. We shall show that $W = \varnothing$ i.e., each connected component C of the partial graph (X, W) reduces to a single vertex. If not, C is a tree, and from Theorem (2, Ch. 3), C necessarily contains a pendant vertex a. If, for example, the pendant arc is incident into a, then

$$0 \neq \sum_{i \in \omega^-(a)} \varphi'_i = \sum_{i \in \omega^+(a)} \varphi'_i = 0 ,$$

which is a contradiction.

Q.E.D.

Corollary. *A necessary and sufficient condition that a vector φ be a flow is that it is of the form*

$$\varphi = s_1 \mu^1 + s_2 \mu^2 + \cdots + s_k \mu^k ,$$

where $s_1, s_2, ..., s_k \in \mathbb{Z}$ and $\mu^1, \mu^2, ..., \mu^k$ are elementary cycles.

Since a cycle is a flow, this linear combination of cycles is a flow. The converse follows immediately from Theorem 3, because graph G may be assumed to be connected (otherwise, each connected component could be considered separately).

Q.E.D.

In particular, Theorem 3 shows that a cycle μ can be obtained by the addition of all the cycles μ^i whose "out of tree" arc is used by μ. In this addition, a cycle is preceded by a $+$ sign if the cycle is in the direction of μ. Otherwise, the cycle is preceded by a $-$ sign. Hence, Theorem 3 reduces the number of unknowns for the determination of a flow from m to $m - n + 1$.

Theorem 4. *A necessary and sufficient condition that a vector* φ *be a flow with no negative components is that it is of the form*

$$\varphi = s_1 \, \mu^1 + s_2 \, \mu^2 + \cdots + s_k \, \mu^k \, ,$$

where $s_1, s_2, \ldots, s_k \in \mathbb{Z}$, $s_1, s_2, \ldots, s_k \geqslant 0$ *and* $\mu^1, \mu^2, \ldots, \mu^k$ *are circuits.*

Clearly, a vector φ of the indicated form is a flow $\geqslant 0$.

Conversely, consider a non-zero flow $\varphi \geqslant 0$, and let $\bar{G}$ be the graph obtained from G by removing all arcs i with $\varphi_i = 0$. Graph $\bar{G}$ contains no cocircuits, and therefore by the arc colouring Lemma, it contains at least one circuit μ^1. Let $s_1 > 0$ be the smallest flow in an arc of μ^1. The vector

$$\varphi^1 = \varphi - s_1 \, \mu^1$$

is a flow $\geqslant 0$ that has more zero components than φ. If φ^1 is not a zero vector, repeat this process, etc. ... Finally, a zero flow of the form

$$\varphi^k = \varphi - s_1 \, \mu^1 - s_2 \, \mu^2 - \cdots - s_k \, \mu^k = 0 \, ,$$

is obtained with $s_1, s_2, \ldots, s_k \in \mathbb{Z}$ and $s_1, s_2, \ldots, s_k \geqslant 0$.

Q.E.D.

A *tension* (or *potential difference*) is defined to be a vector $\theta = (\theta_1, \theta_2, \ldots, \theta_m) \in \mathbb{Z}^m$ such that, for each elementary cycle μ,

$$\sum_{i \in \mu^+} \theta_i = \sum_{i \in \mu^-} \theta_i \, .$$

This equality can be restated by saying that the *scalar product* $\langle \mu, \theta \rangle = \sum \mu_i \, \theta_i$ is zero.

Let Θ denote the set of all tensions. Note that Θ is a submodule of $\mathbb{Z}^m$, i.e.,

$$\begin{aligned} \theta^1, \theta^2 \in \Theta \quad &\Rightarrow \quad \theta^1 + \theta^2 \in \Theta \, , \\ s \in \mathbb{Z}, \theta \in \Theta \quad &\Rightarrow \quad s\theta \in \Theta \, . \end{aligned}$$

Theorem 5. *A vector* $\theta = (\theta_1, \theta_2, \ldots, \theta_m)$ *is a tension if, and only if, there exists a function* $t(x)$ *defined on the vertex set* X *with values in* $\mathbb{Z}$ *such that, for each arc* $i = (a, b)$, $\theta_i = t(b) - t(a)$.

The function $t(x)$ is called a *potential attached to tension* θ.

1. If θ is a vector defined by a function $t(x)$, then consider the cycle

$$\mu = (i_1, i_2, \ldots, i_k)$$

that successively encounters the vertices a, b, c, ..., z. Then, we may write

$$\mu_{i_1} \theta_{i_1} = t(b) - t(a)$$

$$\mu_{i_2} \theta_{i_2} = t(c) - t(b)$$

$$\cdots\cdots\cdots\cdots\cdots\cdots$$

$$\mu_{i_k} \theta_{i_k} = t(a) - t(z).$$

Adding the above equations yields

$$\sum_{i \in \mu^+} \theta_i - \sum_{i \in \mu^-} \theta_i = 0 .$$

2. It is easy to calculate successively the coefficients $t(x)$ for a given tension θ, by the following rules:

RULE 1. Take any vertex x_0, label x_0 and set

$$t(x_0) = 0 .$$

RULE 2. If x is labelled and y is unlabelled, and if $i = (x, y)$ is an arc, put

$$t(y) = t(x) + \theta_i .$$

If $i = (y, x)$ is an arc, put

$$t(y) = t(x) - \theta_i .$$

In this way, all the vertices of a connected graph will be labelled. (If the graph is not connected, each connected component can be treated separately.)

Each coefficient is *uniquely* defined by this process. Otherwise, there would exist two chains μ^1 and μ^2 from x_0 to x such that

$$\langle \mu^1, \theta \rangle \neq \langle \mu^2, \theta \rangle ,$$

and consequently,

$$\langle \mu^1 - \mu^2, \theta \rangle \neq 0 .$$

Since $\mu^1 - \mu^2$ is a flow, it is a linear combination of elementary cycles by virtue of Theorem 3. Therefore, there exists an elementary cycle μ such that

$$\langle \mu, \theta \rangle \neq 0 ,$$

which contradicts the definition of a tension.

Q.E.D.

Consequence. This theorem clearly shows that a cocycle $\omega(A)$ is a tension, since we may let

$$t(x) = \begin{cases} 0 & \text{if} \quad x \in A \\ 1 & \text{if} \quad x \notin A . \end{cases}$$

Hence, for an arc $i = (a, b)$,

$$t(b) - t(a) = \begin{cases} +1 & \text{if} & i \in \omega^+(A) \\ -1 & \text{if} & i \in \omega^-(A) \\ 0 & \text{if} & i \notin \omega(A) \end{cases} = \omega_i(A) \ .$$

Theorem 6. *Let* $G = (X, U)$ *be a connected graph. Let* $H = (X, V)$ *be a spanning tree of* G *with arcs* 1, 2, ..., l, *let* $\omega^1, \omega^2, ..., \omega^l$ *be the cocycles associated with* H. *A tension* $\theta = (\theta_1, \theta_2, ..., \theta_m)$ *is uniquely defined from the values* $\theta_1, \theta_2, ..., \theta_l$ *by:*

$$\theta = \theta_1 \omega^1 + \theta_2 \omega^2 + \cdots + \theta_l \omega^l \ .$$

The vector

$$\theta' = \theta - \theta_1 \omega^1 - \theta_2 \omega^2 - \cdots - \theta_l \omega^l$$

is a tension that has zero value on each arc of the tree H. If θ' corresponds to a potential $t'(x)$, then

$$t'(x_1) = t'(x_2) = \cdots = t'(x_n) \ .$$

Consequently, $\theta' = 0$.

Q.E.D.

Corollary. *A necessary and sufficient condition that a vector* θ *be a tension is that it is of the form*

$$\theta = s_1 \omega^1 + s_2 \omega^2 + \cdots + s_k \omega^k \ ,$$

where $s_1, s_2, ..., s_k \in \mathbb{Z}$, *and* $\omega^1, \omega^2, ..., \omega^k$ *are elementary cocycles.*

Clearly, each linear combination of elementary cocycles is a tension, and conversely, each tension is a linearly combination of elementary cocycles, from Theorem 6.

Q.E.D.

Theorem 6 shows that a tension

$$\theta = (\theta_1, \theta_2, ..., \theta_m) \ ,$$

can be determined from $\lambda(G) = n - 1$ unknowns.

Theorem 7. *A necessary and sufficient condition that* θ *be a tension* $\geqslant 0$ *is that*

$$\theta = s_1 \omega^1 + s_2 \omega^2 + \cdots + s_k \omega^k \ ,$$

where $s_1, s_2, ..., s_k \in \mathbb{Z}$, $s_1, s_2, ..., s_k \geqslant 0$ *and* $\omega^1, \omega^2, ..., \omega^k$ *are elementary cocircuits.*

Clearly, any linear combination of cocircuits with all coefficients $\geqslant 0$ is a tension $\geqslant 0$.

Conversely, consider a tension $\theta \neq 0$, $\theta \geqslant 0$. We shall show that there exists an elementary cocircuit ω^1 and an $s_1 > 0$ such that the vector

$$\theta - s_1 \omega^1$$

has more zero components than the vector θ.

Let $i = 1$ be an arc with

$$\theta_1 = \min \{ \theta_i / \theta_i \neq 0, \quad 1 \leqslant i \leqslant m \}.$$

Put $\theta_1 = s_1 > 0$. Colour black all arcs i such that $\theta_i > 0$. Colour red all arcs i such that $\theta_i = 0$. There cannot exist a red and black elementary cycle with all black arcs in the same direction that contains arc 1. Thus, by virtue of the Arc Colouring Lemma, arc 1 is contained in a black elementary cocycle ω^1 with all arcs in the same direction. Thus, ω^1 is a cocircuit, and the vector

$$\theta - s_1 \omega^1$$

has more zero components than the vector θ. This vector is also a tension $\geqslant 0$.

If $\theta - s_1 \omega^1 \neq 0$, this reduction process can be repeated until a tension

$$\theta - s_1 \omega^1 - s_2 \omega^2 - \cdots - s_k \omega^k = 0.$$

is obtained.

Q.E.D.

Theorem 8. *A vector $\varphi \in \mathbb{Z}^m$ is a flow if, and only if, it is orthogonal to each vector of Θ. A vector $\theta \in \mathbb{Z}^m$ is a tension if, and only if, it is orthogonal to each vector of Φ.*

(Hence, Θ and Φ are two orthogonal submodules of $\mathbb{Z}^m$.)

1. We shall show that if $\varphi \in \Phi$ and $\theta \in \Theta$, then θ and φ are orthogonal, i.e.,

$$\sum_{i=1}^{m} \varphi_i \theta_i = \langle \varphi, \theta \rangle = 0.$$

For each elementary cycle μ,

$$\langle \mu, \theta \rangle = 0.$$

From Theorem 3, φ is of the form

$$\varphi = \sum_k s_k \mu^k.$$

Thus,

$$\langle \varphi, \theta \rangle = \sum_k s_k \langle \mu^k, \theta \rangle = 0.$$

2. Let φ be a vector such that $\langle \varphi, \theta \rangle = 0$ for all $\theta \in \Theta$. Vector φ is a flow, because if we take $\theta = \omega(x)$ for some vertex x, then

$$\sum_{i \in \omega^+(x)} \varphi_i - \sum_{i \in \omega^-(x)} \varphi_i = \langle \omega(x), \varphi \rangle = 0 .$$

3. Let θ be a vector such that $\langle \varphi, \theta \rangle = 0$ for all $\varphi \in \Phi$. Vector θ is a tension, because for every elementary cycle μ, we have

$$\langle \mu, \theta \rangle = 0 .$$

Q.E.D.

Additional algebraic results for flows and tensions appear in Berge and Ghouila-Houri [1962], and Slepian [1968].

4. The maximum tension problem

Consider a graph G whose arcs are denoted by $1, 2, ..., m$. Let $k_1, k_2, ...,$ $k_m, l_1, l_2, ..., l_m \in \mathbb{Z}$, such that

$$- \infty \leqslant k_i \leqslant l_i \leqslant + \infty .$$

Several problems, similar to the above problems can be stated for tensions.

The compatible tension problem. *For graph G, find a tension θ such that*

$$k_i \leqslant \theta_i \leqslant l_i \qquad (i = 1, 2, ..., m) .$$

The maximum tension problem. *For graph G, find a tension θ such that*

(1) $k_i \leqslant \theta_i \leqslant l_i$ $(i = 1, 2, ..., m)$,
(2) θ_1 *is maximum.*

EXAMPLE 1. The shortest path problem (Ch. 4, § 2) is a special case of the maximum tension problem. Let a and b be two vertices of a graph G. Let $l(x, y)$ denote the *length* of arc (x, y). Add to G an arc $1 = (b, a)$ and let $t(x)$ denote the length $l(\mu[a, x])$ of a shortest elementary path $\mu[a, x]$ from a to x. For each arc (x, y) of G, let

$$\theta(x, y) = t(y) - t(x) .$$

Consequently,

$$- \infty \leqslant \theta(x, y) \leqslant l(x, y) .$$

We wish to maximize

$$\theta(b, a) = t(a) - t(b) = - t(b) ,$$

since $t(b)$ must represent the length of the shortest path from a to b.

EXAMPLE 2. *Sequencing problems*: The construction of a factory requires the performance of various distinct tasks designated by $x_1, x_2, ..., x_n$. Before starting a task x_j, it is usually required that another task x_i be sufficiently under way. Let k_{ij} denote the amount of time by which the start of task i must precede the start of task j.

Furthermore, a task x_i cannot begin before a fixed time k_i' and has a known duration k_i''. Given these constraints, when should each task begin so that the construction project finishes as soon as possible?

Consider a graph G whose vertices correspond to the tasks. An arc (x_i, x_j) is present if task x_j can start only after task x_i is sufficiently under way. Add a source a and an arc (a, x) for each task x that cannot begin before a specified time. Finally, add a sink b and an arc (x, b) for each task x.

Denote by $t(x)$ the starting time of task x. Let

$$t(a) = 0 .$$

We wish to minimize $t(b)$, i.e., to maximize the tension

$$\theta(b, a) = t(a) - t(b)$$

in the "return arc" (b, a). The constraints are

$$k_i' \leqslant t(x_i) - 0 \quad = \theta(a, x_i) \leqslant + \infty \qquad \text{if} \qquad (a, x_i) \in U$$

$$k_{ij} \leqslant t(x_j) - t(x_i) = \theta(x_i, x_j) \leqslant + \infty \qquad \text{if} \qquad (x_i, x_j) \in U$$

$$k_i'' \leqslant t(b) \quad - t(x_i) = \theta(x_i, b) \leqslant + \infty \qquad \text{if} \qquad (x_i, b) \in U$$

The solution of sequencing problems as potential problems has been notably developed by B. Roy [1965].

Lemma. *A necessary condition for the existence of a compatible tension is that for each cycle* μ,

$$\sum_{i \in \mu^-} l_i - \sum_{i \in \mu^+} k_i \geqslant 0 .$$

In fact, if such a tension θ exists, then for each cycle μ

$$0 = \langle \mu, \theta \rangle = \langle \mu^+, \theta \rangle - \langle \mu^-, \theta \rangle \geqslant \sum_{i \in \mu^+} k_i - \sum_{i \in \mu^-} l_i .$$

Q.E.D.

The sufficiency of this condition will be demonstrated later.

Compatible tension algorithm (J. C. Herz [1967])

Let G be a graph with the numbers $k_i \leqslant l_i$ associated with each arc i. For a given tension θ, compatible or not, define the *distance of* θ_i *from the interval*

$[k_i, l_i]$ by

$$d_i(\theta_i) \begin{cases} = 0 & \text{if} & \theta_i \in [k_i, l_i], \\ = k_i - \theta_i & \text{if} & \theta_i < k_i, \\ = \theta_i - l_i & \text{if} & \theta_i > l_i. \end{cases}$$

We shall successively reduce the quantity

$$d(\theta) = \sum_{i=1}^{m} d_i(\theta_i).$$

If $d(\theta) = 0$, the tension θ is compatible, and the procedure stops. Otherwise, there exists an arc i with $d_i(\theta_i) > 0$; let

$$1 = (b, a)$$

and suppose that

$$d_1(\theta_1) = k_1 - \theta_1, \qquad \theta_1 < k_1.$$

To construct a tension θ' such that $d(\theta') < d(\theta)$, successively label the vertices of G in the following way:

RULE 1. Label vertex a, the terminal endpoint of arc 1, with the index $+1$.

RULE 2. If x is labelled and y is unlabelled, label y with the index $+i$ if $(x, y) = i$ and if $\theta_i \leqslant k_i$.

RULE 3. If x is labelled and y is unlabelled, label y with the index $-i$ if $(y, x) = i$ and if $\theta_i \geqslant l_i$.

If vertex b, the initial endpoint of arc 1, cannot be labelled by this procedure, then the tension θ can be improved. In fact, the set A of labelled vertices satisfies

$$a \in A, \quad b \notin A.$$

The tension $\theta' = \theta - \omega(A)$ satisfies $\theta_1' = \theta_1 + 1$, and θ' compatible, because $i \in \omega^+(A)$ implies $\theta_i > k_i$ and $\theta_i' \geqslant k_i$. Similarly, $i \in \omega^-(A)$ implies $\theta_i < l_i$, and $\theta_i' \leqslant l_i$.

If this labelling procedure labels vertex b, then there exists a chain

$$v = [a, a_1, a_2, ..., b],$$

in which each vertex has been labelled from the preceding one. Thus

$$\begin{aligned} i \in v^+ &\Rightarrow & \theta_i \leqslant k_i, \\ i \in v^- &\Rightarrow & \theta_i \geqslant l_i. \end{aligned}$$

$\mu = v[a, b] + [b, a]$ is a cycle. Since $\theta_1 < k_1$, we have

$$0 = \langle \, \boldsymbol{\theta}, \boldsymbol{\mu} \, \rangle = \sum_{i \in \mu^+} \theta_i - \sum_{i \in \mu^-} \theta_i < \sum_{i \in \mu^+} k_i - \sum_{i \in \mu^-} l_i \, .$$

From the preceding lemma, it follows that no compatible tension exists.

The algorithm yields the following result:

Compatible Tension Theorem (Ghouila-Houri [1960]). *Given a graph G and numbers k_i and l_i where $-\infty \leqslant k_i \leqslant l_i \leqslant +\infty$ for $i = 1, 2, ..., m$, a necessary and sufficient condition that there exists a tension $\boldsymbol{\theta} = (\theta_1, \theta_2, ..., \theta_m)$ with $k_i \leqslant \theta_i \leqslant l_i$ for all i, is that for each cycle μ,*

$$\sum_{i \in \mu^+} l_i - \sum_{i \in \mu^-} k_i \geqslant 0 \, .$$

The necessary part of the theorem follows from the preceding lemma.

The condition is sufficient because, then, each tension $\boldsymbol{\theta}$ with $d(\boldsymbol{\theta}) > 0$ can be improved using the preceding algorithm until we obtain tension $\boldsymbol{\theta}'$ with $d(\boldsymbol{\theta}') = 0$.

Q.E.D.

Corollary 1 (Roy, [1962]). *There exists a tension $\boldsymbol{\theta}$ with $\theta_i \geqslant k_i$ for all i, if, and only if, for each circuit μ,*

$$\sum_{i \in \mu} k_i \leqslant 0 \, .$$

The proof is achieved by taking $l_i = +\infty$ for all i.

Corollary 2. *There exists a tension $\boldsymbol{\theta}$ such that $\theta_i \leqslant l_i$ for all i, if, and only if, for each circuit μ,*

$$\sum_{i \in \mu} l_i \geqslant 0 \, .$$

Corollary 3. *A vector*

$$\mathbf{y} = (y_1, y_2, ..., y_m)$$

is called a subtension if there exists a tension $\boldsymbol{\theta}$ such that $y_i \leqslant \theta_i$ for all i. A necessary and sufficient condition that a vector $\mathbf{y}$ is a subtension is that

$$\langle \, \boldsymbol{\varphi}, \mathbf{y} \, \rangle \leqslant 0 \quad (\boldsymbol{\varphi} \in \Phi, \quad \boldsymbol{\varphi} \geqslant \boldsymbol{\theta}) \, .$$

Furthermore, a subtension $\mathbf{y}$ and a flow $\boldsymbol{\varphi} \geqslant \mathbf{0}$ satisfy $\langle \, \boldsymbol{\varphi}, \mathbf{y} \, \rangle = 0$, if, and only if, $\mathbf{y}$ is a tension on the partial graph generated by the arcs i with $\varphi_i > 0$.

1. If $\mathbf{y}$ satisfies the above inequality, then by taking for $\boldsymbol{\varphi}$ a circuit $\boldsymbol{\mu}$, we obtain

$$\langle \, \boldsymbol{\mu}, \mathbf{y} \, \rangle = \sum_{i \in \mu} y_i \leqslant 0.$$

From Corollary 1, $\mathbf{y}$ is therefore a subtension.

2. Conversely, let $\mathbf{y}$ be a subtension. If $\varphi \in \Phi$, $\varphi \geqslant 0$, then

$$\langle \varphi, \mathbf{y} \rangle = \sum_{i=1}^{m} \varphi_i \, y_i \leqslant \sum_{i=1}^{m} \varphi_i \, \theta_i = 0 \, .$$

Furthermore, equality holds if, and only if,

$$\varphi_i > 0 \quad \Rightarrow \quad y_i = \theta_i \, .$$

Clearly, if $\langle \varphi, \mathbf{y} \rangle = 0$, then $\mathbf{y}$ is a tension on the partial graph generated by the arcs i such that $\varphi_i > 0$.

Finally, if $\mathbf{y}$ is a tension on the partial graph generated by the arcs i with $\varphi_i > 0$, then

$$\sum_{\varphi_i > 0} y_i \, \varphi_i = 0 \, ,$$

and, consequently,

$$\langle \varphi, \mathbf{y} \rangle = \sum_{\varphi_i > 0} y_i \, \varphi_i + \sum_{\varphi_i = 0} y_i \, \varphi_i = 0 \, .$$

Q.E.D.

Maximum tension algorithm

We shall construct a tension that maximizes the value θ_1 of the tension in arc $1 = (b, a)$.

Starting with any compatible tension θ, achieves the labelling procedure for the compatible tension Problem. If vertex b cannot be labelled, the set A of labelled vertices satisfies $a \in A$, $b \notin A$, and the tension $\theta' = \theta - \omega(A)$ satisfies $\theta_1' = \theta_1 + 1$. Thus, the value of the tension in arc 1 can be improved.

If vertex b can be labelled, we shall show that the value θ_1 of the tension θ in arc 1 is maximum.

For each compatible tension θ and for each cycle μ that uses arc 1 along its direction, we have

$$0 = \langle \theta, \mu \rangle \geqslant \theta_1 + \sum_{\substack{i \neq 1 \\ i \in \mu^+}} k_i - \sum_{i \in \mu^-} l_i \, ,$$

or

(1)
$$\theta_1 \leqslant \sum_{i \in \mu^-} l_i - \sum_{\substack{i \in \mu^+ \\ i \neq 1}} k_i \, .$$

If the algorithm labels vertex b, then there exists a chain $v[a, b]$ such that

$$i \in v^+ \quad \Rightarrow \quad \theta_i = k_i \, ,$$
$$i \in v^- \quad \Rightarrow \quad \theta_i = l_i \, .$$

Hence, for the cycle $\mu = v[a, b] + [b, a]$, we have

$$\theta_1 = \sum_{i \in \mu^-} \theta_i - \sum_{\substack{i \in \mu^+ \\ i \neq 1}} \theta_i = \sum_{i \in \mu^-} l_i - \sum_{\substack{i \in \mu^+ \\ i \neq 1}} k_i .$$

From inequality (1), we see that θ_1 has reached its maximum value.

Theorem 9. *For a graph G with arcs numbers k_i and l_i, the maximum value of a compatible tension in arc 1 is*

$$\max_{\theta} \theta_1 = \min_{\substack{\mu \\ 1 \in \mu^+}} \left(\sum_{i \in \mu^-} l_i - \sum_{\substack{i \in \mu^+ \\ i \neq 1}} k_i \right).$$

If θ is a compatible tension with

$$\theta_1 < \min_{\substack{\mu \\ 1 \in \mu^+}} \left(\sum_{i \in \mu^-} l_i - \sum_{\substack{i \in \mu^+ \\ i \neq 1}} k_i \right)$$

it can be improved by using the above algorithm. The theorem follows.

Q.E.D.

EXERCISES

1. Let

$$\mathcal{T} = (T_1, T_2, ..., T_n)$$

be a partition of a finite set X, and let

$$\mathcal{S} = (S_1, S_2, ..., S_m)$$

be a family of subsets of X. Show that if every union of k of the T_i contains at most k of the S_j (for $k = 1, 2, ..., n$), then there exist indices $i_1, i_2, ..., i_m$ such that

$$T_{i_p} \cap S_p \neq \varnothing \qquad \text{for} \qquad p = 1, 2, ..., m .$$

Hint: This result is easily shown by constructing the appropriate transportation network.

2. Let

$$\mathcal{T} = (T_1, T_2, ..., T_n)$$

be a partition of a finite set X, and let

$$\mathcal{S} = (S_1, S_2, ..., S_m)$$

be a family of subsets, and let $c_1, c_2, ..., c_n$ be positive integers. Show how to construct a set of representatives

$$A = \{ a_1, a_2, ..., a_m \} .$$

such that $a_1 \in S_1, a_2 \in S_2, ...,$ etc., and such that

$$| A \cap T_j | \leqslant c_j \qquad \text{pour} \qquad j = 1, 2, ..., n .$$

By reducing this problem to a flow problem, show that such a set of representatives exists if

$$\left| \bigcup_{i \in I} S_i \cap \bigcup_{j \in J} T_j \right| \geq |I| + \sum_{j \in J} |T_j| - \sum_{j \in J} c_j$$

for all $I \subset \{1, 2, ..., m\}$ and all $J \subset \{1, 2, ..., n\}$.

3. Let

$$\mathscr{T} = (T_1, T_2, ..., T_n)$$

be a partition of a finite set X, and let

$$\mathscr{S} = (S_1, S_2, ..., S_m)$$

be a family of subsets, and let $b_1, b_2, ..., b_n$ be positive integers. Show how to construct a set of representatives

$$A = \{a_1, a_2, ..., a_m\},$$

with $a_1 \in S_1$, $a_2 \in S_2$, ..., and such that

$$|A \cap T_j| \geq b_j \qquad \text{for} \qquad j = 1, 2, ..., n.$$

Show that such a set of representatives exists if

$$\left| \bigcup_{i \in I} S_i \cap \bigcup_{i \in J} T_j \right| \geq |I| - m + \sum_{j \in J} c_j$$

for all $I \subset \{1, 2, ..., m\}$ and for all $J \subset \{1, 2, ..., n\}$.

4. Use the Compatible Flow Theorem to prove the following:

Let R be a transportation network with source a and sink b. Associated with each source arc $i \in \omega^+(a)$ are two numbers b_i and c_i such that $0 \leq b_i \leq c_i$. Associated with each sink arc $j \in \omega^-(b)$ are two numbers b_j' and c_j' such that $0 \leq b_j' \leq c_j'$. A necessary and sufficient condition that there exists a flow φ satisfying

$$b_i \leq \varphi_i \leq c_i \qquad \text{for} \qquad i \in \omega^+(a),$$
$$b_j' \leq \varphi_j \leq c_j' \qquad \text{for} \qquad j \in \omega^-(b),$$

is that both of the following conditions hold:
(1) There exists a flow φ^1 with

$$\varphi_i^1 \geq b_i \qquad \text{for} \qquad i \in \omega^+(a)$$
$$\varphi_j^1 \leq c_j' \qquad \text{for} \qquad j \in \omega^-(b).$$

(2) There exists a flow φ^2 with

$$\varphi_i^2 \leq c_i \qquad \text{for} \qquad i \in \omega^+(a)$$
$$\varphi_j^2 \geq b_j' \qquad \text{for} \qquad j \in \omega^-(b).$$

CHAPTER 6

Degrees and Demi-Degrees

1. Existence of a p-graph with given demi-degrees

For a graph $G = (X, U)$, the *outer demi-degree* $d_G^+(x)$ of a vertex x is defined to be the number of arcs having x as their initial endpoint, i.e.

$$d_G^+(x) = \sum_{y \in X} m_G^+(x, y).$$

Similarly, the *inner demi-degree* $d_G^-(x)$ of a vertex x to be the number of arcs having x as their terminal endpoint, i.e.

$$d_G^-(x) = \sum_{y \in X} m_G^-(x, y).$$

Finally, we define the *degree* $d_G(x)$ of a vertex x is defined to be the integer

$$d_G(x) = d_G^+(x) + d_G^-(x).$$

Thus, a loop at vertex x increases the degree of vertex x by 2.

A *p-graph* is a graph with $m_G^+(x, y) \leqslant p$ for all vertices x, y. Given integers $r_1, r_2, ..., r_n, s_1, s_2, ..., s_n$, we may ask if there exists a *p*-graph G with vertices $x_1, x_2, ..., x_n$ such that

$$d_G^+(x_k) = r_k \qquad (k = 1, 2, ..., n)$$

$$d_G^-(x_k) = s_k \qquad (k = 1, 2, ..., n).$$

In this case, the pairs (r_k, s_k) are said to *constitute the demi-degrees of a p-graph.*

Theorem 1. *Let* $(r_1, s_1), (r_2, s_2), ..., (r_n, s_n)$ *be pairs of integers with*

$$s_1 \geqslant s_2 \geqslant \cdots \geqslant s_n.$$

The pairs (r_k, s_k) *constitute the demi-degrees of a p-graph if, and only if,*

(1) $$\sum_{i=1}^{n} \min\{r_i, pk,\} \geqslant \sum_{j=1}^{k} s_j \qquad (k = 1, 2, ..., n - 1)$$

(2) $$\sum_{i=1}^{n} r_i = \sum_{j=1}^{n} s_j \,.$$

Construct a transportation network R with vertices $x_1, x_2, \ldots, x_n, \bar{x}_1, \bar{x}_2,$ $\ldots, \bar{x}_n$, and with a source a and a sink b. Join vertices x_i and $\bar{x}_j$ by an arc with capacity $c(x_i, \bar{x}_j) = p$. Join vertices a and x_i by an arc with capacity $c(a, x_i) = r_i$. Join vertices $\bar{x}_j$ and b by an arc with capacity $c(\bar{x}_j, b) = s_j$.

Any flow that saturates the source and sink arcs of R defines a p-graph (X, U) having demi-degrees (r_k, s_k). Conversely, each p-graph (X, U) with demi-degrees (r_k, s_k) defines a flow in R that saturates the source and sink arcs. From Theorem (2, Ch. 5), a necessary and sufficient condition for the existence of such a flow is that condition (2) hold and that, for each set $B = \{ \bar{x}_{i_1}, \bar{x}_{i_2}, \ldots, \bar{x}_{i_k} \}$, the total flow that can enter B is greater than the total demand at B, i.e.,

$$F(\bar{x}_{i_1}, \bar{x}_{i_2}, \ldots, \bar{x}_{i_k}) \geqslant d(\bar{x}_{i_1}, \bar{x}_{i_2}, \ldots, \bar{x}_{i_k}) \qquad (1 \leqslant i_1 < i_2 < \cdots < i_k \leqslant n)\,.$$

This is equivalent to

(1′) $$\sum_{i=1}^{n} \min \{ r_i, pk \} \geqslant s_{i_1} + s_{i_2} + \cdots + s_{i_k} \qquad (1 \leqslant i_1 < i_2 < \cdots < i_k \leqslant n)\,.$$

Clearly (1′) $\Rightarrow$ (1) and, since the s_j are indexed in decreasing order, (1) $\Rightarrow$ (1′).

Q.E.D.

For $p = 1$, these necessary and sufficient conditions can be restated in a simpler form. Let

$$r_1 \geqslant r_2 \geqslant \cdots \geqslant r_n$$

be a n-tuple of integers. Corresponding to this n-tuple, associate the sequence $(r_1^*, r_2^*, \ldots)$ where r_k^* denotes the number of r_i that are greater than or equal to the integer k. Hence,

$$r_1^* \geqslant r_2^* \geqslant r_3^* \geqslant \cdots$$

To visualize the numbers r_k^*, we can construct a diagram, called the *Ferrers diagram*, as in Fig. 6.1.

Hatch the first r_i squares in the i-th column in the positive quadrant. Then it is easily seen that r_j^* equals the number of hatched squares in the j-th row in the positive quadrant. By counting, in two different ways, the number of hatched squares, we obtain:

$$\sum_{i=1}^{n} r_i = \sum_{k \geqslant 1} r_k^* \,.$$

The sequences (r_i) and (r_i^*) are called *conjugates*. Using these definitions, several corollaries follow from Theorem 1.

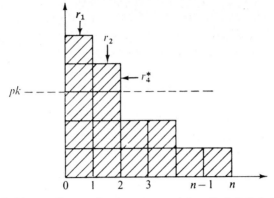

Fig. 6.1. Ferrers diagram for the sequence $(r_i) = (5, 4, 2, 2, 1, 1)$. Here, $(r_i^*) = (6, 4, 2, 2, 1)$

Corollary 1. *Given n pairs (r_i, s_i) of integers such that*

$$s_1 \geqslant s_2 \geqslant \cdots \geqslant s_n,$$

there exists a p-graph with $d_G^+(x_i) = r_i$, $d_G^-(x_i) = s_i$ for $i = 1, 2, ..., n$ if, and only if,

$$\begin{cases} \sum_{i=1}^{pk} r_i^* \geqslant \sum_{j=1}^{k} s_j \qquad (k = 1, 2, ..., n-1) \\ \sum_{i \geqslant 1} r_i^* = \sum_{j=1}^{n} s_j . \end{cases}$$

By counting in two different ways the number of hatched squares in the first pk rows of the associated Ferrers diagram for $(r_1, r_2, ..., r_n)$ we find that

$$\sum_{i=1}^{n} \min \{ r_i, pk \} = \sum_{i=1}^{pk} r_i^* .$$

$$\text{Q.E.D.}$$

Corollary 2 (Ryser [1957]; Gale [1957]). *Consider n pairs (r_i, s_i) of integers such that*

$$s_1 \geqslant s_2 \geqslant \cdots \geqslant s_n .$$

There exists a 1-graph G with $d_G^+(x_i) = r_i$ and $d_G^-(x_i) = s_i$ if, and only if,

$$\begin{cases} \sum_{i=1}^{k} r_i^* \geqslant \sum_{j=1}^{k} s_j \qquad (k = 1, 2, ..., n-1) \\ \sum_{i \geqslant 1} r_i^* = \sum_{j=1}^{n} s_j . \end{cases}$$

This follows immediately from Corollary 1.

Corollary 3. *Given a sequence* $(r_1, r_2, ..., r_n)$, *there exists a* 1-*graph G with* $d_G^+(x_i) = r_i$ *and* $d_G^-(x_i) = 1$ *for all i if, and only if,*

$$\sum_{i=1}^{n} r_i = n .$$

This condition is necessary because if such a graph G exists, then

$$\sum_{i=1}^{n} r_i = \sum_{i=1}^{n} d_G^-(x_i) = n .$$

Conversely, assume that the condition holds; by Theorem 1, we have only to show that, for all $k \leqslant n$,

$$\sum_{i=1}^{n} \min \{ k, r_i \} \geqslant k .$$

If there exists an $r_i \geqslant k$, the above inequality is satisfied. Otherwise, each r_i is less than k, and consequently,

$$\sum_{i=1}^{n} \min \{ k, r_i \} = \sum_{i=1}^{n} r_i = n \geqslant k .$$

Thus there exists a graph G with the desired properties.

Q.E.D.

We shall now present necessary and sufficient conditions for the pairs $(r_1, r_1), (r_2, r_2), ..., (r_n, r_n)$ to constitute the demi-degrees of a symmetric p-graph, i.e. a p-graph G with

$$m_G^+(x, y) = m_G^+(y, x) \qquad (x, y \in X) .$$

Theorem 2. *Let G_0 be a symmetric graph such that each odd cycle contains a vertex with one or more loops, and let $r_1, r_2, ..., r_n$ be positive integers. If G_0 has a partial graph H with $d_H^+(x_i) = d_H^-(x_i) = r_i$ for each vertex x_i, then G_0 has a symmetric partial graph G with*

$$d_G^+(x_i) = d_G^-(x_i) = r_i$$

for all i.

We shall assume that the theorem is true for any graph G_0 of order $< n$, and we shall show that the theorem is true for a graph G_0' of order n. Given a graph H, we shall construct from H the symmetric graph G.

1. Suppose first that there exists a vertex x_0 such that

$$m_H^+(x_0, y) = m_H^+(y, x_0) \qquad (y \in X) .$$

Consider the subgraph $\bar{G}_0$ of order $n - 1$ obtained from G_0 by removing vertex x_0. Graph G_0 has a partial graph $\bar{H}$ with given demi-degrees $\bar{r}_i = r_i - m_H^+(x_0, x_i)$, which can be transformed to a symmetric graph $\bar{G}$ with demi-degrees $\bar{r}_i$ (by the induction hypothesis). This yields graph G.

2. We may now suppose that for each vertex x of G_0, there exists a vertex y adjacent to x with

$$m_H^+(x, y) > m_H^+(y, x) .$$

Since $d_H^+(x) = d_H^-(x)$, there exists a vertex z adjacent to x with

$$m_H^+(x, z) < m_H^+(z, x) .$$

Clearly x, y and z are distinct vertices.

3. Choose any vertex x_1, and let x_2 be any vertex such that

$$m_H^+(x_1, x_2) > m_H^+(x_2, x_1) .$$

Let x_3 be any vertex such that

$$m_H^+(x_2, x_3) > m_H^+(x_3, x_2) .$$

In this way, a sequence $x_1, x_2, x_3, \ldots$ is defined and, since the graph is finite, an elementary cycle

$$\mu = [x_p , x_{p+1}, \ldots, x_{p+k-1}, x_{p+k} = x_p]$$

will be formed.

4. If cycle μ is even, transform H to H' by removing an arc between each of the pairs:

$$(x_p, x_{p+1}), (x_{p+2}, x_{p+3}), \ldots, (x_{p+k-2}, x_{p+k-1})$$

and adding an arc between each of the pairs:

$$(x_{p+2}, x_{p+1}), (x_{p+4}, x_{p+3}), \ldots, (x_{p+k}, x_{p+k-1}).$$

This does not alter the demi-degrees, and produces a graph H' with

$$(1) \qquad \sum_{\substack{x,y \in X \\ x \neq y}} \left| m_{H'}^+(x, y) - m_{H'}^+(y, x) \right| < \sum_{\substack{x,y \in X \\ x \neq y}} \left| m_H^+(x, y) - m_H^+(y, x) \right| .$$

5. If cycle μ is odd, and if H has a loop at one of its vertices, say x_p, then transform H into H' by removing an arc between each pair

$$(x_p, x_p), (x_{p+1}, x_{p+2}), (x_{p+3}, x_{p+4}), \ldots, (x_{p+k-2}, x_{p+k-1}),$$

and adding an arc between each pair

$$(x_{p+1}, x_p), (x_{p+3}, x_{p+2}), \ldots, (x_{p+k}, x_{p+k-1}).$$

This does not alter the demi-degrees, and the graph H' satisfies (1).

6. Finally, if cycle μ is odd and if H has no loops attached to μ, then there exists a vertex of μ, say x_p, that is incident with a loop in G_0. Transform H into H' by adding an arc between each pair

$$(x_p, x_p), (x_{p+2}, x_{p+1}), (x_{p+4}, x_{p+3}), \ldots, (x_{p+k-1}, x_{p+k-2}),$$

and removing an arc between each pair

$$(x_p, x_{p+1}), (x_{p+2}, x_{p+3}), \ldots, (x_{p+k-1}, x_{p+k}).$$

Again, this does not alter the demi-degrees, and graph H' satisfies (1).

After repeating this procedure a finite number of times, a symmetric graph G is obtained.

$$\text{Q.E.D.}$$

Corollary. *Given integers* $r_1 \geqslant r_2 \geqslant \cdots \geqslant r_n$, *a necessary and sufficient condition that there exists a symmetric p-graph with*

$$d_G^+(x_i) = d_G^-(x_i) = r_i$$

for all i, is that

$$\sum_{i=1}^{pk} r_i^* \geqslant \sum_{j=1}^{k} r_j \qquad (k = 1, 2, \ldots, n-1).$$

The result follows by applying Theorem 2 to the p-graph G_0 with vertices $x_1, x_2, \ldots, x_n$ and p arcs going from x_i to x_j for all i and all j, and then, invoking Corollary 1 to Theorem 1.

The following consequence of Theorem 1 is used to characterize tournaments.

Theorem 3. (Landau [1953]; Moon [1963]). *There exists a complete antisymmetric 1-graph with outer demi-degrees*

$$r_1 \leqslant r_2 \leqslant \cdots \leqslant r_n$$

if, and only if,

$$\begin{cases} \sum_{i=1}^{k} r_i \geqslant \binom{k}{2} & (k = 2, 3, \ldots, n-1) \\ \sum_{i=1}^{n} r_i = \binom{n}{2} \end{cases}$$

where $\binom{p}{q}$ *denotes the binomial coefficient* $\dfrac{p!}{q!(p-q)!}$.

The condition is necessary because, in a complete anti-symmetric 1-graph the number of edges joining the vertices of the set $\{x_1, x_2, \ldots, x_k\}$ is less than or equal to the number of arcs leaving the vertices of the set.

The condition is sufficient. To prove *sufficiency*, let

$$s_i = (n - 1) - r_i \qquad (i = 1, 2, ..., n) .$$

Thus,

$$s_1 \geqslant s_2 \geqslant s_3 \geqslant \cdots \geqslant s_n .$$

1. First we shall show that the conditions of Theorem 1 are satisfied for $p = 1$. Note that

$$\sum_{i=1}^{n} s_i = n(n - 1) - \sum_{i=1}^{n} r_i = n(n - 1) - \frac{n(n - 1)}{2} = \frac{n(n - 1)}{2} = \sum_{i=1}^{n} r_i .$$

Furthermore, by letting t_k denote the number of r_i that are $< k$,

$$\sum_{i=1}^{n} \min \{ k, r_i \} = \sum_{i=1}^{t_k} r_i + k(n - t_k) \geqslant \binom{t_k}{2} + k(n - t_k) .$$

Note that, for any integers k and t,

$$2\left[\binom{t}{2} + \binom{k}{2} - k(t - 1)\right] = t(t - 1) + k(k - 1) - 2kt + 2k$$
$$= (k - t)^2 + (k - t) \geqslant 0 .$$

Thus,

$$\sum_{i=1}^{n} \min \{ k, r_i \} \geqslant \binom{t_k}{2} + k(n - t_k) =$$

$$= \binom{t_k}{2} + \binom{k}{2} - \binom{k}{2} - k(t_k - 1) + k(n - 1) \geqslant$$

$$\geqslant k(n - 1) - \binom{k}{2} \geqslant$$

$$\geqslant k(n - 1) - \sum_{i=1}^{k} r_i = \sum_{i=1}^{k} s_i ,$$

and the conditions of Theorem 1 are satisfied.

2. From Part 1, there exists a 1-graph $G = (X, U)$ such that

$$d_G^+(x_i) = r_i , \quad d_G^-(x_i) = s_i \qquad (i = 1, 2, ..., n) .$$

If this 1-graph G has neither loops nor two oppositely directed arcs joining the same pair of vertices (call such arcs *multiple edges*), then G is a complete anti-symmetric 1-graph because the number of arcs is $\sum_{i=1}^{n} r_i = \binom{n}{2}$. In this case, the theorem has been verified.

Otherwise, we shall alter G and decrease the number of loops and multiple edges without changing its demi-degrees. Suppose that at vertex x_i there are $p(x_i)$ loops and $q(x_i)$ multiple edges. The number of vertices that are not adjacent to x_i is

$$(n-1) - | \Gamma_G(x_i) | = (n-1) - [r_i + s_i - q(x_i) - 2p(x_i)] = q(x_i) + 2p(x_i) .$$

Colour red each multiple edge and loop, and insert a green edge between each pair of non-adjacent vertices; then the number of red edges incident to x equals the number of green edges incident to x. Suppose we travel through the coloured edges of the graph without using two edges of the same colour in succession and without using the same edge twice. Then, after arrival at a vertex x, we can always leave x except perhaps if x is the initial vertex of the tour.

In other words, if there exists red edges, then there exists a cycle

$$[y_1, y_2, ..., y_{2k}, y_1] ,$$

with alternately red and green edges such that

$$(y_1, y_2), (y_2, y_1) \in U ,$$
$$(y_2, y_3), (y_3, y_2) \notin U ,$$
$$(y_3, y_4), (y_4, y_3) \in U ,$$
$$.................$$
$$(y_{2k}, y_1), (y_1, y_{2k}) \notin U .$$

Thus G may be altered by removing arcs (y_2, y_1), (y_4, y_3), etc... and by adding arcs (y_2, y_3), (y_4, y_5), ..., (y_{2k}, y_1), without changing any demi-degree. This process decreases the number of red edges. It can be repeated until no more red edges are present.

Q.E.D.

2. Existence of a p-graph without loops and with given demi-degrees

Consider the following problem:

Given a graph $G_0 = (X, U)$, construct a partial graph H with given demi-degrees $d_H^+(x)$ and $d_H^-(x)$.

For $A \subset X$ and $B \subset X$, let $m_{G_0}^+(A, B)$ denote the number of arcs in G_0 whose initial endpoint is in A and whose terminal endpoint is in B. If an integer r_i is associated with each $x_i \in X$ then, for each $A \subset X$, let

$$r(A) = \sum_{x_i \in A} r_i .$$

Theorem 4. *For a graph G_0 with vertices $x_1, x_2, ..., x_n$, and integers r_i, s_i, for $i = 1, 2, ..., n$, a necessary and sufficient condition that G_0 have a partial subgraph H with*

$$d_H^+(x_i) = r_i, \qquad d_H^-(x_i) = s_i \qquad (i = 1, 2, ..., n)$$

is that

(1) $$\sum_{i=1}^{n} \min \{ r_i, m_{G_0}^+(x_i, A) \} \geqslant s(A) \qquad (A \subset X)$$

(2) $$\sum_{i=1}^{n} r_i = \sum_{i=1}^{n} s_i .$$

Consider a bipartite transportation network $R = (X, \bar{X}, \bar{U})$ with vertex sets $X = \{ x_1, x_2, ..., x_n \}$ and $\bar{X} = \{ \bar{x}_1, \bar{x}_2, ..., \bar{x}_n \}$, with a source a and a sink b, and with arcs

$$
\begin{array}{lll}
(x_i, \bar{x}_j) & & \text{with capacity} \quad m_{G_0}^+ (x_i, x_j) \\
(a, x_i) & \text{(for } 1 \leqslant i \leqslant n) & \text{with capacity} \quad r_i \\
(\bar{x}_j, b) & \text{(for } 1 \leqslant j \leqslant n) & \text{with capacity} \quad s_j .
\end{array}
$$

If condition (2) is satisfied, the desired partial graph H exists if, and only if, network R has a maximum flow that saturates the sink arcs; from Theorem (2, Ch. 5), this is equivalent to

(1′) $$F(\bar{A}) \geqslant d(\bar{A}) \qquad (\bar{A} \subset \bar{X}),$$

where $F(\bar{A})$ is the maximum amount of flow that can enter $\bar{A}$, i.e.

$$F(\bar{A}) = \sum_{i=1}^{n} \min \{ r_i, m_{G_0}^+(x_i, \bar{A}) \},$$

and $d(\bar{A})$ is the total demand of set $\bar{A}$, i.e.

$$d(\bar{A}) = \sum_{x_j \in A} s_j = s(\bar{A}) .$$

Clearly, condition (1′) is equivalent to condition (1).

                                                                    Q.E.D.

Corollary 1. *Given pairs of integers* $(r_1, s_1), (r_2, s_2), ..., (r_n, s_n)$, *a necessary and sufficient condition for the pairs to constitute the demi-degrees of a p-graph H without loops, is that*

(1) $$\sum_{i=1}^{n} \min \{ r_i, p \mid A - \{ x_i \} \mid \} \geqslant s(A) \qquad (A \subset X)$$

(2) $$\sum_{i=1}^{n} r_i = \sum_{i=1}^{n} s_i .$$

We apply Theorem 4 to a complete symmetric p-graph G_0 without loops. Clearly,

$$m_{G_0}^+(x_i, A) = p \mid A - \{ x_i \} \mid .$$

The result follows.   Q.E.D.

These conditions simplify considerably when

$$r_1 \geqslant r_2 \geqslant \cdots \geqslant r_n,$$
$$s_1 \geqslant s_2 \geqslant \cdots \geqslant s_n.$$

Consider a sequence $(r_1, r_2, \ldots, r_n)$ of positive integers such that

$$r_1 \geqslant r_2 \geqslant \cdots \geqslant r_n.$$

Let $\bar{r}_k$ denote the number of indices i such that $i < k$ and $r_i \geqslant k - 1$ plus the number of indices i such that $i > k$ and $r_i \geqslant k$. The sequence $(\bar{r}_k)$ is called the *corrected conjugate* of sequence (r_i). The numbers $\bar{r}_k$ can be visualized on the *corrected Ferrers diagram* (Fig. 6.2), formed by dividing the positive quadrant into three parts: hatched, dotted or empty.

All squares on the principal diagonal are dotted. In the i-th column, the first r_i squares not on the diagonal are hatched. All other squares are empty.

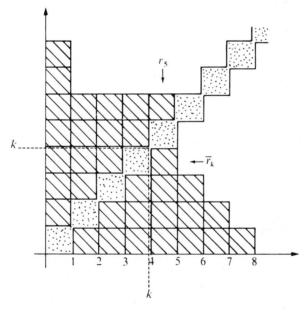

Fig. 6.2. Corrected Ferrers diagram for the sequence $(7, 5, 5, 5, 5, 3, 2, 1)$. The corrected conjugate sequence is $(7, 6, 5, 4, 4, 5, 1, 1)$

Clearly, the number of hatched squares in the k-th row of the diagram equals $\bar{r}_k$, and therefore

$$\sum_{i \geqslant 1} r_i = \sum_{k \geqslant 1} \bar{r}_k.$$

Corollary 2. *Let* (r_i, s_i) *be pairs of integers with*

$$r_1 \geqslant r_2 \geqslant \cdots \geqslant r_n ,$$
$$s_1 \geqslant s_2 \geqslant \cdots \geqslant s_n .$$

Let $(\bar{r}_i)$ *be the corrected conjugate of* (r_i)*. There exists a 1-graph H without loops with* $d_H^+(x_i) = r_i$ *and* $d_H^-(x_i) = s_i$ *for all i, and only if,*

(1)
$$\sum_{i=1}^{k} \bar{r}_i \geqslant \sum_{i=1}^{k} s_i \qquad (k = 1, 2, \ldots, n - 1) ,$$

(2)
$$\sum_{i \geqslant 1} \bar{r}_i = \sum_{i=1}^{n} s_i .$$

1. If such a 1-graph exists, then

$$\sum_{i=1}^{n} \min \{ r_i, | A - \{ x_i \} | \} \geqslant s(A) \qquad (A \subset X) .$$

By taking $A = \{ x_1, x_2, \ldots, x_k \}$, condition (1) follows. Condition (2) obviously holds.

2. Conversely, suppose that conditions (1) and (2) are satisfied. For any set A of cardinality k,

$$\sum_{i=1}^{n} \min \{ r_i, | A - \{ x_i \} | \} =$$

$$= \sum_{x_i \in A} \min \{ r_i, k - 1 \} + \sum_{x_i \in X - A} \min \{ r_i, k \} \geqslant \sum_{i=1}^{k} \bar{r}_i \geqslant \sum_{i=1}^{k} s_i \geqslant s(A) .$$

From Corollary 1, there exists a graph H without loops with $d_H^+(x_i) = r_i$ and $d_H^-(x_i) = s_i$ for all i.

Q.E.D.

We shall now consider necessary and sufficient conditions for the pairs $(r_1, r_1), (r_2, r_2), \ldots, (r_n, r_n)$ to constitute the demi-degrees of a *symmetric p-graph* without loops, i.e. a *p*-graph with:

$$m_G^+(x, y) = m_G^+(y, x) \qquad (x, y \in X) .$$

First, we shall prove the following very general result:

Theorem 5 (Fulkerson, Hoffman, McAndrew [1965]). *Let* $G_0 = (X, U)$ *be a symmetric graph of order n without loops, such that any two vertex-disjoint elementary cycles of odd length are joined by an edge. Let* $r_1, r_2, \ldots, r_n$ *be integers whose sum is even.*

If G_0 *has a partial graph H with* $d_H^+(x_i) = d_H^-(x_i) = r_i$ *for all i, then* G_0 *has a symmetric partial graph G with*

$$d_G^+(x_i) = d_G^-(x_i) = r_i \qquad (i = 1, 2, \ldots, n) \, .$$

1. Let

$$f_1(x, y) = m_H^+(x, y) + m_H^+(y, x) \, .$$

Then,

(1) $$f_1(x, y) = f_1(y, x)$$

(2) $$0 \leqslant f_1(x, y) \leqslant 2 \, m_{G_0}^+(x, y)$$

(3) $$\sum_{y \in X} f_1(x_i, y) = 2 \, r_i \, .$$

If the numbers $f(x, y)$ are all even, then the graph G defined by

$$m_G^+(x, y) = \frac{1}{2} f(x, y)$$

is the required graph.

Otherwise, $f_1(x, y)$ determines a non-empty set $E_1 \subset \mathscr{P}_2(X)$ defined by

$$E_1 = \left\{ [x, y] \, / \, [x, y] \in \mathscr{P}_2(X), f_1(x, y) \equiv 1 \qquad (\text{mod. } 2) \right\} \, .$$

Let $H_1 = (X, E_1)$ denote the simple graph having E_1 as its edge set. Note that in H_1 all degrees are even, because

$$d_{H_1}(x_k) \equiv \sum_{y \in \Gamma_{H_1}(x_k)} f_1(x_k, y) = 2 \, r_k \equiv 0 \qquad (\text{mod. } 2) \, .$$

2. We shall now successively construct a sequence $f_1, f_2, \ldots, f_q$ of functions of two variables that satisfy (1), (2), and (3) and define (as above) sets of edges

$$E_1 \supset E_2 \supset \cdots \supset E_q$$

and simple graphs $H_1, H_2, \ldots, H_q$. When we encounter a set $E_q = \varnothing$ (i.e., a function $f_q(x, y)$ with even values that satisfy conditions (1), (2), and (3)), we shall obtain the required graph G, defined by

$$m_G^+(x, y) = \frac{1}{2} f_q(x, y) \, .$$

3. If the simple graph H_1 has an edge, then it has a cycle (because all its degrees are even). Let μ denote such a cycle.

If cycle μ is even, define a function $f_2(x, y)$, that differs from $f_1(x, y)$ only on the edges of μ, by adding alternately $+ 1$ and $- 1$ to $f_1(x, y)$ while traversing cycle μ.

Since cycle μ is even, $f_2(x, y)$ also satisfies conditions (1), (2) and (3), and determines a new set of edges $E_2 = E_1 - \mu$. The simple graph $H_2 = (X, E_2)$

defined in this way has only even degrees. If H_1 contains an even cycle μ', we can similarly define $H_3 = (X, E_3)$, etc. Repeat this process until a simple graph $H_p = (X, E_p)$ without any even cycles has been found.

4. If the graph H_p has an edge, it contains a cycle μ (because all its degrees are even). This cycle μ is necessarily elementary (because otherwise the edges of μ would contain an even cycle which contradicts the definition of H_p) and odd. Let

$$\mu = [a_1, a_2, ..., a_{2k+1}, a_{2k+2} = a_1]$$

be this elementary odd cycle of H_p. Then

$$\sum_{i=1}^{2k+1} f_p(a_i, a_{i+1}) \equiv 1 \qquad (\text{mod. } 2)$$

and

$$\sum_{\substack{i,j \\ i<j}} f_p(x_i, x_j) = \frac{1}{2} \sum_i \sum_j f_p(x_i, x_j) =$$

$$= \sum_{i=1}^{n} r_i \equiv 0 \qquad (\text{mod. } 2) \ .$$

This shows that f_p has at least one odd value on an edge not in μ, and that H_p contains an edge not in μ. Thus H_p has another odd elementary cycle $v \neq \mu$, and v has no common vertices with μ (since, otherwise, they would form an even cycle).

From the assumptions on G_0, these two cycles are joined by an arc of G_0. For example, let

$$\begin{cases} \mu = [a_1, a_2, ..., a_{2k+1}, a_1] \\ v = [b_1, b_2, ..., b_{2l+1}, b_1] \\ (a_1, b_1) \in U \ . \end{cases}$$

Edge $[a_1, b_1]$, which is not in H_p, satisfies

$$f_p(a_1, b_1) \equiv 0 \qquad (\text{mod. } 2) \ .$$

If $f_p(a_1, b_1) = 0$, the function f_{p+1} is obtained by letting $f_{p+1}(a_1, b_1) = 2$, and by adding alternatively -1 and $+1$ to the edges $[a_1, a_2], ..., [a_{2k+1}, a_1]$, and also to the edges $[b_1, b_2], ..., [b_{2l+1}, b_1]$.

If $f_p(a_1, b_1) \neq 0$, let $f_{p+1}(a_1, b_1) = f_p(a_1, b_1) - 2$ and add alternately $+1$ and -1 to the edges $[a_1, a_2], ..., [a_{2k+1}, a_1]$ and to the edges $[b_1, b_2], ..., [b_{2l+1}, b_1]$.

Function f_{p+1} also satisfies conditions (1), (2) and (3), and it yields the simple graph $H_{p+1} = (X, E_{p+1})$, where $E_{p+1} = E_p - (\mu \cup v)$. Since E is finite, we obtain in a finite number of steps a set $E_q = \varnothing$.

$$Q.E.D.$$

Corollary. *Given the integers $p \geqslant 1$, $r_1 \geqslant r_2 \geqslant \cdots \geqslant r_n \geqslant 1$ such that $\sum r_i$ is even, a necessary and sufficient condition for the existence of a symmetric p-graph G without loops whose vertices satisfy $d_G^+(x_i) = d_G^-(x_i) = r_i$, is that*

$$\sum_{i=1}^{n} \min \{ r_i, p \mid A - \{ x_i \} \mid \} \geqslant \sum_{x_i \in A} r_i \quad (A \subset X) .$$

The proof follows by applying Theorem 5 to a graph G_0 with vertices $x_1, x_2, \ldots, x_n$ and with p arcs from x_i to x_j for each pair (x_i, x_j) with $x_i \neq x_j$.

3. Existence of a simple graph with given degrees

This section describes necessary and sufficient conditions for a sequence of integers $d_1 \geqslant d_2 \geqslant \cdots \geqslant d_n$ to constitute the degrees of a simple graph.

Theorem 6 (Erdös, Gallai [1960]). *Let $d_1 \geqslant d_2 \geqslant \cdots \geqslant d_n$ be a decreasing sequence of n integers with $\sum d_i$ even, and let $(\bar{d}_i)$ denote its corrected conjugate sequence. The following conditions are equivalent:*

(1) *There exists a simple graph G whose vertices x_i satisfy $d_G(x_i) = d_i$;*

(2) $\quad \displaystyle\sum_{i=1}^{k} d_i \leqslant \sum_{i=1}^{k} \bar{d}_i \qquad\qquad (k = 1, 2, \ldots, n) ;$

(3) $\quad \displaystyle\sum_{i=1}^{k} d_i \leqslant k(k-1) + \sum_{j=k+1}^{n} \min \{ k, d_j \} \qquad (k = 1, 2, \ldots, n) .$

(1) $\Rightarrow$ (2) Condition (1) implies the existence of a 1-graph G^* without loops such that $d_{G^*}^+(x_i) = d_{G^*}^-(x_i) = d_i$, for all i; this implies (2), by Corollary 2, Theorem 4.

(2) $\Rightarrow$ (1) Condition (2) implies the existence of a 1-graph H without loops such that $d_H^+(x_i) = d_H^-(x_i) = d_i$, from Corollary 2, Theorem 4. Since $\sum d_i$ is even, this implies from Theorem 5 the existence of a symmetric 1-graph G^* without loops such that

$$d_{G^*}^+(x_i) = d_{G^*}^-(x_i) = d_i .$$

(2) $\Rightarrow$ (3) Consider the corrected Ferrers diagram for the sequence (d_i), and denote by α_k the number of empty squares in $[0, k] \times [0, k]$; condition (3) is equivalent to

(3′) $\qquad\qquad\qquad \displaystyle\sum_{i=1}^{k} d_i \leqslant \sum_{i=1}^{k} \bar{d}_i + \alpha_k .$

Since $\alpha_k \geqslant 0$, condition (2) implies condition (3'), which implies condition (3).

(3) $\Rightarrow$ (2)   Suppose condition (3') is satisfied and there exists an integer k with

$$\sum_{i=1}^{k} d_i > \sum_{i=1}^{k} \bar{d}_i .$$

We shall show that this results in a contradiction.

Clearly $k > 1$. Since only one square in $[0, 1] \times [0, 1]$ is dotted, we have $\alpha_1 = 0$ and, from condition (3'), $d_1 \leqslant \bar{d}_1 + \alpha_1 = \bar{d}_1$. Let q be the largest integer such that $d_q \geqslant k - 1$. Then $q < k$ (since $\alpha_k > 0$). Hence

$$\sum_{i=1}^{k} d_i > \sum_{i=1}^{k} \bar{d}_i = q(k - 1) + \sum_{i=q+1}^{k} d_i + \sum_{i=k+1}^{n} d_i .$$

Thus

(i) $$\sum_{i=1}^{q} d_i > q(k - 1) + \sum_{i=k+1}^{n} d_i .$$

On the other hand, from condition (3), we have

$$\sum_{i=1}^{q} d_i \leqslant q(q - 1) + \sum_{i=q+1}^{n} \min \{ d_i, q \} \leqslant$$

$$\leqslant q(q - 1) + \sum_{i=q+1}^{k} \min \{ d_i, q \} + \sum_{i=k+1}^{n} d_i \leqslant$$

$$\leqslant q(q - 1) + (k - q) q + \sum_{i=k+1}^{n} d_i .$$

Hence

(ii) $$\sum_{i=1}^{q} d_i \leqslant q(k - 1) + \sum_{i=k+1}^{n} d_i .$$

Comparing (i) and (ii), we obtain the desired contradiction.

Q.E.D.

We shall now describe conditions for a sequence (d_i) to constitute the degrees of various types of simple graph.

Theorem 7.   *Consider two sequences $r_1 \geqslant r_2 \geqslant \cdots \geqslant r_p$ and $s_1 \geqslant s_2 \geqslant \ldots \geqslant s_q$, with $p \leqslant q$. There exists a simple bipartite graph $G = (X, Y, E)$ on the sets*

$$X = \{\, x_1, ..., x_p \,\} \qquad and \qquad Y = \{\, y_1, ..., y_q \,\},$$

such that

$$d_G(x_i) = r_i \qquad (i = 1, 2, ..., p),$$
$$d_G(y_j) = s_j \qquad (j = 1, 2, ..., q),$$

if, and only if,

(1) $$\sum_{i=1}^{k} r_i^* \geqslant \sum_{j=1}^{k} s_j \qquad (k = 1, 2, ..., q - 1),$$

(2) $$\sum_{i \geqslant 1} r_i^* = \sum_{j=1}^{q} s_j .$$

Clearly, a necessary and sufficient condition that there exists such a graph G is that there exists a 1-graph whose demi-degrees are given by the pairs

$$(0, s_1), (0, s_2), ..., (0, s_q) \ (r_1, s_{q+1}), (r_2, s_{q+2}), ..., (r_p, s_{q+p}),$$

where

$$s_{q+1} = s_{q+2} = \cdots = s_{q+p} = 0 .$$

From Corollary 2, Theorem 1, such a 1-graph exists if, and only if, we have both

(1') $$\sum_{i=1}^{k} r_i^* \geqslant \sum_{j=1}^{k} s_j \qquad (k = 1, 2, ..., q + p), \text{ and}$$

(2') $$\sum_{i \geqslant 1} r_i^* = \sum_{j=1}^{q+p} s_j .$$

These conditions are equivalent to conditions (1) and (2) above.

Q.E.D.

Theorem 8. *The numbers $d_1, d_2, ..., d_n$ constitute the degrees of a tree if, and only if,*

(1) $$d_i \geqslant 1 \qquad (i = 1, 2, ..., n),$$

(2) $$\sum_{i=1}^{n} d_i = 2(n - 1) .$$

From Theorem (19, Ch. 3), these conditions are equivalent to

$$T(n ; d_1, d_2, ..., d_n) \neq 0 .$$

Q.E.D.

Theorem 9. *Let $d_1 \geqslant d_2 \geqslant \cdots \geqslant d_n$ be a sequence of integers, $n \geqslant 2$. A*

necessary and sufficient condition for the existence of a simple connected graph
G *with degrees* $d_G(x_i) = d_i$, *is that*

(1) $$d_n \geqslant 1$$

(2) $$\sum_{i=1}^{n} d_i \geqslant 2(n-1)$$

(3) $$\sum_{i=1}^{n} d_i \text{ is even}$$

(4) $$\sum_{i=1}^{k} d_i \leqslant \sum_{i=1}^{k} \bar{d}_i \qquad (k = 1, 2, ..., n).$$

Suppose these conditions are satisfied; then from conditions (3) and (4) there exists a simple graph with the given degrees. From (2), it has at least $n - 1$ edges, and therefore, if this graph is not connected, it has a cycle (Theorem 1, Ch. 2). Let $[x, y]$ be an edge of this cycle, and let $[a, b]$ be an edge of a different connected component. This exists, from condition (1).

If edges $[x, y]$ and $[a, b]$ are replaced by two new edges $[x, a]$ and $[y, b]$, the number of connected components is reduced without changing any degrees. By repeating this operation as many times as needed, we obtain a connected graph.

Conversely, suppose there exists a simple connected graph G with degrees

$$d_G(x_i) = d_i.$$

Then conditions (1), (3) and (4) are satisfied. Furthermore, if m denotes the number of edges in G,

$$\sum_{i=1}^{n} d_i = \sum_{i=1}^{n} d_G(x_i) = 2m \geqslant 2(n-1)$$

(since the cyclomatic number satisfies $v(G) = m - n + 1 \geqslant 0$). Thus condition (2) is also satisfied.

Q.E.D.

Recall that a graph of order $n > 2$ is defined to be 2-*connected* if it is connected and if it has no articulation vertex (a vertex whose removal disconnects the graph).

Theorem 10. *Let* $d_1 \geqslant d_2 \geqslant \cdots \geqslant d_n$ *be a sequence of integers,* $n > 2$. *A necessary and sufficient condition for the existence of a simple 2-connected graph G with degrees* $d_G(x_i) = d_i$ *is that*

(1) $$d_n \geqslant 2$$

(2)
$$\sum_{i=1}^{n} d_i \geqslant 2(n + d_1 - 2)$$

(3)
$$\sum_{i=1}^{n} d_i \text{ is even}$$

(4)
$$\sum_{i=1}^{k} d_i \leqslant \sum_{i=1}^{k} \bar{d_i} \qquad (k = 1, 2, ..., n).$$

1. *Assume that conditions* (1), (2), (3) *and* (4) *are satisfied.* Then there exists from Theorem 9 a simple connected graph G with $d_G(x_i) = d_i$ for all i. If x_k is an articulation point, the subgraph G' generated by $X - \{x_k\}$ has $p' \geqslant 2$ connected components. At least one of these connected components has a cycle, since

$$v(G') = m' - n' + p' = m - d_k - (n - 1) + p' \geqslant$$

$$\geqslant \frac{1}{2} \sum_{i=1}^{n} d_i - d_1 - n + 1 + p' \geqslant p' - 1 \geqslant 1 .$$

Let $[y, z]$ be an edge of this cycle, and let $[t, u]$ be an edge of another connected component of G' (which exists from condition (1)). If the edges $[y, z]$ and $[t, u]$ are removed, and two new edges $[y, t]$ and $[z, u]$ are added, the degrees are not altered, but the number of connected components of G' is reduced. Furthermore, for each vertex x, this operation does not increase the number of connected components of the subgraph $G_{x-\{x\}}$ (supposing that $[t, u]$ has been selected in a cycle if its component in G' is not a tree).

Repeating this operation as many times as needed, we obtain a graph H such that $H_{x-\{x\}}$ has only one connected component (for each vertex x). Graph H is the required graph.

2. *Conversely, the degrees of a 2-connected graph G satisfy the required conditions*, because $G_{x-\{x_1\}}$ has $m - d_1$ edges, $n - 1$ vertices, and

$$0 \leqslant v(G_{x - \{x_1\}}) = m' - n' + 1 = m - d_1 - n + 2 .$$

Thus

$$\sum_i d_i = 2m \geqslant 2(n + d_1 - 2) .$$

Hence, condition (2) holds.

Conditions (1), (3) and (4) clearly hold.

Q.E.D.

EXERCISES

1. For a multigraph G without multiple edges (but possibly with loops) let $\delta_G(x) = d_G(x)$ if vertex x has no loops. Let $\delta_G(x) = d_G(x) - 1$ if vertex x has a loop. In other words, a loop increases by 1 (not 2) the "corrected degree" $\delta_G(x)$.

Show from Theorem 2 that for a sequence $d_1 \geqslant d_2 \geqslant \cdots \geqslant d_n$, there exists a multigraph G (without multiple edges) with n vertices $x_1, x_2, \ldots, x_n$, such that $\delta_G(x_i) = d_i$ for $i = 1, 2, \ldots, n$, if and only if

$$\sum_{i=1}^{k} d_i^* \geqslant \sum_{j=1}^{k} d_j \qquad (k = 1, 2, \ldots, n) .$$

(Ramachandra Rao [1969])

2. Show that the pairs (r_i, s_i) are the demi-degrees of an arborescence with root x_1 if, and only if,

$$r_1 = 0$$
$$r_i = 1 \qquad (i \neq 1)$$
$$\sum_{i=1}^{n} s_i = n - 1 .$$

3. Show that the pairs (r_i, s_i) are the demi-degrees of a strongly connected functional graph if, and only if,

$$r_i = s_i = 1 \qquad (i = 1, 2, \ldots, n) .$$

4. Show that a sequence $d_1, d_2, \ldots, d_n$ constitutes the degrees of a multigraph if, and only if, $\sum d_i$ is even.   (Senior [1951])

5. A simple graph G is said to be *k-edge-connected* if it is not disconnected by the removal of less than k edges. Show that if $n > 1$ and $k > 1$, then a sequence $d_1, d_2, \ldots, d_n$ constitutes the degrees of a k-edge-connected graph if, and only if,
(1) the d_i are the degrees of a simple graph,
(2) $d_i \geqslant k$ for all i.   (J. Edmonds [1964])

6. Let $d_1 \geqslant d_2 \geqslant \cdots \geqslant d_n$ be a sequence of integers with $\sum d_i$ even. Show that there exists a multigraph G with multiplicity p without loops with vertices x_i of degree $d_G(x_i) = d_i$ if, and only if,

$$\sum_{i=1}^{k} d_i \leqslant \min_{\substack{l \geqslant k \\ l \leqslant n}} \left(\sum_{i=l+1}^{n} d_i + pkl - pk \right) \qquad (k = 1, 2, \ldots, n) .$$

Hint: This can be shown from the Corollary to Theorem 5.

7. Let $d_1 \geqslant d_2 \geqslant \cdots \geqslant d_n$ be a sequence of integers. From its corrected conjugate sequence $(\bar{d}_1, \bar{d}_2, \bar{d}_3, \ldots)$.

Show that $\bar{d}_1 \geqslant \bar{d}_2 \geqslant \bar{d}_3 \geqslant \cdots$, or that there exists an integer k such that

$$\bar{d}_1 \geqslant \bar{d}_2 \geqslant \cdots \geqslant \bar{d}_k ,$$

$$\bar{d}_{k+1} = \bar{d}_k + 1 \geqslant \bar{d}_{k+2} \geqslant \bar{d}_{k+3} \geqslant \cdots .$$

8. Let $d_1 \geqslant d_2 \geqslant \cdots$ be a decreasing sequence, and let $(\bar{d}_i)$ be its corrected conjugate sequence. Consider an integer $k \leqslant n$, and let l_0 be the largest integer l such that the number of hatched or dotted squares in the l-th column of the corrected Ferrers diagram is $\geqslant k$.

For $l_0 \geqslant k$, show that

$$\sum_{i=1}^{k} \bar{d}_i = \min_{l \geqslant k} \left(\sum_{i=l+1}^{n} d_i + kl - k \right) = \sum_{i=l_0+1}^{n} d_i + kl_0 - k .$$

Let α_k denote the number of empty squares in the square $[0, k] \times [0, k]$ of the corrected Ferrers diagram. For $l_0 < k$, show that

$$\sum_{i=1}^{k} \bar{d}_i = \min_{l \geqslant k} \left(\sum_{i=l+1}^{n} d_i + kl - k \right) - \alpha_k .$$

9. Show that if G is a 3-connected graph with degrees $d_1 \leqslant d_2 \leqslant \cdots \leqslant d_n$, then

(1) $\qquad\qquad d_k \geqslant 3$

(2) $\qquad\qquad \sum_i d_i \geqslant 2(n - 4 + d_n + d_{n-1})$

(3) $\qquad\qquad \sum_i d_i$ is even

(4) $\qquad\qquad \sum_{i=1'}^{k} d_i \leqslant \sum_{i=1}^{k} \bar{d}_i .$

(S. B. Rao and A. Ramachandra Rao [1969] have shown that these necessary conditions are also sufficient for the existence of a 3-connected graph with degrees d_i.)

10. Let $d_1 \geqslant d_2 \geqslant \cdots \geqslant d_n$ be a sequence of integers with $\sum d_i$ even. Show that there exists a multigraph G with multiplicity p without loops with vertices x_i of degree $d_G(x_i) = d_i$ if, and only if,

$$\sum_{i=1}^{k} d_i \leqslant pk(k - 1) + \sum_{i=k+1}^{n} \min \{ pk, d_i \} \quad (k = 1, 2, ..., n)$$

(V. Chungphaisan)

11. Show, for the case $k = 1$, that there exists a simple graph with degree sequence $(d_1, .. , d_n)$ containing a regular partial graph of degree k if, and only if,
 (i) there is a simple graph with degree sequence $(d_1, ..., d_n)$,
(ii) there is a simple graph with degree sequence $(d_1 - k, ..., d_n - k)$.
(V. Chungphaisan has verified this conjecture of B. Grünbaum and S. Kundu has shown that the above statement is true for all k.)

12. Show that the sequence $d_1 \geqslant d_2 \geqslant \cdots \geqslant d_n$ is the degree sequence of a simple graph G if and only if, for all $i = 1, 2, ...,$ max $\{ i / d_i \geqslant i - 1 \}$, the number

$$D_i = \sum_{k=1}^{i} d_k - i(i - 1) - \sum_{k=i+1}^{n} \min \{ d_k, i \}$$

is $\leqslant 0$. Show also that G is unique if and only if $D_i = 0$ for all i.

(Shuo Yen [1975])

CHAPTER 7

Matchings

1. The maximum matching problem

Given a simple graph $G = (X, E)$, a *matching* is defined to be a set E_0 of edges such that no two edges of E_0 are adjacent. If E_0 is a matching, and if $E_1 \subset E_0$, then E_1 is also a matching.

We shall study the following problem: *Find a matching E_0 such that $| E_0 |$ is maximum.*

A vertex x is said to be *saturated* by a matching E_0 if an edge of E_0 is attached to x. Let $S(E_0)$ denote the set of all saturated vertices. A matching that saturates all vertices of G is called a *perfect matching*. Clearly, a perfect matching is a maximum matching. In this chapter, we shall use dark lines to denote the edges of E_0, and light lines to denote the edges of $E - E_0$.

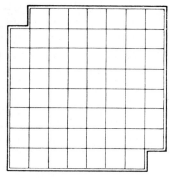

Truncated chessboard

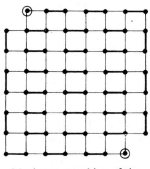

Maximum matching of the corresponding graph

Fig. 7.1

EXAMPLE 1. *Problem of the truncated chessboard.* Consider an 8×8 chessboard whose upper left and lower right corner squares have been removed. (See Fig. 7.1.) We have 31 dominoes, each domino covering exactly two adjacent squares of the chessboard. Can we cover the 62 squares of the chessboard with the 31 dominoes?

This problem is equivalent to finding a maximum matching in a graph whose vertices correspond to the squares of the truncated chessboard. In this

120

graph, two vertices are adjacent if they represent adjacent squares in the chessboard. (See Fig. 7.1.) It is easy to see that the matching in Fig. 7.1 is not perfect. A simple argument that does not use matching theory shows that no
perfect matching is possible: Colour the squares of the chessboard black and
white (as usual). Note that the truncated chessboard does not have the same
number of black and white squares, since the two missing squares must have
the same colour. Clearly, each arrangement of the dominoes covers the same
number of black and white squares. Hence, no perfect matching can exist.

If the chessboard had the same number of black and white squares, the
number of dominoes needed to cover it would be difficult to calculate without
using matching theory.

EXAMPLE 2. *The Battle of Britain.* In 1941, the Royal Air Force consisted
of planes requiring two pilots. However, certain pilots could not fly together
because of language differences or training deficiencies. Given these restrictions, what is the greatest number of planes that can be airborne simultaneously? This problem is solved by finding the maximum matching in a
graph whose vertices correspond to the pilots and whose edges join pilots
who can fly together.

EXAMPLE 3. *Personnel assignment problem.* An office has p secretaries
$x_1, x_2, ..., x_p$ and q jobs $y_1, y_2, ..., y_q$. Each secretary is trained to perform
at least one job. Is it possible to assign each secretary to a job for which she is
qualified? Let $\Gamma(x_i)$ denote the set of jobs for which secretary x_i is qualified.
The problem reduces to finding a matching that saturates all the vertices of X
in a bipartite graph (X, Y, Γ).

EXAMPLE 4. *Dating problem.* In a co-educational American college, each
girl has k boy-friends, and each boy has k girl-friends. Is it possible to have a
dance in which all students simultaneously dance with one of their friends?
Later, it will be shown that this is possible.

Consider a matching E_0. An *alternating chain* is defined as a simple chain
(i.e. a chain that does not use the same edge twice) whose edges are alternately in E_0 and in $F_0 = E - E_0$, i.e. alternating dark and light lines.

Lemma. *Let $G = (X, E)$ be a simple graph, and let E_0 and E_1 be two
matchings in G. Consider the partial graph G' with edge set*

$$(E_0 - E_1) \cup (E_1 - E_0)$$

Each connected component of G' is of one of the following types:
Type 1. Isolated vertex.
Type 2. Even elementary cycle whose edges are alternately in E_0 and E_1.

Type 3. Elementary chain whose edges are alternately in E_0 and E_1 and whose endpoints are distinct and are both unsaturated in one of the two matchings.

Let $a \in X$. We have three cases:

CASE 1. If $a \notin S(E_0 - E_1)$ and $a \notin S(E_1 - E_0)$, then a is an isolated vertex.

CASE 2. If $a \in S(E_0 - E_1)$ and $a \notin S(E_1 - E_0)$, then a is the endpoint of an edge in $E_0 - E_1$. No other edge of $E_0 - E_1$ is attached to a (because E_0 is a matching); no edge of $E_1 - E_0$ is attached to a (because $a \notin S(E_1 - E_0)$). Furthermore, $a \notin S(E_1)$ (because, otherwise, an edge of E_1 attached to a would belong to $E_1 - E_0$).

CASE 3. If $a \in S(E_0 - E_1)$ and $a \in S(E_1 - E_0)$, there exists a unique edge of $E_0 - E_1$ attached to a and a unique edge of $E_1 - E_0$ attached to a.

Since these three cases are exhaustive, the maximum degree of the partial graph

$$(X, (E_0 - E_1) \cup (E_1 - E_0))$$

is 2. This shows that the connected components must be of one of the types described above.

Q.E.D.

Theorem 1 (Berge [1957]). *A matching E_0 is maximum if, and only if, there exists no alternating chain between any two distinct unsaturated vertices.*

1. If E_0 is a matching for which there exists an alternating chain between two unsaturated vertices, then by interchanging the dark and light edges along this chain, we obtain a new matching E_1 with $|E_1| = |E_0| + 1$. Thus, matching E_0 was not maximum.

2. Suppose matching E_0 satisfies the condition of the theorem, and let E_1 be a maximum matching. From 1, we know that matching E_1 satisfies the condition of the theorem. Thus, $|E_0 - E_1| = |E_1 - E_0|$ since the elementary chains of the partial graph $(X, (E_0 - E_1) \cup (E_1 - E_0))$ are necessarily even. Thus

$$|E_0| = |E_1|,$$

and, therefore, E_0 is a maximum matching.

Q.E.D.

Corollary 1. *Let E_0 be a maximum matching and consider the alternating chain $\mu = (e_1, f_1, e_2, f_2, ...)$, where $e_i \in E_0$ and $f_i \in F_0 = E - E_0$. Let the* *operation which interchanges dark and light edges in μ be called a "transfer" on μ.*

Each maximum matching E_1 can be obtained from E_0 by a sequence of transfers along vertex-disjoint alternating chains that are, for the matching E_0, either alternating elementary cycles or alternating elementary even chains starting at an unsaturated vertex.

It suffices to make the transfers along the connected components of the partial graph generated by $(E_0 - E_1) \cup (E_1 - E_0)$, since the connected components are of the above two types by virtue of the lemma.

Q.E.D.

Corollary 2. *An edge is called "free" if it belongs to a maximum matching but does not belong to all maximum matchings. An edge e is free if, and only if, for an arbitrary maximum matching E_0, edge e belongs to an even alternating chain beginning at an unsaturated vertex or to an alternating cycle.*

If e belongs to an alternating chain of this type, then clearly, e is free.

Conversely, if e is free, suppose, for example, that $e \in E_0$ and $e \notin E_1$ for some maximum matching E_1. Thus $e \in (E_0 - E_1) \cup (E_1 - E_0)$ and e belongs to a connected component of the partial graph generated by $(E_0 - E_1) \cup (E_1 - E_0)$. Hence, for the matching E_0, e belongs to an even alternating chain beginning at an unsaturated vertex or to an alternating cycle.

Q.E.D.

Theorem 2 (Erdös, Gallai [1959]). *The maximum number of edges in a simple graph of order n with a maximum matching of q edges ($n \geq 2q > 0$) is*

$$
\begin{cases}
\dbinom{2q}{2} & \text{if} \quad n = 2q \, ; \\[2ex]
\dbinom{2q + 1}{2} & \text{if} \quad 2q < n \leq \dfrac{5q + 3}{2} \, ; \\[2ex]
\dbinom{q}{2} + q(n - q) & \text{if} \quad n > \dfrac{5q + 3}{2} \, .
\end{cases}
$$

Note that in the case $n = 2q$, the graph K_{2q}, a clique with $2q$ vertices, is clearly a graph of order n with $m = \dbinom{2q}{2}$ edges and with a maximum matching of cardinality q.

In the second case, the graph formed by the union of a $(2q + 1)$-clique

K_{2q+1} and a set $S_{n-(2q+1)}$ of $n - (2q + 1)$ isolated vertices is clearly a graph of order n with a maximum matching of cardinality q and with $m = \binom{2q+1}{2}$ edges.

Finally, in the third case, take a q-clique K_q and a stable set S_{n-q} and join in all possible ways the vertices of K_q with the vertices of S_{n-q}.

Clearly, this graph of order n has a maximum matching of cardinality q since $n - q > q$ and has $m = \binom{q}{2} + q(n - q)$ edges.

We shall now show that the given numbers represent the maximum possible number of edges. For the first case ($n = 2q$), this is evident.

Suppose that

$$n \geqslant 2q + 1 .$$

Let $\bar{S}(E_0)$ denote the set of unsaturated vertices in a maximum matching E_0 where $|E_0| = q$. Since $n > 2q$, we have $\bar{S}(E_0) \neq \varnothing$. Let E_1 denote the edges in matching E_0 that have one endpoint adjacent to *several* vertices of $\bar{S}(E_0)$. By Theorem 1, the other endpoint of such an edge cannot be adjacent to $\bar{S}(E_0)$, because then there would exist an alternating chain between two distinct unsaturated vertices, and E_0 would not be a maximum matching.

Let $E_2 = E_0 - E_1$, and let $q_1 = |E_1|$ and $q_2 = |E_2|$. Thus, $q_1 + q_2 = q$. For $i = 1, 2$, let X_i denote the set of endpoints of the edges of E_i. Thus,

$$X_1 \cap X_2 = \varnothing, \quad X_1 \cap \bar{S}(E_0) = \varnothing, \quad X_2 \cap \bar{S}(E_0) = \varnothing .$$

1. Two edges of E_1 cannot generate a 4-clique because then there would be an alternating chain joining two vertices of $\bar{S}(E_0)$. Thus, the number of edges of G joining two vertices of X_1 satisfies

$$m_G(X_1, X_1) \leqslant \binom{2q_1}{2} - \binom{q_1}{2} .$$

2. The number of edges of G joining X_1 and $\bar{S}(E_0)$ satisfies

$$m_G(X_1, \bar{S}(E_0)) \leqslant q_1(n - 2q) .$$

3. Let $[x_2, y_2]$ be an edge of E_2. If neither x_2 nor y_2 is adjacent to the set X_1' of vertices of X_1 that are non-adjacent to $\bar{S}(E_0)$, then

$$m_G(\{x_2, y_2\}, X - X_2) \leqslant 2q_1 + 2 \leqslant 3q_1 + 2 .$$

If the edge $[x_2, y_2]$ has an endpoint x_2 adjacent to X_1', (see Fig. 7.2), its other endpoint y_2 is not adjacent to $\bar{S}(E_0)$. Similarly, endpoint y_2 cannot be adjacent to two vertices of X_1'. Thus,

$$m_G(x_2, X - X_2) \leqslant 2\,q_1 + 1,$$
$$m_G(y_2, X - X_2) \leqslant q_1 + 1,$$

and, hence,

$$m_G(\{\,x_2, y_2\,\}, X - X_2) \leqslant 3\,q_1 + 2.$$

Finally, we obtain

$$m_G(X_2, X - X_2) \leqslant 3\,q_2\,q_1 + 2\,q_2.$$

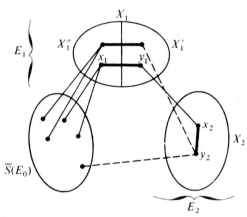

Fig. 7.2

Consequently,

$$m = m_G(X_2, X - X_2) + m_G(X_2, X_2) + m_G(X_1, \overline{S}(E_0)) + m_G(X_1, X_1)$$

$$\leqslant 3\,q_2\,q_1 + 2\,q_2 + \binom{2\,q_2}{2} + q_1(n - 2\,q) + \binom{2\,q_1}{2} - \binom{q_1}{2} =$$

$$= \binom{2\,q + 1}{2} + q_1\left(n - 3\,q + \frac{q_1 - 3}{2}\right)$$

If $n \leqslant \dfrac{5\,q + 3}{2}$, then $n - 3\,q + \dfrac{q_1 - 3}{2} = \left(n - \dfrac{5\,q + 3}{2}\right) + \dfrac{q_1 - q}{2} \leqslant 0.$

Hence $m \leqslant \dbinom{2\,q + 1}{2}$.

If $n > \dfrac{5\,q + 3}{2}$, then

$$m \leqslant \binom{q}{2} + q(n - q) + (q - q_1)\left(\frac{5\,q + 3}{2} - n\right) - \frac{1}{2}q_1(q - q_1),$$

or

$$m \leqslant \binom{q}{2} + q(n - q).$$

Combining the above inequalities for all cases yields,

$$m \leqslant \max \left\{ \binom{2\,q + 1}{2}, \binom{q}{2} + q(n - q) \right\}.$$

Note that

$$\binom{2\,q + 1}{2} \geqslant \binom{q}{2} + q(n - q)$$

is equivalent to

$$2\,q(2\,q + 1) - q(q - 1) + 2\,q^2 \geqslant 2\,qn,$$

or

$$n \leqslant \frac{5\,q + 3}{2}.$$

Q.E.D.

Maximum matching algorithm

Consider a simple graph $G = (X, E)$ with a matching E_0, and associate with it a 1-graph $\overline{G} = (X, U)$, where $(x, y) \in U$ if there exists a vertex such that $[x, z] \in E - E_0$ and $[z, y] \in E_0$. For each even alternating chain μ in G,

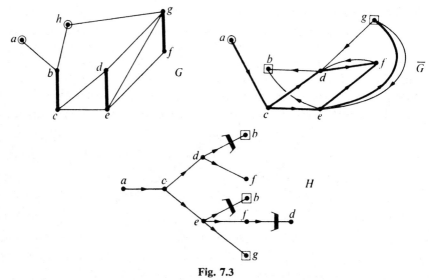

Fig. 7.3

that starts from an unsaturated vertex, there corresponds in $\overline{G}$ a unique elementary path $\overline{\mu}$. For example in Fig. 7.3,

$$\mu = [a,b,c,e,d] \quad \text{corresponds to} \quad \overline{\mu} = [a,c,d] \, .$$

For each elementary path $\overline{\mu}$ in $\overline{G}$, there corresponds a unique even chain μ of G, but this chain is not necessarily simple. For example,

$$\overline{\mu} = [a,c,d,b] \quad \text{corresponds to} \quad \mu = [a,b,c,e,d,c,b]$$

(edge bc appears twice).

A path in $\overline{G}$ will be called *legal* if it corresponds to a simple chain in G. Otherwise, it will be called *illegal*.

Hence, the matching problem reduces to finding a legal path in $\overline{G}$ that connects an unsaturated vertex (e.g. vertex a) to the neighbours of another unsaturated point (e.g. vertices b or g). One could use known algorithms for finding all elementary paths in $\overline{G}$ (see Ch. 4, § 1) and then each path of G that is illegal could be eliminated. In Fig. 7.3, the chain $\overline{\mu} = [a, c, e, g]$ yields the desired alternating chain $\mu = [a, b, c, d, e, f, g, h]$ in G.

As noted by Jack Edmonds ([1962], [1965]), it is not necessary to explore *each* elementary path of $\overline{G}$ starting at vertex a in order to reach an unsaturated vertex by a legal path. Modifications of Edmonds' algorithm have been suggested (C. Witzgall and C. T. Zahn [1965]; M. Balinski [1970]; B. Roy [1969]).

2. The minimum covering problem

Given a simple graph $G = (X, E)$, a *covering* is defined to be a family $F \subset E$ such that each vertex $x \in X$ is the endpoint of at least one edge of F.

The problem of finding a minimum cardinality covering has many similarities to the maximum matching problem. The covering problem is a more general case of a problem known in logic as "Quine's Problem".

EXAMPLE. In the fort shown below (Fig. 7.4), there is a tower at the endpoints

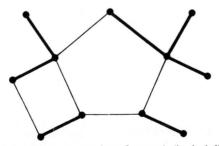

Fig. 7.4. Minimum covering of a graph (in dark lines)

of each wall. A guard stationed at a wall can watch both towers at the end of his wall. What is the minimum number of guards needed to watch all the towers? Since the minimum covering of the corresponding graph is 7 edges, it follows that 7 guards will be required.

Theorem 3 (Norman, Rabin [1959]). *In a simple graph $G = (X, E)$ of order n, a maximum matching E_0 and a minimum covering F_0 satisfy*

$$| E_0 | + | F_0 | = n.$$

Given a maximum matching E_0, a minimum covering

$$F_1 = E_0 \cup \{ \, e_y \mid y \in \bar{S}(E_0) \, \}$$

is obtained by adding to E_0, for each unsaturated vertex y, an edge e_y of G that is incident to y. Given a minimum covering F_0, a maximum matching E_1 is obtained by removing successively from F_0 edges that are adjacent to an unremoved edge.

If E_0 is a maximum matching, the set

$$F_1 = E_0 \cup \{ \, e_y / y \in \bar{S}(E_0) \, \}$$

is clearly a covering, and

$$| F_1 | = | E_0 | + \left(n - 2 \, | E_0 | \right) = n - | E_0 |.$$

Furthermore, if F_0 is a minimum covering, the set E_1 obtained by the successive elimination of edges of F_0 that are adjacent to an (unremoved) edge of F_0 is a matching. Since in G, the edge of F_0 do not form chains of length 3, each removed edge creates exactly one unsaturated vertex of E_1. Hence,

$$| F_0 | - | E_1 | = \left| X - S(E_1) \right| = n - 2 \, | E_1 |$$

and

$$| F_0 | = n - | E_1 |.$$

Since $| E_1 | \leqslant | E_0 |$, it follows that

$$| F_1 | = n - | E_0 | \leqslant n - | E_1 | = | F_0 |.$$

Thus, the covering F_1 is also a minimum covering. Since $| F_1 | = | F_0 |$, it follows that $| E_1 | = | E_0 |$, and, consequently, the matching E_1 is a maximum matching.

Finally, $| E_0 | + | F_0 | = n.$

Q.E.D.

This theorem shows that the minimum covering problem reduces to the maximum matching problem, which we shall study in the next section.

3. Matchings in bipartite graphs

A graph G is said to be *bipartite* if its vertex set can be partitioned into two classes such that no two adjacent vertices belong to the same class.

Theorem 4. *For a graph G, the following conditions are equivalent:*
(1) *G is bipartite,*
(2) *G possesses no elementary cycles of odd length,*
(3) *G possesses no cycles of odd length.*

(1) $\Rightarrow$ (2) because if G is bipartite, we can colour the vertices red and blue such that two adjacent vertices have different colours. If G has an elementary cycle of odd length, then the vertices of the cycle cannot alternate in colour.

(2) $\Rightarrow$ (3) Suppose that G possesses no elementary cycles of odd length, but there exists a cycle $\mu = [x_0, x_1, ..., x_p = x_0]$ of odd length. If there are two vertices x_j and x_k in cycle μ such that $j < k$ and $x_j = x_k$, then the cycle can be decomposed into two cycles $\mu[x_j, x_k]$ and $\mu[x_0, x_j] + \mu[x_k, x_0]$. Furthermore, one of these cycles has odd length (otherwise, μ would have even length).

Clearly, each time that the cycle μ is decomposed in this way, an odd cycle remains. When this decomposition terminates, there will remain an odd elementary cycle, which contradicts (2).

(3) $\Rightarrow$ (1) We shall show that a graph without odd cycles is bipartite. Suppose that the graph is connected (otherwise, each connected component could be considered separately). Successively, colour the vertices using the following rules:

RULE 1. Colour an arbitrary vertex a blue.

RULE 2. If vertex x is blue, colour red all vertices adjacent to x. If vertex y is red, colour blue all vertices adjacent to y.

Since the graph is connected, each vertex is coloured. A vertex x cannot be coloured both red and blue, since then vertices x and a would be contained in a cycle of odd length. This colouring determines a partition of the vertices into two classes and G is bipartite.
Q.E.D.

Henceforth, a bipartite graph with vertex sets X and Y and with edge set E will be denoted by $G = (X, Y, E)$. For any $A \subset X \cup Y$, the set of vertices adjacent to set A is denoted by $\Gamma_G(A)$.

König's Theorem [1931]. *For a bipartite graph* $G = (X, Y, E)$, *the maximum number of edges in a matching equals*

$$\min_{A \subset X} \left(| X - A | + | \Gamma_G(A) | \right)$$

Consider the transportation network with vertices $X \cup Y$ and a source a and a sink b. Source a is joined to each $y_j \in Y$ by an arc of capacity $c(a, y_j) = 1$. Sink b is joined to each $x_i \in X$ by an arc of capacity $c(x_i, b) = 1$. Finally, y_j is joined to x_i by an arc of capacity 1 if $y_j \in \Gamma_G(x_i)$.

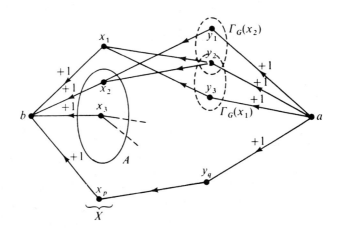

Fig. 7.5

For a set $A \subset X$, the total demand of set A equals $d(A) = | A |$. The maximum amount of flow that can be sent into A equals $F(A) = | \Gamma_G(A) |$.

A flow in this network defines a matching in the graph in which x_i and y_j are matched if a unit of flow traverses arc (y_j, x_i). Conversely, each matching defines a flow.

The cardinality of a maximum matching is, therefore, equal to the value of a maximum flow between a and b. By Theorem (2, Ch. 5),

$$\max_{E_0} | E_0 | = d(X) + \min_{A \subset X} \left(F(A) - d(A) \right) =$$

$$= | X | + \min_{A \subset X} \left(| \Gamma_G(A) | - | A | \right) =$$

$$= \min_{A \subset X} \left(| X - A | + | \Gamma_G(A) | \right).$$

Q.E.D.

For a graph G, a *transversal set* T is defined to be a set of vertices such that each edge has at least one endpoint in T. An equivalent formulation of König's Theorem is:

Corollary 1. *For a bipartite graph G, the maximum number of edges in a matching equals the minimum number of vertices in a transversal set.*

Let E_0 be a maximum matching, and let T_0 be a minimum transversal set. Clearly, $|T_0| \geqslant |E_0|$ since T_0 contains at least one endpoint of each edge in E_0.

Furthermore, for each $A \subset X$, the set $T = (X - A) \cup \Gamma_G(A)$ is a transversal set of G, and from König's theorem,

$$|E_0| = \min_{A \subset X} \big(|X - A| + |\Gamma_G(A)| \big) \geqslant |T_0| \cdot$$

Hence, $|E_0| = |T_0|$.

Q.E.D.

For a graph G, a *stable set* S is defined to be a set of vertices such that no edge has two distinct endpoints in S; another formulation of König's theorem is:

Corollary 2. *For a bipartite graph G, the maximum number of vertices in a stable set equals the minimum number of edges in a covering.*

If $T \subset X$ denotes a transversal set, and if E_0 denotes a matching, then, from Corollary 1,

$$\max_{E_0} |E_0| = \min_{T} |T| \, .$$

If T is a transversal set, its complement $S = (X \cup Y) - T$ is a stable set. If S is a stable set, its complement is a transversal set. Thus,

$$\max_{S} |S| = |X| + |Y| - \min_{T} |T| \, .$$

From Theorem 3,

$$\min_{F} |F| = |X| + |Y| - \max_{E_0} |E_0| \, .$$

Hence,

$$\max_{S} |S| = \min_{F} |F| \, .$$

Q.E.D.

If a bipartite graph $G = (X, Y, E)$ has a matching that saturates all the vertices of X, then we say that X *can be matched into* Y. If this matching also saturates all the vertices of Y, we say that X *can be matched onto* Y.

The following theorem is an easy consequence of the König theorem.

Theorem 5 (P. Hall [1934]; "König-Hall Theorem"). *In a bipartite graph* $G = (X, Y, E)$, X *can be matched into* Y *if, and only if,*

$$\left| \Gamma_G(A) \right| \geqslant |A| \qquad (A \subset X).$$

From the König Theorem, X can be matched into Y if, and only if,

$$|X| = \max_{E_0} |E_0| = \min_{A \subset X} \left(|X - A| + |\Gamma_G(A)| \right).$$

This is equivalent to

$$\min_{A \subset X} \left(|\Gamma_G(A)| - |A| \right) = 0$$

or

$$\left| \Gamma_G(A) \right| - |A| \geqslant 0 \qquad (A \subset X).$$

Q.E.D.

Corollary 1. *In a bipartite multigraph* $G = (X, Y, E)$ *with*

$$|X| = p, \qquad |Y| = q,$$

index the vertices $x_i \in X$ *and* $y_j \in Y$ *such that*

$$d_G(x_1) \leqslant d_G(x_2) \leqslant \cdots \leqslant d_G(x_p),$$
$$d_G(y_1) \geqslant d_G(y_2) \geqslant \cdots \geqslant d_G(y_q).$$

A sufficient condition that X *can be matched into* Y *is that* $q \geqslant p$ *and* $d_G(x_1) > 0$ *and*

$$\sum_{i=1}^{k} d_G(x_i) > \sum_{i=1}^{k-1} d_G(y_i) \qquad (k = 2, 3, \ldots, p).$$

Consider two subsets $A \subset X$ and $B \subset Y$ with k and $k - 1$ elements, respectively. From the above inequality, it follows that

$$m_G(A, Y) = \sum_{x \in A} d_G(x) \geqslant \sum_{i=1}^{k} d_G(x_i) > \sum_{j=1}^{k-1} d_G(y_j) \geqslant$$

$$\geqslant \sum_{y \in B} d_G(y) = m_G(X, B).$$

Thus, the number of edges leaving A is strictly greater than the number of edges entering B. Hence, $\Gamma_G(A) \not\subset B$ for each set B of $k - 1$ elements; consequently $|\Gamma_G(A)| > k - 1 = |A| - 1$. Finally,

$$\left| \Gamma_G(A) \right| \geqslant |A| \qquad (A \subset X).$$

Therefore, X can be matched into Y.

Q.E.D.

Corollary 2. *If, in a bipartite multigraph $G = (X, Y, E)$, we have*

$$\min_{x \in X} d_G(x) \geq \max_{y \in Y} d_G(y) \quad and \quad |Y| \geq |X|,$$

then X can be matched into Y.

Let

$$\min_{x \in X} d_G(x) = d_1, \max_{y \in Y} d_G(y) = d_2.$$

Thus $d_1 \geq d_2$, and by indexing the vertices as described above,

$$d_G(x_1) + d_G(x_2) + \cdots + d_G(x_k) \geq k d_1 > (k-1) d_1 \geq (k-1) d_2 \geq$$
$$\geq d_G(y_1) + d_G(y_2) + \cdots + d_G(y_{k-1}) \quad (k = 2, 3, ..., p).$$

Therefore, X can be matched into Y.

Q.E.D.

Corollary 3. *If $G = (X, Y, E)$ is a bipartite multigraph with no isolated vertices and with $|Y| \geq |X|$, and such that for some vertex $x_1 \in X$,*

$$\min_{\substack{x \neq x_1 \\ x \in X}} d_G(x) \geq \max_{y \in Y} d_G(y),$$

then X can be matched into Y.

We may suppose that x_1 has minimum degree (otherwise, the proof is immediate). Then

$$d_G(x_1) + d_G(x_2) + \cdots + d_G(x_k) \geq d_G(x_1) + (k-1) d_1 >$$
$$> (k-1) d_2 \geq d_G(y_1) + \cdots + d_G(y_{k-1}).$$

Hence, X can be matched into Y.

Q.E.D.

Corollary 4. *In a bipartite multigraph $G = (X, Y, E)$, there exists a matching that saturates all the vertices with maximum degree.*

First, suppose that there exists a bipartite multigraph

$$\overline{G} = (\overline{X}, \overline{Y}, \overline{E}),$$

with $G = (X, Y, E)$ as a subgraph suppose $X \subset \overline{X}$, $Y \subset \overline{Y}$, and $\overline{G}$ is regular of degree

$$h = \max_{z \in X \cup Y} d_G(z).$$

From Corollary 2, $\overline{X}$ can be matched onto $\overline{Y}$ in $\overline{G}$, since $|\overline{X}| = |\overline{Y}|$. This matching saturates each vertex in G of degree h.

We shall construct this graph $\bar{G}$ by taking h replicas of multigraph G. Denote these h replicas by

$$G = (X, Y, E), \quad G' = (X', Y', E'), \quad G'' = (X'', Y'', E''), \ldots$$

$$\ldots, G^{(h-1)} = (X^{(h-1)}, Y^{(h-1)}, E^{(h-1)}).$$

Let $\bar{X}$ consist of $X \cup X' \cup \cdots \cup X^{(h-1)}$ together with some additional vertices. Similarly, let $\bar{Y}$ consist of $Y \cup Y' \cup \cdots \cup Y^{(h-1)}$ together with some additional vertices. These additional vertices are determined as follows: If $x_i \in X$ and $d_G(x_i) < h$, create, in $\bar{Y}$, $h - d_G(x_i)$ additional vertices, and join each of · these vertices to $x_i \in X$, $x_i' \in X'$, $x_i'' \in X''$, ..., $x_i^{(h-1)} \in X^{(h-1)}$, the analogues of x_i. Repeat this construction if $y_j \in Y$ and $d_G(y_j) < h$.

In this way, a multigraph $\bar{G}$ having G as a subgraph is constructed.

Q.E.D.

Note that this result allows us to give an affirmative answer to the Dating Problem (Example 4, § 1).

Corollary 5. *In a bipartite graph $G = (X, Y, E)$, X can be matched into Y if, and only if,*

$$\big| X - \Gamma_G(B) \big| \leqslant | Y - B | \qquad (B \subset Y).$$

(1) If X can be matched into Y, and if $B \subset Y$, then from Theorem 5,

$$\big| X - \Gamma_G(B) \big| \leqslant \big| \Gamma_G(X - \Gamma_G(B)) \big|.$$

No vertex $x \in X - \Gamma_G(B)$ is adjacent to B. Thus,

$$\Gamma_G(x) \subset Y - B,$$

and, hence,

$$\big| X - \Gamma_G(B) \big| \leqslant \big| \Gamma_G(X - \Gamma_G(B)) \big| \leqslant | Y - B |.$$

(2) If X cannot be matched into Y, there exists a set $A \subset X$ such that $| A | > | \Gamma_G(A) |$. Let $B = Y - \Gamma_G(A)$. Since no vertex of B is adjacent to A,

$$A \subset X - \Gamma_G(B).$$

Hence,

$$\big| X - \Gamma_G(B) \big| \geqslant | A | > \big| \Gamma_G(A) \big| = | Y - B |.$$

Q.E.D.

This corollary is in fact a reformulation of the König-Hall theorem that will be needed later.

The next result is a reformulation of Bernstein's theorem.

Theorem 6. *In a bipartite graph $G = (X, Y, E)$, a necessary and sufficient* *condition that there exists a matching that simultaneously saturates* $A \subset X$ *and* $B \subset Y$ *is that*

(1) *A can be matched into Y, i.e.*

$$| \Gamma_G(S) | \geqslant | S | \qquad (S \subset A) ;$$

(2) *B can be matched into A, i.e.*

$$| \Gamma_G(T) | \geqslant | T | \qquad (T \subset B).$$

Clearly, conditions (1) and (2) are necessary, We shall show that if there exists a matching E_0 from B into X, and that if there exists a matching E_1 from A into Y, then there exists a matching saturating both A and B.

We shall now construct from E_1 a matching E_1' in which the saturated vertices of A remain saturated, and an unsaturated vertex $b \in B$ in E_1 becomes saturated in E_1'. Since $b \in S(E_0)$, b is the end-point of a chain μ of the form:

$$\mu[b, z] = (e_1, e_2, e_3, e_4, ...)$$

with

$$e_1, e_3, ... \in E_0 - E_1 ,$$
$$e_2, e_4, ... \in E_1 - E_0 .$$

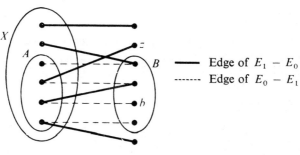

$$\text{—— Edge of } E_1 - E_0$$
$$\text{------ Edge of } E_0 - E_1$$

Fig. 7.6

Suppose that μ is as long as possible. Then, the last vertex z of this chain belongs to Y (otherwise, z belongs to A, and z is an endpoint of an edge of $E_1 - E_0$ that can extend chain μ). If $z \in Y$, then $z \notin B$ (otherwise z is an end-point of an edge of $E_0 - E_1$ that can extend chain μ). Thus, $z \in Y - B$.

Thus, $E_1' = E_1 \cup (\mu \cap E_0) - (\mu \cap E_1)$ is a matching saturating b; besides, each vertex of $A \cup B$ that is saturated in E_1 remains saturated in E_1'.

By repeating this procedure as many times as needed, a matching saturating both A and B can be obtained.

Q.E.D.

Corollary. *In a bipartite graph* $G = (X, Y, E)$, *a necessary and sufficient condition that there exists a matching simultaneously saturating* X *and* $B \subset Y$ *is that*

$$\min \{ \, | \, \Gamma_G(S) \, |, \, | X | - | B - \Gamma_G(S) \, | \} \geqslant | S | \qquad (S \subset X) \, .$$

Using Corollary 5 to Theorem 5, the two conditions of the above theorem can be written as

(1) $\left| \, \Gamma_G(S) \, \right| \geqslant | S |$ $(S \subset X)$

(2') $\left| \, B - \Gamma_G(S) \, \right| \leqslant | X - S |$ $(S \subset X)$

or

(2'') $| X | - \left| \, B - \Gamma_G(S) \, \right| \geqslant | S |$ $(S \subset X) \, .$

$$\text{Q.E.D.}$$

This condition was obtained independently by linear programming methods by Hoffman and Kuhn [1956].

Lemma (Folkman, Fulkerson [1967]). *Let* $G = (X, Y, E)$ *be a bipartite multigraph with maximum degree* $\leqslant h$ *and with* $| E | = m$. *Let* m', m'', h' *and* h'' *be positive integers with*

$$m' + m'' = m, \qquad h' + h'' = h \, .$$

The edges of G *can be partitioned into two classes* E' *and* E'' (*that form two partial multigraphs* G' *and* G''), *such that*

$$| E' | = m' \, , \quad | E'' | = m'' \, , \quad \max_{z \in X \cup Y} d_{G'}(z) \leqslant h' \, , \quad \max_{z \in X \cup Y} d_{G''}(z) \leqslant h'' \, ,$$

if, and only if,

$$m' - h' | X - A | - h' | Y - B | \leqslant m_G(A, B) \, ,$$
$$m'' - h'' | X - A | - h'' | Y - B | \leqslant m_G(A, B) \, ,$$

for all $A \subset X$ *and for all* $B \subset Y$.

1. *Necessity.* If there exist two partial multigraphs G' and G'' satisfying the above conditions, we have

$$m' = | E' | \leqslant m_{G'}(X - A, Y) + m_{G'}(X, Y - B) + m_{G'}(A, B)$$

$$\leqslant h' | X - A | + h' | Y - B | + m_G(A, B) \, .$$

2. *Sufficiency.* Consider the bipartite transportation network R obtained by joining each $x_i \in X$ to a source a and by joining each $y_j \in Y$ to a sink b, and by adding a return arc (b, a). Let

$$\alpha_i = \max \{ \, 0, d_G(x_i) - h'' \, \} \, , \quad \beta_j = \max \{ \, 0, d_G(y_j) - h'' \, \} \, ;$$

Let the interval of permitted flow values in arc u be

$$[b(u), c(u)] = \begin{cases} [\alpha_i, h'] & \text{if} \quad u = (a, x_i) \\ [\beta_j, h'] & \text{if} \quad u = (y_j, b) \\ [0, m_G(x_i, y_j)] & \text{if} \quad u = (x_i, y_j) \\ [m', m'] & \text{if} \quad u = (b, a) \, . \end{cases}$$

Each flow in R that is compatible with these intervals determines a partial graph G' with $|E'| = m'$ and with

$$\max \{ 0, d_G(x_i) - h'' \} \leqslant d_{G'}(x_i) \leqslant h'$$

$$\max \{ 0, d_G(y_j) - h'' \} \leqslant d_{G'}(y_j) \leqslant h'$$

The partial graph $G'' = G - G'$ satisfies $|E''| = m''$ and

$$d_{G''}(x_i) = d_G(x_i) - d_{G'}(x_i) \leqslant h''$$

$$d_{G''}(y_j) = d_G(y_j) - d_{G'}(y_i) \leqslant h''$$

Thus the required partition of the edges of G has been found. Conversely, such a partition determines in graph R a compatible flow within the permitted intervals. From the Compatible Flow Theorem (Ch. 5, § 2), a necessary and sufficient condition for the existence of a compatible flow in R is that

$$b(\omega^-(S)) \leqslant c(\omega^+(S)) \qquad (S \subset X \cup Y \cup \{ a, b \}) \, .$$

It is left to the reader to verify that the above condition is equivalent to that of the lemma.

Q.E.D.

Theorem 7. (Dulmage, Mendelsohn [1961]). *For a bipartite multigraph $G = (X, Y, E)$ with $|E| = m$ and with maximum degree $\leqslant h$, let*

$$\sigma = \max_{\substack{A \subset X \\ B \subset Y}} \{ m - m_G(A, B) - (h - 1)(|X - A| + |Y - B|) \} \, ,$$

$$\rho = \min_{\substack{A \subset X \\ B \subset Y}} \{ m_G(A, B) + |X - A| + |Y - B| \} \, .$$

Then,

$$\sigma \leqslant \left[\frac{m}{h} \right] \leqslant \rho \, .$$

Furthermore, for each integer m' such that $\sigma \leqslant m' \leqslant \rho$, there exists a matching $E' \subset E$ with $|E'| = m'$ whose removal results in a multigraph with maximum degree $\leqslant h - 1$.

1. If $\left[\dfrac{m}{h}\right] > \rho$, then there exist sets $A \subset X$ and $B \subset Y$ such that

$$m > h\big(m_G(A, B) + |X - A| + |Y - B|\big) \geqslant$$

$$\geqslant m_G(A, B) + m_G(X - A, Y) + m_G(A, Y - B) = m$$

which is a contradiction. Thus,

$$\left[\frac{m}{h}\right] \leqslant \rho \,.$$

2. If $\left[\dfrac{m}{h}\right] < \sigma$, then there exist sets $A \subset X$ and $B \subset Y$ such that

$$\frac{m}{h} < m - m_G(A, B) - (h - 1)\big(|X - A| + |Y - B|\big) \,.$$

Hence,

$$m - \frac{m}{h} > m_G(A, B) + (h - 1)\big(|X - A| + |Y - B|\big) \geqslant$$

$$\geqslant \frac{h - 1}{h}\, m_G(A, B) + \frac{h - 1}{h}\big(h\,|X - A| + h\,|Y - B|\big) \geqslant$$

$$\geqslant \frac{h - 1}{h}\big(m_G(A, B) + m_G(X - A, Y) + m_G(X, Y - B)\big) \geqslant$$

$$\geqslant \frac{h - 1}{h}\, m \,,$$

which is a contradiction.

(3) If m' satisfies $\sigma \leqslant m' \leqslant \rho$, then, for all $A \subset X$ and for all $B \subset Y$, we have

$$\begin{cases} m' \geqslant m - m_G(A, B) - \text{'}(h - 1)\big(|X - A| + |Y - B|\big) \\ m' \leqslant m_G(A, B) + |X - A| + |Y - B| \,. \end{cases}$$

Let $h' = 1$, $h'' = h - 1$, $m'' = m - m'$. The above inequalities become

$$\begin{cases} m'' - h''\,|X - A| - h''\,|Y - B| \leqslant m_G(A, B) \\ m' - h'\,|X - A| - h'\,|Y - B| \leqslant m_G(A, B) \,. \end{cases}$$

From the lemma, G can be decomposed into two partial bipartite multigraphs

$$G' = (X, E') \qquad \text{and} \qquad G'' = (X, E - E') \,,$$

with

$$|E'| = m', \quad \max d_{G'}(z) \leqslant 1, \quad \max d_{G''}(z) \leqslant h - 1 \,.$$

Clearly, E' is the required matching.

$$\text{Q.E.D.}$$

4. An extension of the König theorem

In this section, we shall consider a multigraph with the properties:
(1) there are no loops,
(2) if μ and μ' are two odd cycles without a common vertex, there exist two adjacent vertices $x \in \mu$ and $x' \in \mu'$.

Such a graph is called *semi-bipartite* (see Ch. 6, § 2). We shall now study conditions for which a semi-bipartite graph possesses a *perfect matching* (i.e. a matching that saturates all vertices).

Lemma 1. *Let* $G = (X, \Gamma)$ *be a symmetric semi-bipartite 1-graph with* $| X |$ *even; there exists a perfect matching if, and only if, there exists a partial graph* H *of* G *with*

$$d_H^+(x) = d_H^-(x) = 1 \qquad (x \in X) .$$

The proof follows from Theorem (5, Ch. 6).

Lemma 2. *Let* $G = (X, \Gamma)$ *be a 1-graph; there exists a set of elementary circuits of* G *that partition* X *if, and only if,*

$$| \Gamma(A) | \geqslant | A | \qquad (A \subset X) .$$

Associate with G a bipartite graph $G_1 = (X, \bar{X}, E)$ obtained by taking two replicas X and $\bar{X}$ of set X, and joining $x_i \in X$ to $\bar{x}_j \in \bar{X}$ by an edge if $x_j \in \Gamma(x_i)$.

In graph G, set X can be partitioned into circuits if, and only if, the bipartite graph G_1 has a perfect matching, i.e., if, and only if,

$$| \Gamma_{G_1}(A) | = | \Gamma(A) | \geqslant | A | \qquad (A \subset X) .$$

Q.E.D.

Theorem 8. *A semi-bipartite graph* $G = (X, E)$ *possesses a perfect matching if, and only if,*
(1) $| X |$ *is even,*
(2) $| \Gamma_G(A) | \geqslant | A | \qquad (A \subset X)$.

Apply Lemmas 1 and 2 to the symmetric semi-bipartite 1-graph G^* obtained from G by replacing each edge by two oppositely directed arcs.

Corollary. *If* G *is a semi-bipartite regular multigraph of degree* h *that has an even number of vertices, then* G *possesses a perfect matching.*

Make two replicas X and $\bar{X}$ of the vertex set of G, and construct a bipartite multigraph $H = (X, \bar{X}, E)$ with $m_H(x, \bar{y}) = m_G(x, y)$. For each $A \subset X$, we have

$$h | A | = m_H(A, \Gamma_H(A)) \leqslant m_H(X, \Gamma_H(A)) = h | \Gamma_G(A) | .$$

Thus $| \Gamma_G(A) | \geq | A |$ and, from Theorem 8, the simple graph G' obtained from G by collecting the multiple edges has a perfect matching, which is also a perfect matching in G.

Q.E.D.

5. Counting perfect matchings

For certain graphs (particularly planar graphs) simple methods are available to count the number of distinct matchings. These methods are the result of work done independently by Fisher [1961] and by Kasteleyn [1961]. Kasteleyn's proof has been simplified by Rényi [1966] and by Pla [1970]. We shall use Pla's proof.

First, let us review some definitions from matrix algebra. If $A = ((a_j^i))$ is a square matrix of order n, the *determinant* of A is written as

$$\text{Det } A = \sum_\sigma \varepsilon(\sigma)\, a_{\sigma(1)}^1\, a_{\sigma(2)}^2 \cdots a_{\sigma(n)}^n$$

where

$$\sigma = \begin{pmatrix} 1 & 2 & \cdots & n \\ \sigma(1) & \sigma(2) & \cdots & \sigma(n) \end{pmatrix}$$

is a permutation of degree n, and $\varepsilon(\sigma) = +1$ or -1 if the permutation is even or odd, respectively. The *permanent* of A is defined to be the number

$$\text{Perm } A = \sum_\sigma a_{\sigma(1)}^1\, a_{\sigma(2)}^2 \cdots a_{\sigma(n)}^n \,,$$

Proposition 1. *Let $G = (X, U)$ be a 1-graph with vertices $x_1, x_2, \ldots, x_n$, and let $A = ((a_j^i))$ be a square matrix of order n defined by $a_j^i = 1$ if $(x_i, x_j) \in U$, and $a_j^i = 0$ if $(x_i, x_j) \notin U$. Then the permanent of A equals the number of pairwise disjoint systems of circuits that partition X.*

Each non-zero term of the expansion of the permanent corresponds to such a system of circuits, and conversely.

Q.E.D.

A skew-symmetric matrix $B = ((b_j^i))$ can be associated with an anti-symmetric 1-graph $G = (X, U)$ by letting

$$b_j^i = \begin{cases} +1 & \text{if } (x_i, x_j) \in U\,, \\ -1 & \text{if } (x_j, x_i) \in U\,, \\ 0 & \text{otherwise}\,. \end{cases}$$

Matrix B is called the *adjoint matrix* of G.

If $B = ((b_j^i))$ is a skew-symmetric matrix of even order $n = 2k$, the *pfaffian* of B is defined to be

$$\text{Pf } B = \sum_\pi \varepsilon(\sigma_\pi)\, b_{i_2}^{i_1}\, b_{i_4}^{i_3} \cdots b_{2k}^{i_{2k-1}} \,,$$

where $\pi = [i_1, i_2] \cdot [i_3, i_4] \ldots [i_{2k-1}, i_{2k}]$ is a permutation of degree n that decomposes into k cycles of length 2 (i.e., $\pi(i) = j$ implies $j \neq i$ and $\pi(j) = i$), such that

$$i_1 < i_3 < \cdots < i_{2k-1}$$
$$i_1 < i_2, i_3 < i_4, \ldots, i_{2k-1} < i_{2k},$$

and where σ_π is the permutation

$$\sigma_\pi = \begin{pmatrix} 1 & 2 & \cdots & n \\ i_1 & i_2 & \cdots & i_n \end{pmatrix}.$$

Proposition 2. *If $G = (X, U)$ is an anti-symmetric 1-graph of even order $n = 2k$, and if $B = ((b_j^i))$ is its adjoint matrix, then $|$ Pf $B |$ is less than or equal to the number of perfect matchings in G. Furthermore, the number of perfect matchings in G equals $|$ Pf $B |$ if, and only if, each term in the expansion of Pf B has the same sign.*

Each term in the expansion of the pfaffian corresponds to a perfect matching, and conversely. If two non-zero terms have opposite signs, then they cancel out one another.

Q.E.D.

Note that the pfaffian is easily calculated using the following well-known theorem from linear algebra:

Proposition 3. *If B is a square skew-symmetric matrix of order n, then*

$$\text{Det } B = (\text{Pf } B)^2 \qquad \text{if } n \text{ is even},$$
$$= 0 \qquad \text{if } n \text{ is odd}.$$

The proof can be found in most comprehensive texts on linear algebra.

Consider an anti-symmetric 1-graph $G = (X, U)$ with a perfect matching $W_0 \subset U$. If μ is a cycle of G, let μ^+ denote the arcs of μ that are directed in the direction of travel through the cycle, and let $| \mu^+ |$ denote the cardinality of this arc set. If, for a family $M = (\mu_i \, / \, i \in I)$ of cycles, the numbers $| \mu_i^+ |$ are all odd, G is said to be *well directed with M*. Finally, cycle μ is said to be *alternable* if there exists a perfect matching in which μ is an alternating cycle.

Theorem 9 (Kasteleyn [1961]). *Let $G = (X, U)$ be an anti-symmetric 1-graph of even order $2k$; let B be its adjoint matrix, and let W_0 be a perfect matching in G. The following three statements are equivalent:*
(1) *All the non-zero term in the expansion of Pf B have the same sign,*
(2) *G is well directed for the family of its alternable cycles,*
(3) *G is well directed with the family of the alternating cycles for W_0.*

If W_0 is the only perfect matching in G, the result follows since G has no other alternable cycle. Thus we may assume that G has several perfect matchings.

(1) $\Rightarrow$ (2) For example, let $\mu = [x_1, x_2, ..., x_{2p}, x_1]$ be an alternable cycle, and let W and W' be two perfect matchings in G such that

$$(W - W') \cup (W' - W) = \mu .$$

Suppose that the set of arcs common to W and W' is

$$W \cap W' = \{ (x_{2p+1}, x_{2p+2}), ..., (x_{2k-1}, x_{2k}) \} .$$

The terms corresponding to W and W' in the expansion of Pf B are respectively

$$\theta = \varepsilon(\sigma)\ b_2^1\ b_4^3\ ...\ b_{2p}^{2p-1}\ b_{2p+2}^{2p+1}\ ...\ b_{2k}^{2k-1}$$

and

$$\theta' = \varepsilon(\sigma')\ b_{2p}^1\ b_3^2\ b_5^4\ ...\qquad b_{2p+2}^{2p+1}\ ...\ b_{2k}^{2k-1} ,$$

where

$$\sigma = \begin{pmatrix} 1 & 2 & ... & 2k \\ 1 & 2 & ... & 2k \end{pmatrix} ,$$

$$\sigma' = \begin{pmatrix} 1 & 2 & 3 & ... & 2p & 2p+1 & ... & 2k \\ 1 & 2p & 2 & ... & 2p-1 & 2p+1 & ... & 2k \end{pmatrix} .$$

Note that $\varepsilon(\sigma)\varepsilon(\sigma') = +1$, because $2p - 2$ transpositions are required to pass from σ to σ'. Thus, from (1),

$$+ 1 = \theta\ \theta' = - b_2^1\ b_3^2\ ...\ b_{2p}^{2p-1}\ b_1^{2p} .$$

Consequently,

$$| \mu^+ | \equiv | \mu^- | \equiv 1 \qquad (\mathrm{mod}\ 2),$$

and for the alternable cycle μ, graph G is well directed.

(2) $\Rightarrow$ (3) This follows because each alternating cycle of W_0 is an alternable cycle.

(3) $\Rightarrow$ (1) Let W be a perfect matching different from W_0. From Corollary 1 to Theorem 1, a sequence of perfect matchings $W_0, W_1, ..., W_q$ can be constructed so that for each i the arcs of

$$(W_i - W_{i+1}) \cup (W_{i+1} - W_i)$$

define an elementary cycle μ_i that is alternating in W_0. Thus, from (3), $| \mu_i^- |$ is odd.

As above, we can show that the terms $\theta(W_0)$ and $\theta(W)$ corresponding to matchings W_0 and W satisfy $\theta(W_0)\theta(W) = +1$. Consequently, all non-zero terms in the expansion of Pf B have the same sign.

Q.E.D.

Theorem 10. *If $G = (X, U)$ is an anti-symmetric 1-graph, and if $M = (\mu_i \mid i \in I)$ is a family of linearly independent cycles, then, by reversing the direction of certain arcs, a 1-graph $G' = (X, U')$ that is well directed for M can be obtained.*

Let

s_i = number of arcs of G directed in the direction of travel in μ_i,

s' = number of arcs of G directed against the direction of travel in μ_i.

Let

$$c_{ij} \begin{cases} = 1 & \text{if} \quad u_j \in \mu_i \\ = 0 & \text{if} \quad u_j \notin \mu_i . \end{cases}$$

Finally, let

$$z_j \begin{cases} = 1 & \text{if the direction of arc } u_j \text{ should remain unchanged} \\ & \quad \text{to obtain graph } G', \\ = 0 & \text{otherwise.} \end{cases}$$

Graph G' will be well directed for M if, and only if, for each i,

that is

$$s'_i \equiv 1 \quad (\text{mod } 2),$$

$$s_i + \sum_{j=1}^{m} c_{ij} z_j \equiv 1 \quad (\text{mod } 2).$$

Hence, G' is obtained from a system of equations of m variables $z_1, z_2, \ldots, z_m$ in the field of integers modulo 2:

$$\sum_{j=1}^{m} c_{ij} z_j \equiv 1 + s_i \quad (i \in I) .$$

If family M consists of independent cycles, then this system consists of principal equations and, therefore, it has a solution $(z_1, z_2, \ldots, z_m)$ in the field of integers modulo 2.

Q.E.D.

For example, consider the non-planar graph in Fig. 7.7 with the perfect matching W_0 shown in dark lines. The alternating cycles are linearly independent since each possesses a distinct edge (marked with a cross in Fig. 7.8). Thus, from Theorem 10, it is possible, by directing the edges, to obtain a graph $G = (X, U)$ in which the alternating cycles are well directed. The adjoint matrix of graph $G = (X, U)$ is:

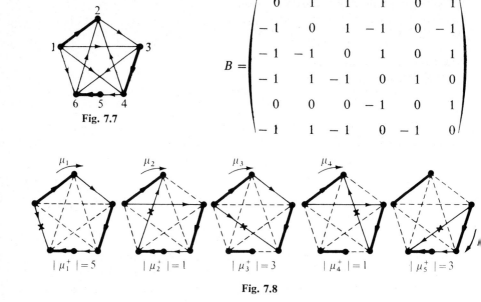

Fig. 7.7

$$B = \begin{pmatrix} 0 & 1 & 1 & 1 & 0 & 1 \\ -1 & 0 & 1 & -1 & 0 & -1 \\ -1 & -1 & 0 & 1 & 0 & 1 \\ -1 & 1 & -1 & 0 & 1 & 0 \\ 0 & 0 & 0 & -1 & 0 & 1 \\ -1 & 1 & -1 & 0 & -1 & 0 \end{pmatrix}$$

$| \mu_1^+ | = 5$ $| \mu_2^+ | = 1$ $| \mu_3^+ | = 3$ $| \mu_4^+ | = 1$ $| \mu_5^+ | = 3$

Fig. 7.8

From Theorem 9, the number of distinct perfect matchings is

$$| \text{Pf } B | = \sqrt{\text{Dét } B} = \sqrt{3 \cdot 3 \cdot 2 \cdot 2} = 6$$

We shall show that this method always works for planar graphs.

Theorem 11 (Kasteleyn [1961]). *Let $G = (X, U)$ be a connected planar graph with a perfect matching W_0 that is well directed for the family of contours of its bounded faces. (From Theorem 10, one such orientation always exists.) Then G is well oriented for the family of its alternating cycles.*

Let μ be an alternating cycle for the perfect matching W_0. Cycle μ surrounds an even number of vertices because the vertices in the interior μ are matched together by W_0.

Let H be the subgraph of G generated by vertices situated on μ or in the interior of μ. Clearly, H is planar and connected. Suppose H has n vertices, m arcs and f finite faces, $v_1, v_2, ..., v_f$.

Successively traverse the cycles $v_1, v_2, ..., v_f$ along their direction and then traverse μ against its direction. While doing this, count the total number ξ of arcs traversed along their direction. It is clear that each arc of H will be traversed once in each direction; thus, $\xi = m$.

Next, by summing over the cycles, we have

$$\xi = \sum_{i=1}^{f} | v_i^+ | + \left(l(\mu) - | \mu^+ | \right) ,$$

where $l(\mu)$ is the length of μ.
Since G is well directed for the cycles v_i, we have

$$m \equiv f + l(\mu) + | \mu^+ | .$$

The numbers n, m and f satisfy the Euler relation for planar graphs (see Corollary 1, Theorem 2, Ch. 2), $f = m - n + 1$. Hence

$$| \mu^+ | \equiv m + f + l(\mu) \equiv n + 1 + l(\mu)$$
$$\equiv 1 + \left(n - l(\mu) \right) \equiv 1 ,$$

since $n - l(\mu)$ is the number of vertices in the interior of μ, which is even. Thus, G is well directed for μ.

Q.E.D.

Remark. To calculate the number of distinct perfect matchings in a simple planar graph G, it suffices to give the edges of G a suitable anti-symmetric orientation, to determine the adjoint matrix B and to calculate $\sqrt{\text{Det } B}$.

EXERCISES

1. Suppose that $G = (X, E)$ is a connected graph without isthmi (an edge whose removal disconnects the graph), and each vertex of G has degree 3. Consider a maximum matching E_0. Show that there exists a chain whose edges belong alternately to E_0 and $E - E_0$ that uses exactly once each edge of $E - E_0$ and uses each edge of E_0 twice.

(P. Medgyessy [1950])

2. In a bipartite graph $G = (X, Y, E)$, let

$$\delta(A) = | A | - | \Gamma_G(A) |, \qquad \delta_0 = \max_{A \subset X} \delta(A) .$$

Show that

(1) $\delta(A_1 \cup A_2) + \delta(A_1 \cap A_2) \geq \delta(A_1) + \delta(A_2)$.
(2) Using the above inequality, show that the family

$$\mathcal{A} = \{ A \mid \delta(A) = \delta_0 \}$$

satisfies

$$A_1, A_2 \in \mathcal{A} \quad \Rightarrow \quad A_1 \cup A_2, \ A_1 \cap A_2 \in \mathcal{A} .$$

(3) If $\delta_0 > 0$, the set of vertices of X not saturated by at least one maximum matching is

$$A_0 = \bigcap_{A \in \mathcal{A}} A .$$

3. Consider a bipartite graph $G = (X, Y, E)$ for which there exists a matching of X into Y. Show that there exists an $x_0 \in X$ such that, for each $y \in \Gamma_G(x_0)$, at least one maximum matching uses the edge $[x_0, y]$.

Hint: If $| \Gamma_G(S) | > | S |$ for each $S \neq \varnothing$, any $x_0 \in X$ can be chosen.

(M. Hall [1948])

4. Deduce from the preceding exercise that if the minimum degree $d_G(x)$ for $x \in X$ is equal to k, then there exist at least k ! distinct maximum matchings.

(M. Hall [1948])

5. In a bipartite graph (X, Y, Γ) with $| X | = | Y |$, let

$$k = \max_k \left\{ d_G(x_1) + d_G(x_2) + \cdots + d_G(x_k) - d_G(y_1) - d_G(y_2) - \cdots - d_G(y_{k-1}) \right\} .$$

Show that there exist k disjoint matchings of X into Y.

(O. Ore [1955])

6. Show that in a graph G with n vertices and minimum degree k, a maximum matching V satisfies

$$| V | \geqslant \min \left\{ k, \left[\frac{n}{2} \right] \right\} .$$

(P. Erdös, L. Pósa [1962])

7. Show that if V is a matching and T is a transversal set, then

$$\min | T | \leqslant 2 \max | V | .$$

Show that the equality holds if, and only if, the connected components of the graphs are all cliques of odd cardinality.

8. If $k = \min d_G(x)$, if G is connected, and if $\max | V | < \dfrac{n + 2}{3}$, then

$$\min | T | \leqslant 2 \max | V | - k .$$

(Erdös, Gallai [1961])

9. If a graph remains connected after any $k - 1$ of its vertices are removed, and if $\max | V | < \dfrac{n - 1}{2}$, then $k \leqslant \max | V |$, and

$$\min | T | \leqslant 2 \max | V | - k$$

(Erdös, Gallai [1961])

Show that this bound is attained by a graph G formed from a k-clique K_k and $l > k + 1$ cliques $K_{2n_i + 1}$, with each vertex of $K_{2n_i + 1}$ being joined to each vertex of K_k.

10. Show that

$$2 \min | V | \leqslant m + \max | V | .$$

Also show by an induction on m that the equality holds if, and only if, the connected components of the graph are 2-cliques or 3-cliques.

11. Show that

$$4 \min | T | \leqslant 2 n + m - \max | V | .$$

If G is a 2-clique, 3-clique, 4-clique or two triangles joined by one edge, then the equality holds. Are there other connected graphs for which this equality holds?

(Erdös, Gallai [1961])

12. Let a tree with diameter $\leqslant 3$ be called a *double star*. Let $f(G)$ denote the minimum number of double stars needed to cover all the edges of a graph G with n vertices.

(1) If r is the cardinality of a maximum matching of G, show that

$$f(G) \leqslant 2 r .$$

(2) Show that $f(G) \leqslant n - 2r$.

(3) Using (1) and (2), show that $f(G) \leqslant \left[\dfrac{2n}{3} \right] .$

(L. Lovász [1968])

13. In a simple graph $G = (X, E)$, consider a family of sets
$$\mathscr{C} = \{ C_1, C_2, ..., C_p \},$$
with
(1) $| C_i |$ odd $(i = 1, 2, ..., p)$.
(2) For each $[x, y] \in E$, there exists an index i such that
$$| C_i \cap \{ x, y \} | = \min \{ | C_i |, | \{ x, y \} | \}.$$
If $| C_i | \leq 2k + 1$, let $c(C_i) = \max \{ k, 1 \}$, and let
$$c(\mathscr{C}) = \sum_{i=1}^{p} c(C_i).$$
Show that for each matching E_0,
$$| E_0 | \leq \min_{\mathscr{C}} c(\mathscr{C}).$$
Also show that
$$\max_{E_0} | E_0 | = \min_{\mathscr{C}} c(\mathscr{C}).$$
(J. Edmonds [1964])

14. Show that if G is a connected graph of even order with no subgraph isomorphic to $K_{1,3}$, then G has a perfect matching. Thus, there exists an edge $[x, y]$ such that $G_{X - \{x,y\}}$ is connected.
(Las Vergnas [1975])

15. Let G be a simple graph, and let $f(x)$ and $g(x)$ be even integers such that $g(x) \leq f(x)$ and $f(x) \neq 0$. Show that by removing or doubling various edges, it is possible to obtain a multigraph H such that
$$g(x) \leq d_H(x) \leq f(x) \qquad \text{for all } x,$$
if and only if each stable set S of G satisfies $f(\Gamma S) \geq g(S)$.
(Berge, Las Vergnas [1978])

16. Using Corollary 1 of Theorem 5, show that if G is a multigraph of order n, with no loops, and is regular with degree h odd, $h \geq n + 1$, then G has a perfect matching.

17. Let $\mathscr{A} = (A_1, A_2, ..., A_n)$ be a family of sets such that $A' < A$, $A \in \mathscr{A}$ implies $A' \in \mathscr{A}$. Let G be the graph whose vertices correspond to the sets A_i, two vertices being adjacent whenever the corresponding sets are disjoint. Show that $G - \emptyset$ has a perfect matching.
(Berge [1975])

CHAPTER 8

c-Matchings

1. The maximum c-matching problem

Consider a multigraph $G = (X, E)$ with vertices $x_1, x_2, ..., x_n$ and a n-tuple $\mathbf{c} = (c_1, c_2, ..., c_n)$ of integers with

$$0 \leqslant c_i \leqslant d_G(x_i) \qquad (i = 1, 2, ..., n) .$$

The set $E_0 \subset E$ is called a $\mathbf{c}$-*matching* if for each i, the set $E_0(x_i)$ of edges of E_0 incident to x_i satisfies

$$\left| E_0(x_i) \right| \leqslant c_i .$$

A vertex x_i is said to be *saturated* in the $\mathbf{c}$-matching E_0 if $\left| E_0(x_i) \right| = c_i$.

In this section, we shall study the problem of constructing a maximum $\mathbf{c}$-matching. The maximum matching problem (Chap. 7) is a special case of the maximum $\mathbf{c}$-matching problem for $\mathbf{c} = (1, 1, ..., 1)$.

To each multigraph G, there corresponds a simple graph $\bar{G}$ defined as follows:

For each $x_i \in X$, define two disjoint sets

$$A_i = \left\{ a_i^e \mid e \in E(x_i) \right\} \quad \text{and} \quad B_i = \left\{ b_i^k \mid k = 1, 2, ..., d_G(x_i) - c_i \right\}$$

Let the vertex set of $\bar{G}$ be the union of $\bigcup_{i=1}^n A_i$ and of $\bigcup_{i=1}^n B_i$. For each i, join each vertex of A_i to each vertex of B_i. For each edge $e = [x_i, x_j]$ of G, construct an edge $\bar{e} = [a_j^e, a_j^e]$ in $\bar{G}$.

Theorem 1. *A maximum matching $\bar{E}_0$ in graph $\bar{G}$ that saturates $\bigcup_{i=1}^n B_i$ induces a maximum $\mathbf{c}$-matching E_0 in graph G, and conversely.*

1. Let $\bar{E}_0$ be a maximum matching of $\bar{G}$. We may assume that $\bar{E}_0$ saturates each b_i^k (by interchanging if necessary the edges of $\bar{E}_0$ and of $\bar{E} - \bar{E}_0$ along a chain $[b_i^k, a_i^e, a_j^e]$ of length 2). Matching $\bar{E}_0$ in G defines a set of edges E_0 in G, and this set E_0 satisfy

$$\left| E_0(x_i) \right| \leqslant d_G(x_i) - \left| B_i \right| = d_G(x_i) - \left(d_G(x_i) - c_i \right) = c_i .$$

Thus, E_0 is a $\mathbf{c}$-matching in G; furthermore,

$$\left| E_0 \right| = \left| \bar{E}_0 \right| - \sum_{i=1}^n \left| B_i \right| .$$

148

2. Now consider a maximum c-matching $E_1 \subset E$ of graph G. Set E_1 defines in $\overline{G}$ a matching $\overline{E}_1$ that saturates each b_i^k, as shown in Fig. 8.1.

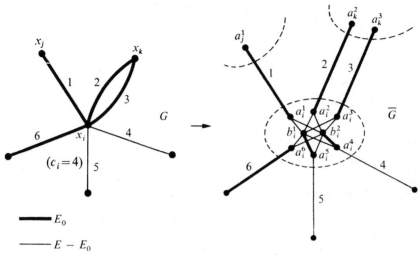

Fig. 8.1

Consequently,

$$| \overline{E}_1 | = | E_1 | + \sum_{i=1}^{n} | B_i | .$$

Since E_0 is a c-matching in G, $| E_0 | \leqslant | E_1 |$, and therefore

$$| \overline{E}_1 | = | E_1 | + \sum | B_i | \geqslant | E_0 | + \sum | B_i | = | \overline{E}_0 | .$$

Since $\overline{E}_0$ is a maximum matching, $| \overline{E}_1 | = | \overline{E}_0 |$. Consequently, $| E_1 | = | E_0 |$. Hence, E_0 is a maximum c-matching of G, and $\overline{E}_1$ is a maximum matching of $\overline{G}$.

$$\text{Q.E.D.}$$

Remark. This theorem demonstrates that a maximum c-matching can be constructed by determining the maximum matching $\overline{E}_0$ in graph $\overline{G}$ and then saturating each vertex b_i^k by an interchange along alternating chains of length 2.

Consider a multigraph G with a c-matching E_0. From G, construct a multigraph $R(E_0)$ by adding to G a vertex x_0 called the *origin* that is joined to each vertex x_i by c_i edges, for all i. Multigraph $R(E_0)$ is called a *transfer network*.

Let the edges of E_0 be represented by dark lines, and let $c_i - | E_0(x_i) |$ edges from x_0 to x_i for all i be also represented by dark lines. All other edges of $R(E_0)$ are represented by light lines.

An *alternating chain* is defined as a chain of $R(E_0)$ with edges alternately dark and light that repeats no edge. A *transfer* along an alternating chain μ is defined as the interchange of the dark and light colouring along μ.

Theorem 2. (Berge [1958]). *A c-matching E_0 of a multigraph $G = (X, E)$ is maximum if, and only if, there exists no alternating chain in $R(E_0)$ that joins x_0 to itself and has dark initial and terminal edges.*

As in Theorem 1, construct a graph $\overline{G}$ corresponding to the multigraph G and a matching $\overline{E}_0$ in $\overline{G}$ corresponding to E_0 that saturates each vertex in $\bigcup B_i$. An alternating chain with the above properties in $R(E_0)$ corresponds in $\overline{G}$ to an alternating chain that connects two distinct unsaturated vertices (if $\overline{E}_0$ is properly chosen), and vice versa. In graph $\overline{G}$, such a chain exists if, and only if, the matching $\overline{E}_0$ in $\overline{G}$ is not maximum (Chapter 7, Theorem 1), or, from Theorem 1, if, and only if, the c-matching E_0 in G is not maximum.

Q.E.D.

Theorem 3. *If E_0 and E_1 are two maximum c-matchings of a multigraph $G = (X, E)$, then E_1 can be obtained from E_0 by transfers along alternating cycles of $R(E_0)$ that are pairwise edge-disjoint (but not necessarily elementary).*

From Corollary 1 to Theorem 1 (Ch. 7), each maximum matching $\overline{E}_1$ in $\overline{G}$ can be obtained from the maximum matching $\overline{E}_0$ by a series of transfers along vertex disjoint alternating chains. Each of these chains is either an alternating elementary cycle or an even elementary chain starting at an unsaturated vertex. In either case, this alternating chain corresponds in $R(E_0)$ to an alternating cycle that is not necessarily elementary.

Q.E.D.

Theorem 4. *A "free edge" is defined to be any edge of $G = (X, E)$ that is contained in some maximum c-matching but not in every maximum c-matching.*
An edge e is free if, and only if, given a maximum c-matching E_0 of G, e lies on an alternating cycle of $R(E_0)$.

1. If e is contained in an alternating cycle of $R(E_0)$, then e is evidently a free edge, since a transfer can be made along this alternating cycle.

2. Let e be a free edge, and suppose at first that e is contained in the maximum c-matching E_0. There exists a maximum c-matching E_1 that does not contain e. Thus, from Theorem 3, e is contained in an alternating cycle of $R(E_0)$.

If e is not contained in E_0, then e is contained in a maximum c-matching

$\bar{E}_1$ obtained from E_0 by a transfer along an alternating cycle of $R(E_0)$. Again, e is contained in an alternating cycle of $R(E_0)$.

Q.E.D.

2. Transfers

In this section we shall consider a simple graph G and show how to obtain from G all the graphs with the same degrees as G by a sequence of transfers of a particular form.

Lemma. *If $G = (X, E)$ is a simple graph such that each even cycle of length > 4 has a chord that divides the cycle into two even cycles. Let $E_0 \subset E$ be a set of dark edges. Then any transfer along an alternating cycle μ can be obtained by a sequence of transfers along alternating cycles of length 4.*

If the length of cycle μ equals 4, the result is trivial. If the result is true for cycles of length $< 2k$, then we shall show that it is also true for an alternating cycle μ of length $2k$, say

$$\mu = [a_1, a_2, ..., a_{2k}, a_1] \ .$$

Since this cycle has a chord of the form $[a_i, a_{i+2p+1}]$, we have two even cycles:

$$\mu_1 = [a_i, a_{i+1}, ..., a_{i+2p+1}, a_i] \ ,$$
$$\mu_2 = [a_1, ..., a_i, a_{i+2p+1}, a_{i+2p+2}, ..., a_{2k}, a_1] \ .$$

Relative to E_0, only one of these two cycles is alternating, say μ_1. Since the length of μ_1 is $< 2k$, the transfer along μ_1 can be accomplished by a sequence of transfers along alternating cycles of length 4:

$$E_0 \quad \rightarrow \quad E_0' = E_0 - (\mu_1 \cap E_0) \cup (\mu_1 - E_0) \ .$$

Relative to E_0, cycle μ_2 is an alternating cycle of length $< 2k$. The transfer

$$E_0' \quad \rightarrow \quad E_0'' = E_0' - (\mu_2 \cap E_0') \cup (\mu_2 - E_0')$$

can be accomplished by a sequence of transfers along alternating cycles of length 4. Thus, the transfer

$$E_0 \quad \rightarrow \quad E_0'' = E_0 - (\mu \cap E_0) \cup (\mu - E_0)$$

has been obtained.

Q.E.D.

Theorem 5. *Let $H_0 = (X, E_0)$ and $H_1 = (X, E_1)$ be two simple graphs, with the same vertex set, such that*

$$d_{H_0}(x_i) = d_{H_1}(x_i) = d_i \quad (i = 1, 2, ..., n) \ .$$

Let a, b, c, d be any four vertices in X with $ac \in E_0$, $bd \in E_0$, $ad \notin E_0$ and $bc \notin E_0$. A "direct transfer on H_0" is defined as an operation that removes edges ac and bd and adds edges ad and bc. Then, graph H_1 can be obtained from graph H_0 by a sequence of direct transfers.

Clearly, H_0 and H_1 are two partial graphs of a complete graph of the form $G = (X, E)$. In G each even cycle $[a_1, a_2, ..., a_1]$ has a chord of the form $[a_i, a_i + 3]$. From the above lemma and from Theorem 3, H_0 can be transformed into H_1 by transfers along alternating cycles of length 4. These transfers are direct transfers.

Q.E.D.

Theorem 6. Let $H_0 = (X, Y, E_0)$ and $H_1 = (X, Y, E_1)$ be two bipartite graphs such that

$$d_{H_0}(x_i) = d_{H_1}(x_i) = r_i \qquad (i = 1, 2, ..., p),$$

$$d_{H_0}(y_j) = d_{H_1}(y_j) = s_j \qquad (j = 1, 2, ..., q).$$

Let $x_1, x_2 \in X$ be two distinct vertices, and let $y_1, y_2 \in Y$ be two distinct vertices with

$$x_1 y_1 \in E_0, \qquad x_2 y_2 \in E_0, \qquad x_1 y_2 \notin E_0, \qquad x_2 y_1 \notin E_0.$$

A "bipartite transfer on H_0" is defined to be an operation that removes edges $x_1 y_1$ and $x_2 y_2$ and adds edges $x_1 y_2$ and $x_2 y_1$.

Graph H_1 can be obtained from graph H_0 by a sequence of bipartite transfers.

Clearly, H_0 and H_1 are partial graphs of a complete bipartite graph of the form $G = (X, Y, E)$ in which each $x \in X$ is joined to each $y \in Y$. From the lemma and from Theorem 3, H_0 can be transformed into H_1 by a sequence of transfers along alternating cycles of length 4. These transfers are bipartite transfers.

Q.E.D.

Theorem 7. Let $G = (X, U)$ and $G' = (X, U)$ be l-graphs such that

$$d_G^+(x_i) = d_{G'}^+(x_i) = r_i \qquad (i = 1, 2, ..., n),$$

$$d_G^-(x_i) = d_{G'}^-(x_i) = s_i \qquad (i = 1, 2, ..., n).$$

Let a, b, x, y be vertices of X with $a \neq b$, $x \neq y$, $(a, x) \in U$, $(b, y) \in U$, $(a, y) \notin U$, $(b, x) \notin U$. An "oriented transfer" is defined to be an operation that removes arcs (a, x) and (b, y) and adds arcs (a, y) and (b, x).

Graph G' can be obtained from Graph G by a sequence of oriented transfers.

Graph $G = (X, U)$ corresponds to a bipartite graph $H = (X, \bar{X}, E)$ where $[x_i, \bar{x}_j] \in E$ if, and only if, $(x_i, x_j) \in U$. The proof follows when Theorem 6 is applied to the bipartite graph H.

Q.E.D.

Remark. Consider a l-graph G such that $d_G^+(x) = d_G^-(x)$ for each vertex x. Such a graph defines a permutation of degree n, and vice versa. Thus, Theorem 7 generalizes the well known theorem from algebra: *Every permutation is a product of transpositions.*

3. Maximum cardinality of a c-matching

Consider a multigraph $G = (X, E)$ with a c-matching $E_0 \subset E$, and the corresponding multigraph $R(E_0)$ with dark and light edges (see Section 1). Recall that the origin x_0 is joined to some of the vertices of X by dark and light lines, the edges of E_0 are dark, and the edges of $F_0 = E - E_0$ are light. Consider the chains starting at the x_0 and beginning with a dark edge.

If there exists an alternating chain μ starting at x_0 and going to $x \in X$, then orient each edge in μ toward x. It is possible that an edge will be given two opposite directions. If vertex x is the endpoint of a dark edge oriented toward x and is not the endpoint of any light edge oriented toward x, then vertex x is called *dark*. The set of all dark vertices is denoted by X^d.

If vertex x is the endpoint of a light edge directed toward x and not the endpoint of any dark edge directed toward x, then vertex is called *light*. The set of all light vertices is denoted by X^l.

If a vertex x is the endpoint of a light edge directed toward x and also the endpoint of a dark edge directed toward x, then x is called a *mixed* vertex. The set of all mixed vertices is denoted by X^m. Finally, if a vertex x is not the endpoint of any edge directed toward x, then vertex x is called *inaccessible*. The set of all inaccessible vertices is denoted by X^i.

Vertex x_0, the origin of the network $R(E_0)$, is so far unclassified: it will be defined to be *light* if there exist no alternating chains from x_0 to x_0 with dark edges at each endpoint; otherwise, x_0 is defined to be *mixed*.

Each vertex of $R(E_0)$ is either light, dark, mixed or inaccessible.

Theorem 8. *Two dark vertices can only be joined by a dark edge. Two light vertices can only be joined by a light edge. A mixed vertex and an inaccessible vertex cannot be joined by an edge. A dark vertex and an inaccessible vertex can only be joined by a dark edge. A light vertex and an inaccessible vertex can only be joined by a light edge.*

This follows immediately from the definitions.

These results are summarized below:

Type of vertex	Dark	Light	Mixed	Inaccessible
Dark	Dark edge			Dark edge
Light		Light edge		Light edge
Mixed				
Inaccessible	Dark edge	Light edge		

Consider the subgraph of $R(E_0)$ generated by the set X^m. Denote the connected components of this graph by $M_1, M_2, \ldots$. Chain μ is said to *enter* M_1 *via edge* $[a, b]$ if μ is of the form

$$\mu = [x_0, a_1, a_2, \ldots, a_k = a, \quad b_1 = b, b_2, \ldots, b_l]$$

with

$$x_0, a_1, a_2, \ldots, a_k \notin M_1$$

$$b_1, b_2, b_3, \ldots, b_l \in M_1.$$

Lemma 1. *Let $x_0 \notin M_1$ and $x \in M_1$. If an alternating chain $\mu[x_0, x]$ enters M_1 via an edge $[a, b]$ and terminates with a dark (respectively, light) edge, then there exists an alternating chain $\mu'[x_0, x]$ entering M_1 via $[a, b]$ and terminating with a light (respectively, dark) edge.*

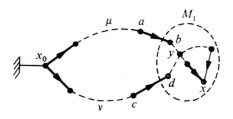

Fig. 8.2

If x is a mixed vertex, then there exists an alternating chain $v[x_0, x]$ that terminates with a light edge. Since $x_0 \notin M_1$ and $x \in M_1$, there exist in $v[x_0, x]$ edges with one endpoint in $X - M_1$ and the other endpoint in M_1. Let $[c, d]$ be the last edge of this type in $v[x_0, x]$.

Finally, let y be the first vertex of μ that is on $v[d, x]$ (such a vertex always exists). If $\mu[b, y]$ and $v[d, y]$ terminate with edges of the same type, the theorem is true because of the alternating chain:

$$\mu' = \mu[x_0, y] + v[y, x] \,.$$

If $\mu[b, y]$ and $v[d, y]$ terminate with edges of different types, then $[a, b] = [c, d]$ since, otherwise, $\mu[x_0, y] + v[y, c]$ would be a simple alternating chain, and c would be mixed, which contradicts $c \notin M_1$.

Thus, the chain

$$\mu' = \mu[x_0, b] + v[d, x]$$

is alternating and satisfies the requirements of the theorem.

Q.E.D.

Lemma 2. *Let $x_0 \notin M_1$, $x \in M_1$; let $[a, b]$ be an edge incident to M_1 and directed into M_1. There exists an alternating chain $\mu[x_0, x]$ from x_0 to x that enters M_1 via edge $[a, b]$.*

Let Y be the set of vertices of M_1 that are accessible by an alternating chain entering via $[a, b]$. Let Z be the set of other vertices of M_1. Since $b \in Y$, we have $Y \neq \varnothing$. Suppose that $Z \neq \varnothing$, then there exists an edge $[y, z]$ with $y \in Y$ and $z \in Z$. We shall show that this leads to a contradiction.

From Lemma 1, y is accessible by two alternating chains $\mu_1[x_0, y]$ and $\mu_2[x_0, y]$ entering M_1 via edge $[a, b]$ and terminating respectively with a dark and light edge. Thus, z is accessible to an alternating chain entering via $[a, b]$. This contradicts $z \in Z$.

Q.E.D.

Theorem 9. (Gallai [1950]). *Let E_0 be a set of dark edges in G_1. Let M_1 be a component of the subgraph of $R(E_0)$ generated by the mixed vertices. If $x_0 \notin M_1$, there exists exactly one edge incident to M_1 and directed into M_1. If $x_0 \in M_1$, there exists no edge incident to M_1 and directed into M_1.*

1. If $x_0 \notin M_1$, let μ be an alternating chain going from x_0 to M_1. Let b be the first vertex of μ in M_1. The alternating chain $\mu[x_0, b]$ enters M_1 via an edge $[a, b]$ that is incident to M_1 and is directed into M_1.

Let $[c, d]$ be an edge other than $[a, b]$ that is also incident to M_1 and directed into M_1. From Lemma 2, we know that $d \in M_1$ is accessible by an alternating chain entering via $[a, b]$; from Lemma 1, we may assume that this chain terminates with an edge of a type different from $[c, d]$. Thus c is a mixed vertex, which contradicts $c \notin M_1$.

Thus $[a, b]$ is the only edge that is directed into M_1.

2. If $x_0 \in M_1$, then we can return to the preceding case simply by adding vertices a_0 and b_0, a dark edge $[a_0, b_0]$ and a light edge $[b_0, x_0]$. Let a_0 replace x_0 as the origin. The only edge entering M_1 is $[b_0, x_0]$. Thus there exists no edge of the original graph that enters M_1.

Q.E.D.

Theorem 10. *Let E_0 be a maximum matching, and let M_1 be a component of the subgraph generated by the mixed vertices in the multigraph $R(E_0)$. There is a dark edge incident to M_1 and directed exclusively into M_1. All other edges incident to M_1 are light and exclusively directed out of M_1.*

1. Since the matching E_0 is maximum, x_0 is light, and $x_0 \notin M_1$. From Theorem 9, there exists exactly one edge directed into M_1. Denote this edge by $[c, x]$, where $c \notin M_1$, $x \in M_1$.

Suppose $[c, x]$ is light: since x is mixed, there exists an alternating chain μ from x_0 to x that terminates with a dark edge. Chain μ necessarily uses edge $[c, x]$ (which is the only possible entrance edge to M_1), followed by a dark edge incident to x and terminates with another dark edge incident to x. But this is impossible, since there is only one dark edge incident to x. Thus, edge $[c, x]$ is dark.

2. Let $[y, b]$ be another edge incident to M_1 with $y \in M_1$ and $b \notin M_1$. From Theorem 9, $[y, b]$ is necessarily directed out of M_1. Furthermore, $[y, b]$ cannot be dark because y, being a mixed vertex, can be reached by an alternating chain terminating with a dark edge.

Q.E.D.

Corollary 1. *If E_0 is a maximum matching, and if M_1 is a component of the subgraph generated by X^m, then $\mid M_1 \mid \geqslant 3$ and $\mid M_1 \mid$ is odd.*

Let $[c, x]$ be a dark edge incident to M_1 with $x \in M_1$. Then, $\mid M_1 \mid \geqslant 3$, because, otherwise, x cannot be reached by an alternating chain terminating with a light edge. Furthermore, since M_1 consists of vertex x and pairs of vertices joined by dark edges, $\mid M_1 \mid$ is odd.

Q.E.D.

Corollary 2. *If E_0 is a maximum matching, each non-mixed vertex that is adjacent to a mixed vertex is a light vertex.*

If $[c, x]$ is a dark edge incident to a component M_1 with $c \notin M_1$, and $x \in M_1$, then $c \notin X^i$, $c \notin X^d$ and $c \notin X^m$. Consequently, $c \in X^l$.

If $[a, b]$ is a light edge incident to M_1 with $a \notin M_1$, $b \in M_1$, then similarly, $a \notin X^i$, $a \notin X^d$, and $a \notin X^m$. Thus, $a \in X^l$.

Q.E.D.

Corollary 3. *If E_0 is a maximum matching, each vertex adjacent to a dark vertex is light.*

Let $a \in X^d$. If $[x, a]$ is a dark edge, then $x \notin X^i$ (since x is the endpoint of a directed edge); $x \notin X^d$ (because, otherwise, a would be mixed), and $x \notin X^m$ (from Corollary 2). Thus $x \in X^l$.

If $[x, a]$ is a light edge, then $x \notin X^i$, $x \notin X^d$, and $x \notin X^m$ (from Corollary 2). Thus, $x \in X^l$.

$$Q.E.D.$$

Theorem 11. *Let E_0 be a maximum matching and let I_1 be a connected component of the subgraph generated by the inaccessible vertices. Each edge incident to I_1 is light and undirected. Each vertex $x \notin I_1$ that is adjacent to a vertex of I_1 is light. If graph G has no isolated vertices, then $| I_1 |$ is even and $\geqslant 2$.*

1. If $[c, x]$ is an edge incident to I_1 with $x \in I_1$, then $c \notin X^m$, $c \notin X^i$, and $c \notin X^d$ (from Corollary 3). Thus, $c \in X^l$, and $[c, x]$ is light.

2. Since I_1 does not contain x_0 nor unsaturated vertices, it contains only pairs of vertices joined by dark edges. Thus, $| I_1 |$ is even and $\geqslant 2$.

$$Q.E.D.$$

Theorem 12. (Berge, [1958]). *Given a simple connected graph G and a subset S of the vertices, let $p_i(S)$ denote the number of components of odd order in the subgraph generated by $X - S$. The number of unsaturated vertices in a maximum matching is*

$$\xi = \max_{S \subset X} \left(p_i(S) - | S | \right).$$

1. Consider a set $S \subset X$, and let $C_1, C_2, ..., C_p$ be the components of odd order in the subgraph generated by $X - S$. If a component C_k has no unsaturated vertices, there is at least one dark edge going from C_k to a vertex $s_k \in S$ because $| C_k |$ is odd. Two distinct components C_k correspond to two distinct vertices s_k. Thus, n_0, the number of unsaturated vertices, satisfies

$$p_i(S) - n_0 \leqslant (\text{number of } C_k \text{ without unsaturated vertices}) \leqslant | S |.$$

Hence,

$$p_i(S) - | S | \leqslant n_0 \qquad (S \subset X).$$

2. We shall show that S can be chosen so that $p_i(S) - S = n_0$ (this establishes the theorem).

Let E_0 be a maximum matching, and let X^l be the set of light vertices. From Theorem 10 and Corollaries 1, 2, and 3,

$$p_i(X^l) = (\text{number of components in } G_{X^m}) + | X^d |.$$

Furthermore, the dark edges of G define a bijection between the set X^l and the components M_k that contains no unsaturated vertices and no dark saturated vertices. Thus

$$| X^l | = (\text{number of components in } G_{X^m}) + | X^d | - n_0.$$

Hence

$$n_0 = p_i(X^l) - | X^l |.$$

Q.E.D.

Corollary 1. *In a simple connected graph $G = (X, E)$ of order n, the number of edges in a maximum matching equals*

$$\frac{1}{2}(n - \xi),$$

where

$$\xi = \max_{S \subset X} \big[p_i(S) - | S | \big].$$

The proof is immediate from Theorem 12.

Corollary 2. (Tutte [1947]). *A necessary and sufficient condition for a connected graph to possess a perfect matching is that*

$$p_i(S) \leqslant | S |^\ulcorner \qquad (S \subset X).$$

The proof is immediate.

The following result generalizes the Petersen theorem [1891] for regular graphs of degree 3.

Theorem 13. *Let $G = (X, E)$ be a connected multigraph that is regular of degree h, with an even number of vertices, without loops, and with*

$$m_G(S, X - S) \geqslant h - 1 \qquad (S \subset X; \quad S \neq \varnothing, X).$$

Then G possesses a perfect matching. Furthermore, each edge of G is free (i.e. each edge of G belongs to at least one perfect matching but does not belong to all perfect matchings).

1. By using Corollary 2, we shall show that there exists a perfect matching.

Let S be a non-empty subset of X, where $S \neq X$, and let $C_1, C_2, \ldots$ be the connected components of odd order in the subgraph generated by $X - S$. By hypothesis, $m_G(S, C_1) \geqslant h - 1$. However,

$$m_G(S, C_1) > h - 1,$$

because, otherwise, the number of edges in C_1 would equal

$$m_G(C_1, C_1) = \frac{1}{2}\big[h | C_1 | - (h - 1) \big] = \frac{1}{2}\big[h(| C_1 | - 1) + 1 \big]$$

which is not an integer, since $|C_1|$ is odd. Thus,

$$m_G(S, C_1) \geqslant h .$$

Hence,

$$h\,|\,S\,| \geqslant m_G(S, X - S) \geqslant m_G(S, C_1 \cup C_2 \cup \cdots) = \sum_k m_G(S, C_k) \geqslant h p_i(S) .$$

Thus, $p_i(S) \leqslant |S|$ for all $S \neq X$, $S \neq \varnothing$. This inequality remains valid when $S = X$ because $p_i(X) = 0$, or when $S = \varnothing$ because $p_i(\varnothing) = 0$ (since the graph has an even number of vertices and is connected). Thus, from Corollary 2, there exists a perfect matching E_0.

2. To show that an edge $e \in E$ is free, it is sufficient from Theorem 4 to show that e is on an alternating cycle relative to a perfect matching E_0.

We may assume that $e \in E - E_0$ (because if $e \in E_0$ and if some edge of $E - E_0$ that is adjacent to e is on an alternating cycle, then edge e will also be contained in an alternating cycle).

We shall suppose that e is a light edge of G that does not belong to any alternating cycle, and we shall show that this leads to a contradiction.

Contract the endpoints of edge e into a single vertex x_0 and define a transfer network $R(E_0)$ with origin x_0. Since edge e appears in no alternating cycle, vertex x_0 is light.

Assume that there exist mixed vertices; let C_1 be a connected component of the subgraph of $R(E_0)$ generated by the mixed vertices. Clearly, $x_0 \notin C_1$. Since $|C_1|$ is odd from Corollary 1 to Theorem 10, we know from Part 1 of this proof that

$$m_G(X - C_1, C_1) \geqslant h .$$

Furthermore, from Theorem 10, there is a single dark edge of $R(E_0)$ that leaves C_1, and thus there are at least $h - 1$ light edges that leave C_1.

Now consider the network $\bar{R}(E_0)$ obtained from $R(E_0)$ by contracting each component C_k of mixed vertices. Thus, C_1 becomes a dark vertex c_1, and no edge has been oriented in two directions. From Theorem 10, c_1 is incident to a dark edge directed into c_1 and to light edges each directed out of c_1.

Let $\bar{X}^d$ and $\bar{X}^l$ respectively denote the sets of dark vertices and light vertices in network $\bar{R}(E_0)$. Let $d_l^+(x)$ denote the number of light edges of $R(E_0)$ directed out of x. Consequently,

$$d_l^+(x) \geqslant h - 1 \qquad (x \in \bar{X}^d) .$$

Besides, the number $d_l^-(x)$ of light edges directed into x satisfies

$$d_l^-(x) \leqslant d_G(x) - 1 = h - 1 \qquad (x \in \bar{X}^l - \{x_0\}) .$$

Thus

$$| \overline{X}^d | (h - 1) \leqslant \sum_{x \in \overline{X}^d} d_I^+(x) = \sum_{x \in \overline{X}^l} d_I^-(x) \leqslant (h - 1) | \overline{X}^l - \{ x_0 \} | + d_I^-(x_0)$$

$$\leqslant (h - 1) | \overline{X}^l | - (h - 1) + 2(h - 2).$$

Since the theorem is obvious for $h = 1$, we may assume that $h > 1$; thus,

$$| \overline{X}^d | \leqslant | \overline{X}^l | + 1 - \frac{2}{h - 1}.$$

By counting in two different ways the number of dark edges of $\overline{R}(E_0)$ that have exactly one direction, we obtain

$$| \overline{X}^d | = | \overline{X}^l | + 1.$$

Since we have obtained two incompatible relations, the proof is achieved.

Q.E.D.

Corollary (Errera [1922]). *If $G = (X, E)$ is a simple connected graph regular of degree 3 such that all the isthmi of G are on the same elementary chain, then G possesses a perfect matching.*

Recall that an "isthmus" is an edge $[a, b]$ whose removal disconnects the graph. In a connected graph, the removal of an isthmus creates exactly two connected components.

Since $3 | X | = 2 | E |$, the number of vertices of G is even. Consider the different connected components created by the removal of all of the isthmi. From the hypothesis, each of these connected components is joined to the rest of the graph G by one or two isthmi.

Suppose a connected component C_i is joined to the rest of the graph by two isthmi $[x, c]$ and $[y, c']$, where $c, c' \in C_i$ and $x, y \notin C_i$. Consider the graph G_i obtained from graph G_{C_i} by joining vertices c and c'. Graph G_i is connected, regular of degree 3 and has no isthmi. Therefore, from Theorem 13, graph G_i possesses a perfect matching that contains edge $[c, c']$. This matching corresponds to a matching $E_0(C_i)$ of graph G_{C_i} that saturates all vertices except c and c'.

Suppose a connected component D_j is joined to the rest of the graph by only one edge $[d, x]$ with $d \in D_j$, $x \notin D_j$. Consider the graph H_j obtained from G_{D_j} by removing vertex d and by joining the vertices d_1 and d_2 of D that are adjacent to vertex d in graph G. From Theorem 13, graph H_j has a perfect matching that does not contain edge $[d_1, d_2]$, and that corresponds in graph G_{D_j} to a matching $E_0(D_j)$ that saturates every vertex except $d \in D_j$.

Thus, a perfect matching for G can be formed from the union of sets $\bigcup E_0(C_i)$, $\bigcup E_0(D_j)$, and the set of isthmi of graph G.

Q.E.D.

EXERCISE

1. Show the following result:

Given a multigraph $G = (X, E)$, there exists a partial multigraph H with $d_H(x_i) = d(i)$ if, and only if, for each pair S, T of disjoint sets, the number $q(S, T)$ of components C of G_S with $d(C) + m_G(C, T) \equiv 0$ modulo 2 satisfies

$$q(S, T) \leqslant d(X - S - T) - d(T) + d_{G_{S \cup T}}]$$

Hint: Apply Theorem 5 to the graph $\bar{G}$ constructed from G as shown in Theorem 1.

(W. T. Tutte [1954])

CHAPTER 9

Connectivity

1. *h*-Connected graphs

The *connectivity* $\kappa(G)$ of a connected graph G is defined to be the minimum number of vertices whose removal disconnects G or reduces G to a single vertex.

If G is not a clique, there exist two non-adjacent vertices a and b; thus, $X - \{a, b\}$ is a set whose removal disconnects G, and, consequently,

$$\kappa(G) \leqslant | X - \{a, b\} | = n - 2 .$$

On the other hand, if G is the n-clique K_n, we have

$$\kappa(K_n) = n - 1 .$$

Graph G is said to be *h-connected* if its connectivity $\kappa(G)$ is $\geqslant h$. An *articulation set* of G is any set of vertices of G whose removal disconnects G. Let $\mathscr{A}$ denote the family of all articulation sets of G. A graph $G \neq K_n$ is *h*-connected if, and only if,

$$\kappa(G) = \min_{A \in \mathscr{A}} | A | \geqslant h .$$

Thus G is *h-connected if, and only if,*

(1) $h \leqslant n - 1$, and
(2) *there is no articulation set A of G with $| A | = h - 1$.*

Theorem 1. *For a simple connected graph G,*

$$\kappa(G) \leqslant \min_{x \in X} d_G(x) .$$

If G has n vertices, then $\kappa(G) \leqslant n - 1$. Let x_0 be a vertex of minimum degree h, and suppose that $h < \kappa(G)$. The set $\Gamma_G(x_0)$ has cardinality $h < n - 1$. Thus G is not a clique, and

$$\kappa(G) = \min_{A \in \mathscr{A}} | A | \leqslant | \Gamma_G(x_0) | = h < \kappa(G) ,$$

162

which is a contradiction.

Q.E.D.

Theorem 2 (Harary [1962]). *For each m, n with $0 \leqslant n - 1 \leqslant m \leqslant \binom{n}{2}$, the maximum connectivity for a simple connected graph with n vertices and m edges is* $\left[\dfrac{2\,m}{n}\right]$.

1. From Theorem 1, a connected graph G satisfies

$$2\,m = \sum_{x \in X} d_G(x) \geqslant n \min_{x \in X} d_G(x) \geqslant n \,/\, \kappa(G) \,.$$

Hence,

$$\kappa(G) \leqslant \frac{2\,m}{n} \,,$$

and so

$$\kappa(G) \leqslant \left[\frac{2\,m}{n}\right] \,.$$

2. Note that, if G is a simple connected graph, then

$$n - 1 \leqslant m \leqslant \binom{n}{2} \,.$$

It now remains to show that for each n, m, satisfying these inequalities there exists a simple connected graph $G(n, m)$ such that

$$\kappa\big(G(n, m)\big) = \left[\frac{2\,m}{n}\right] \,.$$

CASE 1. Suppose $\dfrac{2\,m}{n}$ is an integer. Consider two subcases:

SUBCASE 1'. Suppose $\dfrac{2\,m}{n} = 2k$ is an even integer. Construct a graph $G_{n,2k}$ with vertices $x_0, x_1, \ldots, x_{n-1}$. Join each vertex x_i to the vertices of

$$\{\, x_j \,/\, j \equiv i \pm k' \;(\text{mod. } n) \;;\; 1 < k' < k \,\} \,.$$

Since each vertex has degree $2\,k$, the number of edges in graph $G_{n,2k}$ is

$$m(G_{n,2k}) = \frac{1}{2} \sum_i d_{G_{n,2k}}(x_i) = kn = m \,.$$

It can be easily shown that this graph is $(2\,k)$-connected.[1] Thus, take

$$G(n, m) = G_{n,2k} .$$

SUBCASE 1''. Suppose that $\dfrac{2\,m}{n} = 2\,k + 1$ is an odd integer. Then $n(2\,k + 1) = 2\,m$ implies that n is even. Form graph $G(n, m)$ by taking graph $G_{n,2k}$ and by joining vertex x_i to vertex x_j if

$$j \equiv i + \frac{n}{2} \quad (\text{mod. } n),$$

$$j = 0, 1, ..., \frac{n}{2} - 1 .$$

Thus, for $G = G(n, m)$,

$$m(G) = \frac{1}{2} \sum d_G(x_i) = \frac{1}{2} (2\,k + 1)\, n = m .$$

It can be easily shown that this graph is $(2\,k + 1)$-connected.[2]

CASE 2. Suppose that $\dfrac{2\,m}{n}$ is not an integer. Let $q = \left[\dfrac{2\,m}{n} \right] .$

SUBCASE 2'. Suppose that n or q is an even number. Form graph $G(n, m)$ by taking graph $G_{n,q}$ and adding $m - nq$ edges arbitrarily.

---

[1] We shall show that $G_{n,2k}$ is $(2\,k)$-connected. From the symmetry of G, it is sufficient to show that x_0 and x_α, $\alpha = 1, 2, ..., n - 1$, cannot be disconnected by less than $2\,k$ vertices. Suppose that x_0 and x can be disconnected by $2\,k - 1$ vertices, $x_{i_1}, x_{i_2}, ... = x_{i_{2k-1}}$ where

$$1 \leqslant i_1 < i_2 < \cdots < i_{2k-1} \leqslant n - 1 .$$

One of the two intervals $[0, \alpha]$, $[\alpha, n]$ contains at most $k - 1$ of these indices. Suppose it is interval $[0, \alpha]$. Then, two consecutive vertices of the sequence obtained by removing $x_{i_1}, x_{i_2} ...$ from x_0, $x_1, ..., x_\alpha$, are joined by an edge (because the difference between their indices is $\leqslant k - 1$). Consequently, there is a chain from x_0 to x_α, which is a contradiction.

[2] We shall show that $G_{n, 2k+1}$ is $(2\,k + 1)$ connected. Suppose that x_0 and x_α are disconnected by $2\,k$ vertices $x_{i_1}, x_{i_2}, ..., x_{i_{2k}}$ where $1 < i_1 < i_2 \leqslant ... i_{2k} \leqslant n - 1$. If one of the two intervals $[0, \alpha]$ $[\alpha, n]$ does not contain k consecutive indices, then a chain from x_0 to x_α can be constructed as shown above. Therefore, suppose that x_0 and x_α are disconnected by $x_i, x_{i+1}, ..., x_{i+k-1}$ where $1 \leqslant i < \alpha - k + 1$, and

$$x_{\alpha+j}, x_{\alpha+j+1}, ..., x_{\alpha+j+k-1} ,$$

where

$$1 \leqslant j \leqslant n - k - \alpha .$$

Let

$$\beta = \left[\frac{\alpha + i + j + k - n - 1}{2} \right], \quad \beta' = \beta + \frac{n}{2} \quad (\text{indices mod. } n) .$$

Then,

$$\beta \in [\alpha + j + k, n + i - 1], \quad \beta' \in [i + k, \alpha + j - 1] .$$

There is a chain from x_0 to x_β and from $x_{\beta'}$ to x_α. Hence, there is a chain from x_0 to x_α since $[x_\beta, x_{\beta'}]$ is an edge of $G_{n, 2k+1}$.

SUBCASE 2″. Suppose that both n and q are odd integers. Let $q = 2k + 1$. Form the graph $G(n, m)$ by adding to $G_{n,2k}$ n edges of the form $[x_i, x_j]$, where $j \equiv i + \dfrac{n-1}{2}$ modulo n, and also $m - nq$ arbitrary edges. The total number of edges now equals $(m - nq) + n + 2kn = m$. Similarly, it can be shown that this graph $G(n, m)$ is q-connected.

Q.E.D.

The following theorem for h-connected graphs was discovered first by K. Menger [1926] and rediscovered by H. Whitney [1932].

Lemma *If a and b are two non-adjacent vertices of a simple graph G, then the maximum number of vertex-disjoint chains between a and b equals*

$$c_G^0(a, b) = \min_A |A|,$$

where the minimization is taken over all vertex sets A that do not contain either a or b and whose removal disconnects a from b.

To prove the lemma, we shall construct a transportation network R with source a and sink b in which the flow φ that maximizes φ_1 determines φ_1 vertex-disjoint chains from a to b in G.

First, replace each edge in G by two oppositely directed arcs with the same endpoints. Replace each vertex $x \neq a, b$ by two vertices x' and x'' joined by an arc (x', x'') with capacity 1. Replace each arc entering x by an arc entering x' with capacity 1, and replace each arc leaving x by an arc leaving x'' with capacity 1.

Thus, the maximum flow between a and b represents the maximum number of vertex-disjoint paths between a and b. The minimum capacity of a cut between a and b represents the minimum number of vertices needed to disconnect a from b. From Theorem (1, Ch. 5), these quantities are equal.

Q.E.D.

Theorem 3 (Menger [1926]). *A necessary and sufficient condition for a simple graph to be h-connected is that any two distinct vertices a and b can be joined by h vertex-disjoint chains.*

Let h be a positive integer. Since the proof is obvious for $h = 1$, suppose $h \geqslant 2$.

Necessity. Let $G = (X, E)$ be a h-connected graph. The lemma shows that two non-adjacent vertices can be joined by h vertex-disjoint chains. We shall show that this is also true for two adjacent vertices.

Suppose this were not true. Let a and b be two adjacent vertices that cannot be joined by h vertex disjoint chains. Consider the graph G' obtained from G by removing edge $[a, b]$. Vertices a and b cannot be joined in G' by $h - 1$ vertex-disjoint chains. Thus, from the lemma, there exists in G' a set $A \subset X - \{a, b\}$ of cardinality $\leq h - 2$ that disconnects a and b. Then

$$| X - A | = | X | - | A | \geq (h + 1) - (h - 2) = 3 ,$$

and there is a vertex c distinct from a and b in $X - A$.

We shall show that it is possible to join vertices c and a in graph G' by a chain that avoids A. If c and a are adjacent, clearly this is possible. If c and a are not adjacent, they are joined by h disjoint chains in G, and therefore, they are joined by $h - 1$ disjoint chains in G'; one of these chains avoids A, because $| A | < h - 1$.

Similarly, vertices c and b can be joined by a chain that avoids A. Finally, vertices a and b in G' can be joined by a chain that avoids A, which contradicts the definition of A.

Sufficiency. If two arbitrary vertices are joined by h disjoint chains, graph G is connected. At most one of these chains has length 1, and the union of the vertices of the $h - 1$ other chains contains at least $h - 1$ distinct vertices other than a and b. Hence,

$$n \geq (h - 1) + 2 > h .$$

Thus, $h \leq n - 1$.

Furthermore, G has no articulation set A with $| A | < h$ because, otherwise, we could choose two vertices a and b in two distinct components of the subgraph generated by $X - A$; each chain joining a and b passes through set A, but there cannot be h distinct vertices in A.

Thus G is h-connected.   Q.E.D.

Corollary 1. *If G is h-connected, the partial graph obtained by removing an edge is $(h - 1)$-connected.*

If two vertices can be joined by h vertex-disjoint chains in G, they can be joined by at least $h - 1$ vertex-disjoint chains in the new graph.

Q.E.D.

Corollary 2. *If G is h-connected, the subgraph obtained by removing a vertex is $(h - 1)$-connected.*

If two vertices of the new graph can be joined by h vertex-disjoint chains in G, then they can be joined by at least $h - 1$ vertex-disjoint chains in the new graph.

Q.E.D.

Corollary 3. *Let G be a simple h-connected graph. Let $B = \{ b_1, b_2, ..., b_h \}$ be a set of vertices with $|B| = h$. If $a \in X - B$, there exist h vertex-disjoint elementary chains $\mu_i[a, b_i]$ joining a and B.*

We may assume that none of the b_i are adjacent to a because, if k vertices of B are adjacent to a, then h can be replaced by $h - k$ in the theorem.

Let graph G' be the graph obtained from graph G by adding a vertex z and joining it to each vertex $b_i \in B$. We shall show that graph G' is h-connected. Let S be a subset of X with $|S| \leqslant h - 1$; we have then to show that the subgraph generated by $X \cup \{z\} - S$ is connected. This is clearly true if $z \in S$, since G_{X-S} is connected. If $z \notin S$, then, since $|S| < h$, one of the b_i does not belong to S, and since this b_i is adjacent to z, the subgraph under consideration is connected.

From the lemma, there exist, then, h vertex-disjoint chains in G' between a and z. These chains induce in G the required chains $\mu_i[a, b_i]$.

Q.E.D.

Corollary 4. *Let G be a simple h-connected graph with $h \geqslant 2$. An elementary cycle passes through an arbitrary set of two edges e_1 and e_2 and $h - 2$ vertices $a_1, a_2, ..., a_{h-2}$.*

1. If $h = 2$, the graph G' obtained from G by adding vertices a and b in the middle of edges e_1 and e_2, respectively, is again 2-connected because no vertex of G' can disconnect G'. From the Menger theorem, there exist two vertex-disjoint chains in G' that join a and b. These chains determine an elementary cycle passing through edges e_1 and e_2, and the proof is achieved.

2. Suppose $h > 2$ and suppose that the theorem is true for all k-connected graphs with $k < h$. We shall show that it is also true for a given h-connected graph G. We may assume that $a_1, a_2, ..., a_{h-2}$ are not the endpoints of either e_1 or e_2 since then the theorem would follow from the induction hypothesis.

The subgraph obtained from G by removing vertex a_{h-2} is $(h-1)$-connected, from Corollary 2. Hence by the induction hypothesis, there is a cycle μ_0 passing through $e_1, e_2, a_1, a_2, ..., a_{h-3}$. Let B be the vertex set of cycle μ_0. Clearly, $|B| \geqslant h$.

From Corollary 3, vertex a_{h-2} and set B can be joined by h vertex-disjoint chains. Suppose that each of these chains encounters B at only one vertex (i.e. does not return to B). Denote these chains by

$$\mu_1[a_{h-2}, b_1], \mu_2[a_{h-2}, b_2], ..., \mu_h[a_{h-2}, b_h] .$$

Suppose, for example, that cycle μ_0 encounters vertices

$$b_1, b_2, ..., b_h \in B$$

in this order. Among the h segments of μ_0 determined by two consecutive vertices of the sequence $b_1, b_2, ..., b_h, b_1$, there is at least one segment that does not contain in its interior any of the $h - 1$ elements $a_1, a_2, ..., a_{h-3}, e_1, e_2$. For example, let $\mu_0[b_1, b_2]$ be this segment. The cycle required for the proof is then:

$$\mu_2[a_{h-2}, b_2] + \mu_0[b_2, b_h] + \mu_0[b_h, b_1] - \mu_1[a_{h-2}, b_1].$$

Q.E.D.

Corollary 5 (Dirac [1960]). *If G is a simple h-connected graph with $h \geqslant 2$, then there is an elementary cycle passing through h arbitrary vertices.*

Select any set of h vertices $a_1, a_2, ..., a_h$. Consider an edge $[a_{h-1}, x]$ and an edge $[a_h, y]$. From Corollary 4, there exists an elementary cycle passing through these two edges and through vertices $a_1, a_2, ..., a_{h-2}$.

Therefore, there exists an elementary cycle containing vertices $a_1, a_2, ..., a_h$.

Q.E.D.

Theorem. (Halin [1969]). *If G is a simple h-connected graph, either there exists an edge such that the partial graph obtained by removing this edge is also h-connected or there exists a vertex of degree h.*

The proof, which is omitted, uses Theorem 3. For the cases $h = 3$ and 4, the theorem was first proved by Las Vergnas [1968].

We shall now study conditions on the degrees of a graph that imply h-connectivity. The following result is a slight generalization of a result of Bondy [1969].

Theorem 4. *Let G be a simple graph with n vertices $x_1, x_2, ..., x_n$ such that*

$$d_G(x_1) \leqslant d_G(x_2) \leqslant \cdots \leqslant d_G(x_n) = d.$$

Let $i_1 < i_2 < \cdots$ be the sequence of indices i such that $d_G(x_i) < i$. For the l-th index of this sequence, let

$$q(l) = \left[\frac{1}{l} \sum_{k=1}^{l} d_G(x_{i_k}) \right].$$

Then, the number p of connected components of G satisfies

$$p \leqslant \max \left\{ l, \left[\frac{n + q(l) - d}{q(l) + 1} \right] \right\}.$$

Let $c_1 \leqslant c_2 \leqslant \cdots \leqslant c_p$ be the cardinalities of the vertex sets of the p connected components of G. We may assume that $p > l$ because, if $p \leqslant l$, the proposed inequality is obvious.

Graph G contains at least c_1 vertices of degree $\leqslant c_1 - 1$, and $c_1 + c_2 + \cdots + c_k$ vertices of degree $\leqslant c_k - 1$. Hence,

$$d_G(x_{c_1 + c_2 + \cdots + c_k}) \leqslant c_k - 1 < c_1 + c_2 + \cdots + c_k \qquad (k = 1, 2, \ldots, p) .$$

For $k = 1, 2, \ldots, p$, the numbers $c_1 + c_2 + \cdots + c_k$ are all different. Hence, $c_1 + c_2 + \cdots + c_k \geqslant i_k$, and so

$$\begin{cases} c_k \geqslant d_G(x_{c_1 + c_2 + \cdots + c_k}) + 1 \geqslant d_G(x_{i_k}) + 1 \qquad (k = 1, 2, \ldots, p - 1) . \\ c_p \geqslant d + 1 . \end{cases}$$

Since $l \leqslant p - 1$, we have

$$n = c_1 + c_2 + \cdots + c_p \geqslant \sum_{k=1}^{l} d_G(x_{i_k}) + \sum_{k=l+1}^{p-1} d_G(x_{i_k}) + d + p .$$

Thus, from the definition of $q(l)$,

$$\sum_{k=1}^{l} d_G(x_{i_k}) \geqslant l q(l) .$$

Furthermore, $k > l$, we have $d_G(x_{i_k}) \geqslant q(l)$. Hence,

$$n \geqslant l q(l) + (p - 1 - l) q(l) + d + p ,$$

and thus

$$p \leqslant \frac{n + q(l) - d}{q(l) + 1} ,$$

which implies that

$$p \leqslant \max \left\{ l, \left[\frac{n + q(l) - d}{q(l) + 1} \right] \right\} .$$

Q.E.D.

EXAMPLE. Consider the graph G in Fig. 9.1 with the following degree sequence; 1, 1, 3, 3, 4, 4, 4, 4, 4. The vertices x_i with $d_G(x_i) < i$ are circled.

For $l = 1$, we have $q(1) = 1$, and

$$p \leqslant \left[\frac{9 + 1 - 4}{2} \right] = 3 .$$

For $l = 2$, we have $q(2) = 2$, and

$$p \leqslant \max \left\{ 2, \left[\frac{9 + 2 - 4}{3} \right] \right\} = 2 .$$

Thus, Theorem 4 shows that G cannot have more than 2 components.

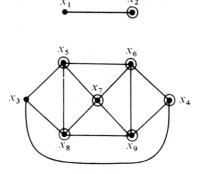

Fig. 9.1

Corollary 1 (Bondy [1969]). *Let G be a simple graph with n vertices $x_1, x_2, \ldots, x_n$ such that*

$$d_G(x_1) \leqslant d_G(x_2) \leqslant \cdots \leqslant d_G(x_n) = d \,.$$

If, for some integer q,

$$d_G(x_k) \geqslant k \qquad (k = 1, 2, \ldots, q) \,,$$

then the number p of connected components of G satisfies

$$p \leqslant \frac{n + q - d}{q + 1} \,.$$

There exists an index i with $d_G(x_i) < i$ because $d_G(x_n) < n$; let i_1 be the smallest such index. We have

$$q(1) = d_G(x_{i_1}) \geqslant d_G(x_q) \geqslant q \,.$$

Hence

$$\frac{n + q(1) - d}{q(1) + 1} \leqslant \frac{n + q - d}{q + 1} \,.$$

The proof follows by letting $l = 1$ in Theorem 4.

                                                                Q.E.D.

The graph in Fig. 9.1 shows that Corollary 1 is weaker than Theorem 4.

Corollary 2. *Let G be a simple graph with n vertices $x_1, x_2, \ldots, x_n$ such that*

$$d_G(x_1) \leqslant d_G(x_2) \leqslant \cdots \leqslant d_G(x_n) = d \,.$$

If $d_G(x_k) \geqslant k$ for all $k \leqslant n - d - 1$, then G is connected.

In Corollary 1, let

$$q = n - d - 1 .$$

We have

$$n + q - d = (q + d + 1) + q - d < 2(q + 1) ,$$

and, from Corollary 1,

$$p \leqslant \frac{n + q - d}{q + 1} < 2 .$$

Thus graph G is connected.

Q.E.D.

Corollary 3 (Bondy [1969]). *Let G be a simple graph with n vertices x_1, $x_2, ..., x_n$ such that*

$$d_G(x_1) \leqslant d_G(x_2) \leqslant \cdots \leqslant d_G(x_n) .$$

If for some integer $h < n$

$$d_G(x_k) \geqslant k + h - 1 \qquad (k \leqslant n - d_G(x_{n-h+1}) - 1)$$

then G is h-connected.

Consider a set $A \subset X$ with cardinality $h - 1$. We shall show that A cannot be an articulation set. Index the vertices x_i' of $G' = G_{X-A}$ so that

$$d_{G'}(x_1') \leqslant d_{G'}(x_2') \leqslant \cdots \leqslant d_{G'}(x_{n-h+1}') = d' .$$

If $k \leqslant |X'| - d' - 1$, then

$$k \leqslant (n - h + 1) - \left(d_G(x_{n-h+1}) - h + 1\right) - 1 = n - d_G(x_{n-h+1}) - 1 .$$

From the hypothesis, this implies that

$$d_G(x_k) \geqslant k + h - 1 ,$$

which implies that

$$d_{G'}(x_k') \geqslant d_G(x_k) - |A| \geqslant k + h - 1 - (h - 1) = k.$$

Corollary 2 shows that G_{X-A} is connected. Since this is proved for each set $A \subset X$ with cardinality $h - 1$, graph G is h-connected.

Q.E.D.

Corollary 4 (Chartrand, Kapoor, Kronk [1968]). *Let G be a simple graph of order n such that, for some integer $h < n$, the following two conditions hold:*

(1) for each $k \leqslant \dfrac{n - h}{2}$, the number of vertices with degree $< k + h - 1$ is less than k;

(2) *the number of vertices with degree* $< \dfrac{n+h}{2} - 1$ *is less than* $n - h + 1$.

Then graph G is h-connected.

After indexing the vertices as in Corollary 3, conditions (1) and (2) become equivalent to

(1') $d_G(x_k) \geqslant k + h - 1 \qquad \left(k \leqslant \dfrac{n-h}{2} \right)$

(2') $d_G(x_{n-h+1}) \geqslant \dfrac{n+h}{2} - 1$.

The inequalities $k > \dfrac{n-h}{2}$ and $k \leqslant n - d_G(x_{n-h+1}) - 1$ cannot be satisfied simultaneously because this would imply that

$$\frac{n-h}{2} < n - d_G(x_{n-h+1}) - 1 ,$$

and hence

$$\frac{n+h}{2} - 1 \leqslant d(x_{n-h+1}) < n - 1 - \frac{n-h}{2} = \frac{n+h}{2} - 1 ,$$

which is a contradiction.

Thus, if $k \leqslant n - d_G(x_{n-h+1}) - 1$, then $k \leqslant \dfrac{n-h}{2}$ and, from condition (1'),

$$d_G(x_k) \geqslant k + h - 1 .$$

Corollary 3 shows that graph G is h-connected.

Q.E.D.

Corollary 5 (Chartrand, Harary [1968]). *Let G be a simple graph of order n such that, for some positive integer $h < n$,*

$$d_G(x) \geqslant \frac{n+h}{2} - 1 \qquad (x \in X) .$$

Then G is h-connected.

To prove the corollary, it is sufficient to show that conditions (1') and (2') of Corollary 4 are satisfied. If $k \leqslant \dfrac{n-h}{2}$, then, for each vertex x of graph G,

$$d_G(x) \geqslant \frac{n-h}{2} + h - 1 \geqslant k + h - 1 .$$

Thus condition (1′) is satisfied; condition (2′) is obviously satisfied.

Q.E.D.

2. Articulation vertices and blocks

A vertex whose removal from the graph increases the number of connected components is called an *articulation vertex*. An edge whose removal from the graph increases the number of connected components of the graph is called an *isthmus*. Using these definitions, we can redefine 2-connectivity: *A graph is 2-connected if, and only if, it is of order $n \geqslant 3$, connected, and has no articulation vertices.*

A set A of vertices in graph G that generates a subgraph G_A that is connected and without articulation vertices and is maximal with respect to this

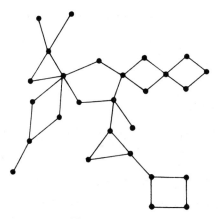

Fig. 9.2. Cactus

property is called a *block*. Thus, subgraph G_A is either 2-connected (if $|A| > 2$), or an isthmus of G (if $|A| = 2$), or an isolated vertex of G (if $|A| = 1$). It is left to the reader to verify that the graph in Fig. 9.2 has 9 articulation vertices, 6 isthmi, and 13 blocks.

A *chord* of an elementary cycle is defined to be an edge that joins two non-consecutive vertices of the cycle. A cycle of length 3 has no chords. A cycle of length 4 can have 0, 1 or 2 chords.

A *cactus* is defined to be a connected graph in which every block is either an isthmus or an elementary cycle without chords (see Fig. 9.2).

The principal characteristics of 2-connected graphs are described in the following theorem.

Theorem 5. *The following properties are equivalent in a graph G of order* $\geqslant 3$:

(1) *G is 2-connected.*
(2) *Every two vertices of G lie on a common elementary cycle.*
(3) *Every vertex and edge of G lie on a common elementary cycle.*
(4) *Every two edges of G lie on a common elementary cycle.*
(5) *Every two edges of G lie on a common elementary cocycle.*
(6) *Every two adjacent edges of G lie on a common elementary cycle.*

(1) $\Rightarrow$ (4) This was established by Corollary 4 to Theorem 3.

(4) $\Rightarrow$ (3) $\Rightarrow$ (2) This proof is obvious.

(2) $\Rightarrow$ (1) This follows from Theorem 3.

(4) $\Rightarrow$ (5) Let $[a, b]$ and $[c, d]$ be two edges, and let

$$[b, a, a_1, a_2, ..., a_k, c, d, d_1, d_2, ..., d_l, b]$$

be an elementary cycle that contains them.

Let $A = \{ a, a_1, ..., a_k, c \}$. Consider the subgraph G_{X-A}, and let B denote the connected component of G_{X-A} that contains b and d. Then cocycle $\omega(B)$ in G is an elementary cocycle that contains edges $[a, b]$ and $[c, d]$.

(5) $\Rightarrow$ (6) Let $[a, b]$ and $[a, c]$ be two adjacent edges. Let $\omega(A)$ be an elementary cocycle that contains these edges and that separates the graph into two connected components A and B. If $a \in A$, then $b \in B$ and $c \in B$. Consequently, there exists an elementary chain $\mu[b, c]$ in B that joins b and c. Hence,

$$\mu[b, c] + [c, a] + [a, b]$$

is an elementary cycle that contains edges $[a, b]$ and $[a, c]$.

(6) $\Rightarrow$ (1) We shall show that G is 2-connected. Otherwise, G has an articulation vertex a. Let B and C denote the two connected components of the subgraph $G_{X-\{a\}}$. Vertex a is joined to B by an edge $[a, b]$ and to C by an edge $[a, c]$; clearly, no elementary cycle can contain both these edges. This contradicts (6).

Q.E.D.

In the following theorem Ramachandra Rao has characterized the simple connected graphs with n vertices and r articulation vertices that have the maximum number of edges.

Lemma. *A connected graph of order* $n \geqslant 2$ *has at least two vertices that are not articulation vertices. There are exactly two such vertices if, and only if, the graph consists of an elementary chain.*

From Chapter 3, Theorem 2, we know that a spanning tree of G has at least two pendant vertices. Clearly, these vertices cannot be articulation vertices of G.

If G has only two vertices that are not articulation vertices, then each spanning tree of G has exactly two pendant vertices. This implies that G is an elementary chain.

Q.E.D.

Theorem 6 (Ramachandra Rao [1968]). *In a simple connected graph of order $n \geq 2$ with r articulation vertices, the maximum number of possible edges is*

$$\binom{n-r}{2} + r .$$

Recall from the lemma that $n - r \geq 2$; thus this formula is meaningful.

If G is such a graph with the maximum number of possible edges, then each block of G is a clique with at least two vertices. If there are p blocks, then $p \geq r + 1$. Let $n_i \geq 2$ be the number of vertices in the i-th block. We see easily, by induction on the number of blocks, that

$$\sum_{i=1}^{p} n_i = n + p - 1 .$$

Thus, the number of edges of G is

$$\max \left\{ \sum_{i=1}^{p} \binom{n_i}{2} \Big/ \sum_{i=1}^{p} n_i = n + p - 1 ; \; n_1, n_2, \ldots, n_p \geq 2 ; \; p \geq r + 1 \right\}$$

$$= \max_{p \geq r+1} \left(p - 1 + \binom{n + p - 1 - 2p + 2}{2} \right)$$

$$= \max_{p \geq r+1} \left(\binom{n - p + 1}{2} + p - 1 \right)$$

$$= \binom{n - r}{2} + r .$$

Note that we can construct a graph with r articulation vertices and with the maximum number of possible edges by taking a clique K_{n-r} with $n - r$ vertices and attaching to one of these vertices an elementary chain of length r. Thus, there are r articulation vertices, and the number of edges is

$$m = \binom{n - r}{2} + r.$$

Q.E.D.

We shall now apply the preceding results to graphs in which *each elementary cycle of even length has at least two chords*. Trees and cacti with only odd cycles are examples of such graphs. Other less trivial examples will be encountered later.

Lemma 1. *If each even elementary cycle has at least two chords, then each even elementary cycle generates a clique.*

If this were not true, then there would exist an even cycle μ of minimum length that does not generate a clique. Let $\mu = [a_1, a_2, ..., a_p, a_1]$.

Clearly, $p \geqslant 6$, since for $p = 4$, a quadrilateral with two diagonals is a clique.

An *even chord* (respectively, *odd chord*) is defined to be a chord that divides μ into two even chains (respectively, odd chains). There are only two cases to consider:

(1) *Cycle μ has an odd chord* $[a_1, a_i]$. Then the cycles

$$\mu_1 = \mu[a_1, a_i] + [a_i, a_1]$$
$$\mu_2 = [a_1, a_i] + \mu[a_i, a_1]$$

are even and of length less than $l(\mu)$, the length of μ. These cycles generate two cliques (see Fig. 9.3).

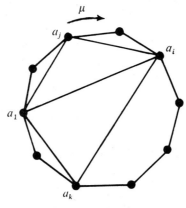

Fig. 9.3

Vertices

$$a_j \in \mu[a_2, a_{i-1}] \quad \text{and} \quad a_k \in \mu[a_{i+1}, a_p]$$

are adjacent because the cycle

$$[a_1, a_j] + [a_j, a_i] + [a_i, a_k] + [a_k, a_1]$$

is a quadrilateral and therefore possesses two chords, one of which is necessarily $[a_j, a_k]$. Thus μ generates a clique.

(2) *Cycle μ has two non-adjacent even chords $[a_1, a_i]$ and $[a_j, a_k]$.* Suppose that these chords cross one another; for example, suppose

$$a_j \in \mu[a_1, a_i], \quad a_k \in \mu[a_i, a_1],$$

as in Fig. 9.4. Consider the following two even cycles:

$$v_1 = \mu[a_1, a_j] + [a_j, a_k] - \mu[a_i, a_k] + [a_i, a_1],$$
$$v_2 = \mu[a_j, a_i] + [a_i, a_1] - \mu[a_k, a_1] + [a_k, a_j].$$

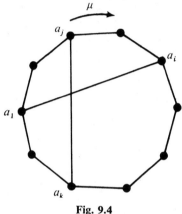

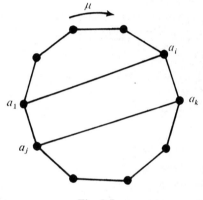

Fig. 9.4   Fig. 9.5

At least one of the cycles v_i has length $l(v_i)$ with

$$l(v_i) \leqslant \frac{1}{2} l(\mu) + 2,$$

since, otherwise,

$$l(\mu) = l(v_1) + l(v_2) - 4 > \frac{1}{2} l(\mu) + 2 + \frac{1}{2} l(\mu) + 2 - 4 = l(\mu).$$

For example, let v_1 be this cycle v_i. Then,

$$l(v_1) \leqslant \frac{1}{2} l(\mu) + 2 < l(\mu).$$

Consequently, v_1 generates a clique, and there exists in μ at least one odd chord. From Part (1) of the proof, μ generates a clique.

If the two chords do not cross one another, suppose, for example, $a_k \in [a_i, a_1]$, and $a_j \in [a_i, a_1]$, where $k < j$ (see Fig. 9.5). Consider the cycle

$$v = [a_1, a_i] + \mu[a_i, a_k] + [a_k, a_j] + \mu[a_j, a_1].$$

Since cycle v is shorter than cycle μ, it generates a clique, and cycle μ contains an odd chord. From part (1) of the proof, μ generates a clique. Hence, in all cases, μ generates a clique. This contradicts the definition of μ.

Q.E.D.

Lemma 2. *If each even cycle has at least two chords, then an odd elementary cycle with at least one chord generates a clique.*

It is sufficient to show that if $\mu = [a_1, a_2, ..., a_p, a_1]$ is an odd cycle and if $[a_1, a_i]$ is one of its chords, then $[a_1, a_{i-1}]$ and $[a_1, a_{i+1}]$ are also chords. It is evident that one of the cycles

$$\mu_1 = \mu[a_1, a_i] + [a_i, a_1] \quad \text{or} \quad \mu_2 = [a_1, a_i] + \mu[a_i, a_1]$$

is even. Suppose μ_1 is even. From Lemma 1, we know that μ_1 generates a clique and this implies that $[a_1, a_{i-1}] \in E$. Since $[a_1, a_{i-1}] \in E$, consider the cycle

$$\mu' = [a_1, a_{i-1}] + \mu[a_{i-1}, a_1].$$

This cycle is even, and so we know from Lemma 1 that $[a_1, a_{i+1}] \in E$.

Q.E.D.

Theorem 7. (Dirac [1960]). *If each even elementary cycle has at least two chords, then each block is either a clique or an odd cycle without chords.*

Let B be a block of G. If B is not a clique, then there exist two non-adjacent vertices b_1 and b_2 in B. Since G_B is 2-connected, there is a cycle μ_0 that contains b_1 and b_2. This cycle μ_0 is odd and has no chords (since, otherwise, b_1 and b_2 would be adjacent, from Lemmas 1 and 2).

Let $B_0 \subset B$ denote the set of vertices on μ_0. Suppose first that $B_0 \neq B$. Let $a \in B - B_0$. From Corollary 3 to Theorem 3, there are two vertex-disjoint elementary chains $\mu_1[a, x]$ and $\mu_2[a, y]$ joining a and B_0, that contain no vertices of B_0 except x and y.

One of the cycles

$$\mu_1[a, x] + \mu_0[x, y] - \mu_2[a, y]$$

$$\mu_1[a, x] - \mu_0[y, x] - \mu_2[a, y]$$

is even. Since this cycle generates a clique (from Lemma 1), vertices x and y are adjacent. Since μ_0 has no chords, x and y are consecutive vertices of the cycle. Therefore, the even cycle contains b_1 and b_2. Thus b_1 and b_2 are adjacent, which is impossible.

Thus $B_0 = B$ is an odd cycle without chords.

Q.E.D.

3. k-Edge-connected graphs

In this section we shall study, for a general graph $G = (X, U)$, the cardinality of a minimum cut between two vertices a and b, i.e. the number:

$$c_G^+(a, b) = \min_{\substack{A \subset X \\ a \in A \\ b \in X - A}} m_G^+(A, X - A),$$

where $m_G^+(A, B)$ denotes the number of arcs going from A to B. We shall also study, for a multigraph $G = (X, E)$, the number:

$$c_G(a, b) = \min_{\substack{A \subset X \\ a \in A \\ b \in X - A}} m_G(A, X - A).$$

Theorem 8. 1. *Let $G = (X, U)$ be a graph; then $c_G^+(a, b)$ equals the maximum number of arc-disjoint paths from a to b.*

2. *Let $G = (X, E)$ be a multigraph; then $c_G(a, b)$ equals the maximum number of edge-disjoint chains from a to b.*

1. If $G = (X, U)$ is a graph, form from G a transportation network R with a as a source a and b as a sink. Let each arc $u \in U$ have capacity $c(u) = 1$, and add a return arc $(b, a) = u_0$ with infinite capacity.

In network R, the maximum flow problem (Ch. 5, § 1) is to find a flow φ, satisfying $0 \leqslant \varphi(u) \leqslant 1$ for all u in U, that maximizes the flow value in the return arc u_0.

Theorem (1, Ch. 5) shows that

$$\max_{\varphi} \varphi(u_0) = \min_{\substack{A \subset X \\ A \ni a \\ A \not\ni b}} m_G^+(A, X - A).$$

The left side of this equality equals the maximum number of arc-disjoint paths from a to b in graph G. The right side is, by definition, $c_G^+(a, b)$. This proves Part 1 of the theorem.

2. If $G = (X, E)$ is a multigraph, consider the graph G^* obtained from G by replacing each edge of G by two oppositely directed arcs. The maximum number of elementary chains in G between a and b with no common edge equals the maximum number of elementary paths in G^* from a to b with no common arc. From part 1, this number equals

$$c_{G^*}^+(a, b) = c_G(a, b).$$

Q.E.D.

Nash-Williams Lemma. *If* $G = (X, E)$ *is a multigraph, it is always possible to construct a set* E' *of new edges that matches the vertices of odd degree, and is such that graph* $G' = (X, E')$ *satisfies*

$$\min_{\substack{A \ni a \\ A \not\ni b}} \left(m_G(A, X - A) - m_{G'}(A, X - A) \right) = 2 \left[\frac{1}{2} c_G(a, b) \right],$$

for all $a \in X$, $b \in X$.

Let Y denote the set of vertices of odd degree, and let Z denote the set of vertices of even degree; then

$$2 \mid E \mid = \sum_{x \in X} d_G(x) = \sum_{x \in Y} d_G(x) + \sum_{x \in Z} d_G(x).$$

Thus $\mid Y \mid$ is an even number, and it is always possible to match together the vertices of Y. The proof of this lemma is long, and the reader is referred to the original presentation (Nash-Williams [1960]).

A connected multigraph G is said to be *k-edge-connected* if it cannot be disconnected by the removal of less than k edges, i.e., if and only if

$$c_G(x, y) \geqslant k \quad \text{for all} \quad x, y \in X, x \neq y.$$

A multigraph is 1-edge-connected if, and only if, it is connected. A multigraph is 2-edge-connected if, and only if, it is connected and has no isthmus. The least k such that G is k-edge-connected is called the edge-connectivity.

Lemma. *In a connected multigraph* G, *an edge is an isthmus if, and only if, no elementary cycle of* G *contains this edge.*

If $[a, b]$ is not an isthmus, the graph remains connected after the removal of $[a, b]$ and there is an elementary chain $\mu[a, b]$ of G that does not contain $[a, b]$. Thus, $[a, b]$ together with $\mu[a, b]$ form the required elementary cycle.

Conversely, if edge $[a, b]$ lies on a cycle, then the removal of $[a, b]$ does not disconnect the graph, and so $[a, b]$ cannot be an isthmus.

Q.E.D.

Theorem 9. *A connected multigraph is* 2-edge-connected *if, and only if, every edge lies on a cycle.*

The proof follows from the lemma.

Theorem 10 (Robbins [1941]). *Given a simple graph* G, *its edges can be directed to form a strongly connected 1-graph* H *if, and only if,* G *is 2-edge-connected.*

1. If such a graph H exists, then G is 2-edge-connected because the removal of an edge cannot disconnect H and, consequently, cannot disconnect G.

2. Conversely, if G is 2-edge-connected, each edge of G lies on an elementary cycle. Let A_1 denote the set of vertices on a first elementary cycle of G. Direct the edges of this cycle to form a circuit, and arbitrarily direct all the edges with both endpoints in A_1. The subgraph generated by A_1 is strongly connected. If $X = A_1$, the proof is achieved.

If $X - A_1 \neq \varnothing$, there is a vertex $a_2 \notin A_1$ that is adjacent to a vertex $a_1 \in A_1$ (because the graph is connected). By hypothesis, edge $[a_1, a_2]$ is on an elementary cycle that contains a chain leaving A_1 at vertex a_1 and returning to A_1 at some vertex b_1, no edge of which is already directed. Direct this chain so that it becomes a path. Let A_2 denote set A_1 augmented by the vertices of this path. Arbitrarily direct all undirected edges with both endpoints in A_2.

The subgraph generated by A_2 is strongly connected. If $X \neq A_2$, this procedure can be repeated, as many times as needed, to obtain a strongly connected graph H.

Q.E.D.

A stronger result is due to Las Vergnas [1975]:

Given a strongly connected 1-graph G, with no isthmus consisting of a double edge, one can obtain a strongly connected 1-graph H by removing one arc from each double edge of G.

Another extension is due to Nash-Williams [1960]:

Given a multigraph $G = (X, E)$, one can obtain a graph $H = (X, U)$ by orienting each edge of G, such that

$$c_H^+(x, y) \geqslant [\tfrac{1}{2}c_G(x, y)] \qquad (x, y \in X, x \neq y).$$

Even stronger results are due to Mader [1978]. See also Thomassen [1978]. The following is an immediate corollary to the theorem of Nash-Williams:

A necessary and sufficient condition for a multigraph G to admit a strongly h-connected orientation H is that:

$$m_G(A, X - A) \geqslant 2h \qquad (A \neq \varnothing, A \neq X).$$

The case where $h = 1$ gives Robbins' theorem. Other extensions are due to Franck [1980, 1982].

The notion of a minimal strongly h-connected graph generalizes that of a minimally h-connected graph. Mader has conjectured:

If G is a minimal strongly h-connected graph, then there exists a vertex x of G such that

$$d_G^+(x) = d_G^-(x) = h.$$

Hamidoune has shown that every minimal strongly h-connected graph contains at least $h + 1$ vertices of demi-degree h. He has also shown that every strongly 2-connected subgraph of a minimal strongly h-connected graph contains a vertex x with

$$d^+(x) = h \quad \text{or} \quad d^-(x) = h.$$

(Hamidoune [1981])

These two results generalize, for the case of directed graphs, two of Mader's theorems on the existence of $h + 1$ vertices of degree h in every minimally h-connected graph, and on the existence of a vertex of degree h in every cycle of a minimally h-connected graph (Mader [1972]).

Recall that Mader has solved this problem for the case of minimal strongly h-arc-connected graphs (Mader [1974]):

Every minimal strongly h-arc-connected graph contains two vertices of degree 2h.

Mader [1971] also showed:

Every graph G with $m(G) > (h - 1)n - \binom{h}{2}$ and $n \geqslant h$ contains an h-edge-connected subgraph.

EXERCISES

1. Show that a k-connected graph is k-edge-connected.

2. Show that a connected graph is 2-connected if, and only if, for every three vertices a, b, x, there exists an elementary chain $\mu[a, b]$ that contains x.

3. Let G be a simple regular graph of degree 3. Show that G is k-connected if, and only if, G is k-edge-connected.

4. Tutte has shown that:

A graph G is 3-connected if, and only if, G is a "wheel" (an elementary cycle μ and a vertex x_0 joined to each vertex in μ) or can be formed from a wheel by a sequence of operations of the two following types:

(1) the addition of a new edge,

(2) the replacement of a vertex x of degree $\geqslant 4$ by two adjacent vertices x' and x'' of degrees $\geqslant 3$ such that a neighbour of x in the original graph is a neighbour of exactly one of x' and x'' in the new graph.

Verify this result with some simple examples.

(Tutte [1961])

5. A simple 2-connected graph is defined to be *minimally 2-connected* if the partial graph resulting from the removal of any edge is not 2-connected.
If G is minimally 2-connected, show that:
(1) No edge is the chord of a cycle.
(2) If $G \neq K_3$, then G contains no triangles.
(3) If $[x, y]$ is an edge and z is an articulation vertex of $G - [x, y]$, then z lies on every cycle that contains $[x, y]$.
(4) Each chain that contains $[x, y]$ and z contains a vertex of degree 2 in its interior.
(5) If $G \neq K_3$, there is a cycle of G that contains two non-adjacent vertices of degree 2.
(6) The set of all the vertices of degree 2 is an articulation set.

(M. Plummer [1968])

6. If the sequence $d_1 \leqslant d_2 \leqslant \cdots \leqslant d_n$ does not satisfy Bondy's condition (Corollary 3 of Theorem 4) for some $h < n$, then there exists an $h + 1$-connected graph whose degree sequence majorises (d_i).

(Boesch [1974])

7. The sequence $d_1 \leqslant d_2 \leqslant \cdots \leqslant d_n$ is the degree sequence of an h-connected graph if and only if:
(1) it is the degree of a simple graph;
(2) $d_i \geqslant h$ for all i;

$$(3) \quad m - \sum_{i=h}^{n} d_i + \binom{h-1}{2} \geqslant n - h.$$

(Wang, Kleitman [1973])

8. A simple graph is said to be k-critical h-connected if it is h-connected, but if the graph obtained by deleting any non-empty set of at most k vertices is not h-connected. Show that:

(1) Every 1-critical h-connected graph has a vertex of degree $\leqslant \left\lceil \dfrac{3h}{2} \right\rceil - 1$.

(Chartrand, Mader)

(2) Every 1-critical 3-connected graph has two vertices of degree 3.

(Entriger, Slater [1978])

(3) Every k-critical h-connected graph, with $k > \dfrac{h}{2}$, has two vertices of degree h.

(Hamidoune)

Give examples of these results.

CHAPTER 10

Hamiltonian Cycles

1. Hamiltonian paths and circuits

In a graph $G = (X, U)$, a *hamiltonian path* is defined to be a path that meets every vertex exactly once. Similarly, a *hamiltonian circuit* is defined to be a circuit that passes through every vertex exactly once. In a simple graph $G = (X, E)$, a *hamiltonian chain* and a *hamiltonian cycle* are defined similarly.

EXAMPLE 1. *Voyage around the world* (Hamilton). Consider 20 cities $a, b, c, ..., t$, represented by the vertices of a regular dodecahedron (polyhedron with twelve pentagonal faces and 20 vertices). How can we travel to every city exactly once and return home using only the edges of the dodecahedron? In other words, can we find a hamiltonian cycle in the graph in Fig. 10.1?

Hamilton solved this problem as follows. When the traveller arrives at the endpoint of an edge, he has the choice of taking the edge to his right (denote this choice by R), or the edge to his left (denote this choice by L), or by staying where he is (denote this choice by 1). In the obvious way, define a *product* of these operations: for example, L^2R denotes the operation of going left twice and then right once. Finally, two operations are defined to be *equal* if after starting from the same vertex they terminate at the same vertex. Note that the product is not commutative (for example $LR \neq RL$), but it is associative (for example $(LL)R = L(LR)$). For the graph in Fig. 10.1,

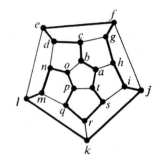

Fig. 10.1

$$\begin{cases} R^5 = L^5 = 1 \\ RL^2\,R = LRL \\ LR^2\,L = RLR \\ RL^3\,R = L^2 \\ LR^3\,L = R^2 \,. \end{cases}$$

184

Hence,

$$1 = R^5 = R^2 R^3 = (LR^3 L) R^3 = (LR^3)^2 = [L(LR^3 L) R]^2 =$$
$$= (L^2 R^3 LR)^2 = [L^2(LR^3 L) RLR]^2 = [L^3 R^3 LRLR]^2 =$$
$$= LLLRRRLRLRLLLRRRLRLR .$$

This sequence contains twenty operations and contains no partial sequence that equals 1. Therefore, it represents a hamiltonian cycle. Another hamiltonian cycle is obtained by reversing the sequence. It is left to the reader to verify that no other cycles exist.

Note that the voyage around the world can begin (and terminate) at any of the twenty cities. Hamilton solved the problem with the additional constraint that the first five cities to be visited had to be a, b, c, d, e in that order. He found four solutions, starting with edge $[a, b]$:

RLRLRLLLRRRLRLRLLLRR
RLRLLLRRRLRLRLLLRRRL
RLRLRRRLLLRLRLRRRLLL
RLRRRLLLRLRLRRRLLLRL

EXAMPLE 2. *Open voyage around the world.* Suppose that the traveller in Example 1 need not return home. In this case, each trip corresponds to a hamilton chain in the graph in Fig. 10.1, and the number of distinct trips available to the traveller increases greatly. All the trips starting with an R are indicated below. All other possible trips can be obtained by interchanging the operations L and R in the sequences given below.

RRRLRLRLRLLLRRRLRLRL
RLRLLLRRRLRLRLLLRR   RLRRRLLLRLRLRRRLLLRL
RLLLRRRLRLRLLLRRRL   RLRLLRRRLRLLLRRLRL
RRRLLLRLRLRRRLLLRL
RLRRRLLLRLRLRRRLLL
RLRLRRRLLLRLRLRRRL   RRRLRRLRLRLRLLRLLL
RRLRLRLLLRRRLRLRLL   RRRLLRLRRRLLLRLRLR
RLRLRLLLRRRLRLRLLL   RRRLRRLRLRLLLRRRLR
RLLLRLRLRRRLLLRLRL
RRLLLRLRLRRRLLLRLR   RLRLRLLLRRRLRLLRRR
                            RLLLRLRLLRRLRLRRRL

RRRLLLRLRLRLRRLRRR
RLRRRLLLRLRLRRLRRR   RRRLRRLRLRLRLLLLRR
RLRLLLRRRLRLLRRRLRL   RLRLRRRLLLRRLRLLLR
RLLLRLRRRLLLRRRLRLR

                            RRRLLLRLRLRLRRRLLL
RLLLRLRLLRRRLRLLLR   RLLLRLRRRLLRLRLLLR
RLLLRLRRRLLLRLRRRL

EXAMPLE 3. *Knight's journey* (Euler). How can a knight be moved on a chessboard so that he visits each square exactly once? This problem of finding a hamiltonian chain has interested many mathematicians: e.g. Euler, de Moivre, Vandermonde, etc. Many methods have been proposed. One method that seems to work in practice is the following: Move the knight to the square from which he will control the smallest number of unvisited squares.

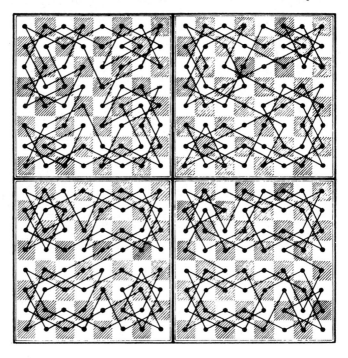

Fig. 10.2. Examples of the knight's journey

Another method is to move the knight on only half of the chessboard and then to have him repeat this pattern on the other half (see Fig. 10.2). This method depends on the special structure of the chessboard and does not work in more general situations.

EXAMPLE 4. *The problem of Mr. No.* Mr. No, a mythical Japanese detective, lives in the upper left square of a chessboard. He wishes to visit Mr. Go who lives in the lower right square of the chessboard. Mr. No can move to any adjacent square but cannot move diagonally. Is it possible for him to visit each square of the chessboard exactly once en route to Mr. Go?

Clearly, the problem of Mr. No is to find a hamiltonian chain between two given vertices. To visit each square exactly once, Mr. No must make 63 moves and 63 colour changes. Clearly, after 63 colour changes Mr. No stops on a square with a colour different from the colour of his home square. Since Mr. No and Mr. Go have home squares of the same colour, no such trip is possible.

Theorem 1. *The number $h(G)$ of hamiltonian paths in a 1-graph $G = (X, U)$ has the same parity as the number of hamiltonian paths in the complementary 1-graph $\overline{G} = (X, X \times X - U)$.*

We shall show that if G and $\overline{G}$ are two complementary 1-graphs, then $h(G) \equiv h(\overline{G})$ modulo 2.

Suppose that the vertices of G are indexed from 1 to n. Given a subset V of the arcs of G, let $h(V)$ denote the number of arrangements $(i_1, i_2, ..., i_n)$ of $(1, 2, ..., n)$ such that each arc of V is of the form (i_k, i_{k+1}). Note that if $h(V)$ is not zero, then

$$| V | \leqslant n - 1 \,,$$

and the arcs of V form a family of pairwise disjoint paths.

Furthermore, if $h(V) \neq 0$ and if $| V | < n - 1$, the connected components of the partial graph (X, V) consist of r disjoint paths, where $r > 1$, and consequently, $h(V) = r! \equiv 0$. Hence,

$$h(V) \not\equiv 0 \quad \Rightarrow \quad | V | = n - 1 \,.$$

Now note that $h(\overline{G})$ equals the number of arrangements $(i_1, i_2, ..., i_n)$ such that no arc of G is of the form (i_k, i_{k+1}). Thus,

$$n\,! - h(\overline{G}) = \sum_{\substack{V \subset U \\ | V | = 1}} h(V) - \sum_{\substack{V \subset U \\ | V | = 2}} h(V) + \cdots + (-1)^{k-1} \sum_{\substack{V \subset U \\ | V | = k}} h(V) + \cdots$$

Therefore

$$h(\overline{G}) \equiv n\,! - h(\overline{G}) \equiv \sum_{\substack{V \subset U \\ | V | = n - 1}} h(V) = h(G) \,.$$

$$\text{Q.E.D.}$$

Hence, $h(\overline{G}) \equiv h(G)$ modulo 2.

Theorem 2 (C. A. B. Smith [1946]). *In a simple regular graph of degree 3, the number of hamiltonian cycles that contain a given edge is even.*

Suppose that there exists a hamiltonian cycle (otherwise, the result is trivially true). This cycle is of even length (since there are $\frac{3}{2} n$ edges, which implies that n is even). Thus the edges of the cycle can alternately be coloured with two colours α and β and the edges not on the cycle can be coloured γ.

Thus, there exists a partition $\{E_\alpha, E_\beta, E_\gamma\}$ of the edge set into three perfect matchings. Such a partition will be called here a 3-*colouration* of the edges. A 3-colouration of the edges with three colours α, β, γ determines a vector $\mu_{\alpha\beta} = (\mu^1, \mu^2, ..., \mu^m)$ with $\mu^i = 1$ if edge e_i is α or β, and $\mu^i = 0$ otherwise. Let $\mu \in M$ if vector μ defines a family of vertex-disjoint even cycles that use all vertices. Each 3-colouration $\{E_\alpha, E_\beta, E_\gamma\}$ determines three vectors $\mu_{\alpha\beta}, \mu_{\alpha\gamma}, \mu_{\beta\gamma} \in M$, with

$$\mu_{\alpha\beta} + \mu_{\alpha\gamma} + \mu_{\beta\gamma} \equiv 0 \qquad (\text{mod. } 2).$$

Furthermore, if a vector $\mu \in M$ consists of a family of $k(\mu)$ pairwise disjoint cycles, then it corresponds to $2^{k(\mu)-1}$ distinct 3-colourations. Summing the above identity over all 3-colourations $\{E_\alpha, E_\beta, E_\gamma\}$ gives

$$\sum_{\mu \in M} 2^{k(\mu)-1} \mu \equiv 0 \qquad (\text{mod. } 2).$$

Hence,

$$\sum_{\mu \,/\, k(\mu)=1} \mu \equiv 0 \qquad (\text{mod. } 2).$$

Since this sum is over all the hamiltonian cycles of G, the number of distinct hamiltonian cycles that contain a given edge is even.

Q.E.D.

Corollary 1. *If a simple regular graph of degree* 3 *has a hamiltonian cycle, it has at least three hamiltonian cycles.*

Let e be an edge of the hamiltonian cycle. At least two distinct hamiltonian cycles μ_1 and μ_2 contain e. Let x denote the first vertex at which these two cycles diverge after passing through e. If $[x, y]$ is an edge of μ_1 and not of μ_2, then at least two hamiltonian cycles pass through $[x, y]$, and they are both different from μ_2.

Q.E.D.

Corollary 2 (N. J. A. Sloane [1969]). *If a graph* $G = (X, E)$ *has two hamiltonian cycles without common edges, it has at least three hamiltonian cycles.*

Let $\mu_1 = [x_1, x_2, ..., x_n, x_1]$ be a hamiltonian cycle of G, and let μ_2 be another hamiltonian cycle of G that does not contain any edge of μ_1.

CASE 1. Suppose n is even. A regular graph of degree 3 can be formed with all the edges of μ_1 and some edges of μ_2. Corollary 1 established that this graph has three hamiltonian cycles. Therefore, the result is true for G.

CASE 2. Suppose n is odd and $[x_i, x_{i+2}] \in E$ for all i. Then, there exists a third hamiltonian cycle, namely:

$$\mu_3 = [x_1, x_n, x_2, x_3, x_4, ..., x_{n-1}, x_1].$$

CASE 3. Suppose n is odd and $[x_n, x_2] \notin E$. Suppose that

$$\mu_2 = [x_i, x_1, x_j, x_{k_1}, x_{k_2}, \ldots, x_{k_{n-3}}, x_i], \qquad i, j \neq n .$$

Form a new graph $\bar{G}$ by adding edge $[x_n, x_2]$ and by removing vertex x_1 and all edges incident to x_1. Graph $\bar{G}$ has a hamiltonian cycle $\bar{\mu}_1 = [x_2, x_3, \ldots, x_n, x_2]$. The edges of cycle $\bar{\mu}_1$ and the edges

$$[x_j, x_{k_1}], [x_{k_2}, x_{k_3}], [x_{k_4}, x_{k_5}], \ldots, [x_{k_{n-3}}, x_i],$$

form a regular graph of degree 3 that has two hamiltonian cycles that use edge $[x_2, x_n]$ by Theorem 2. Therefore G possesses two hamiltonian cycles distinct from μ_2.

$$\text{Q.E.D.}$$

Theorem 3 (Kotzig [1967]). *A simple regular bipartite graph of degree 3 has an even number of hamiltonian cycles.*

Let $G = (X, Y, E)$ be a regular bipartite graph of degree 3 with $| X | = | Y | = \dfrac{n}{2}$, and let $h(G)$ denote the number of hamiltonian cycles in G. For $n \leqslant 6$, the theorem is true since the complete bipartite graph $K_{3,3}$ contains exactly 6 hamiltonian cycles.

Assume that the theorem is true for graphs of order $< n$; we shall show that it is also true for a graph $G = (X, Y, E)$ of order $n > 6$. Since G contains no triangles, an edge $[x_1, y_1]$ is adjacent to four edges $[x_1, y_2], [x_1, y_3], [y_1, x_2], [y_1, x_3]$ such that the vertices $x_1, x_2, x_3, y_1, y_2, y_3$ are all distinct.

Let G' and G'' denote the two graphs obtained from G by the transformations shown in Fig. 10.3.

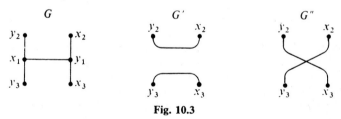

Fig. 10.3

Let h_i, h_i' and h_i'' denote the number of distinct hamiltonian cycles in the graphs G, G', and G'' using the vertices $x_1, x_2, x_3, y_1, y_2, y_3$ as shown in Fig. 10.4. From the induction hypothesis,

$$h(G') = h_1' + h_2' + h_3' + h_4' + h_5' \equiv 0 \qquad (\text{mod. } 2)$$

$$h(G'') = h_1'' + h_2'' + h_3'' + h_4'' + h_5'' \equiv 0 \qquad (\text{mod. } 2) .$$

Furthermore, it is evident that

$$h_1 = h_1'', \quad h_2 = h_2', \quad h_3 = h_3', \quad h_4 = h_4'$$

$$h_5 = h_3'', \quad h_6 = h_4'', \quad h_1' = h_2'', \quad h_5' = h_5''.$$

Hence,

$$h(G) = h_1 + h_2 + h_3 + h_4 + h_5 + h_6 = h_1'' + h_2' + h_3' + h_4' + h_3'' + h_4''$$

$$\equiv h_1'' + h_2' + h_3' + h_4' + h_3'' + h_4'' + (h_1' + h_2'') + (h_5' + h_5'')$$

$$= \sum h_i' + \sum h_i'' \equiv 0 \quad (\text{mod. } 2).$$

Therefore, the number of hamiltonian cycles in G is even.

Q.E.D.

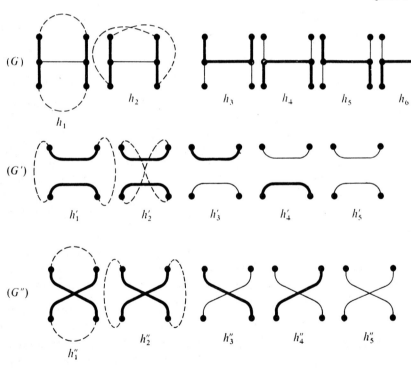

Fig. 10.4

2. Hamiltonian paths in complete graphs

Theorem 4 (Camion [1959]). *If $G = (X, \Gamma)$ is a strongly connected, complete 1-graph, then G has a hamiltonian circuit.*

Let $\mu = [a_1, a_2, ..., a_{h-1}, a_h, a_{h+1} = a_1]$ be a circuit of maximum length h. Suppose that circuit μ does not encounter some vertex b. Then,

$$a_i \in \Gamma(b) \quad \Rightarrow \quad b \notin \Gamma(a_{i-1}) \quad \Rightarrow \quad a_{i-1} \in \Gamma(b)$$
$$b \in \Gamma(a_i) \quad \Rightarrow \quad a_{i+1} \notin \Gamma(b) \quad \Rightarrow \quad b \in \Gamma(a_{i+1})$$

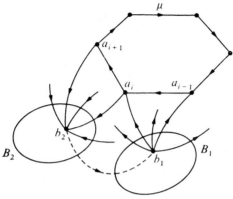

Fig. 10.5

Therefore, the vertices not lying on circuit μ can be divided into two classes B_1 and B_2 as follows: If $b \in B_1$, then each arc joining b and μ is directed towards μ. If $b \in B_2$, then each arc joining b and μ is directed towards b.

By hypothesis, $B_1 \cup B_2 \neq \varnothing$. Since G is strongly connected, $B_1 \neq \varnothing$, $B_2 \neq \varnothing$, and there exists an arc from B_2 to B_1. Denote this arc by (b_2, b_1), where $b_2 \in B_2$ and $b_1 \in B_1$.

Thus the circuit $[a_1, a_2, ..., a_h, b_2, b_1, a_1]$ is longer than μ, which contradicts the maximality of μ.

Q.E.D.

Corollary. *Let $G = (X, \Gamma)$ be a complete 1-graph. There exists a vertex x_0 such that for each vertex $y \neq x_0$ there is a path from x_0 to y. Such a vertex x_0 is called a "root" of G. Each root is the initial endpoint of a hamiltonian path of G.*

1. The existence of a root of G was established in Theorem (4, Ch. 4), since any centre of G is a root.

2. Let x_0 be a root of G. Consider the graph G' obtained from graph G by adding a vertex z, an arc (z, x_0), and the arc (x, z) for every $x \neq x_0$. By Theorem 4, graph G', which is complete and strongly connected, has a hamiltonian circuit, which corresponds in G to a hamiltonian path starting from x_0.

Q.E.D.

Algorithm. This corollary provides a very simple algorithm to construct a hamiltonian path in a complete 1-graph $G = (X, \Gamma)$ without loops:

Let a_1 denote the vertex with the largest outer demi-degree $d_G(a_1)$. Vertex a_1 is a centre (Theorem 4, Ch. 4). In the subgraph generated by $\Gamma(a_1) - \{\, a_1 \,\}$, let a_2 denote the vertex with the largest outer demi-degree. In the subgraph generated by $\Gamma(a_2) - \{\, a_1, a_2 \,\}$, let a_3 denote the vertex with the largest outer demi-degree, etc.

Then $[a_1, a_2, ...,]$ is a hamiltonian path.

Theorem 5. *If* $G = (X, \Gamma)$ *is a complete, anti-symmetric, transitive* 1-*graph, then* G *has exactly one hamiltonian path.*

A hamiltonian path must start from a root. Since G is transitive, vertex x_1 is a root if, and only if,

$$\Gamma(x_1) \cup \{\, x_1 \,\} = X .$$

There is always a root in G, and since G is anti-symmetric, the above equation shows that the root is unique. Thus, the first vertex of a hamiltonian path must necessarily be this root x_1.

By the same argument, the second vertex of the hamiltonian path is the unique root of the subgraph generated by $X - \{\, x_1 \,\}$, etc. ..., and the hamiltonian path is unique.

$$\text{Q.E.D.}$$

Theorem 6 (Rédei [1934]). *If* G *is a complete, anti-symmetric* 1-*graph, then the number* $h(G)$ *of distinct hamiltonian paths in* G *is odd.*

We shall show that if G is a complete, anti-symmetric graph, then reversing the direction of an arc does not change the parity of $h(G)$. Since graph G can be obtained from a complete, anti-symmetric, transitive graph by making successive reversals of arc directions, this fact will establish the theorem.

Let (a, b) be an arc of G, and let G' be the graph obtained from G by reversing the direction of (a, b). Let G_1 be the graph obtained from G by adding arc (b, a). Let G_2 be the graph obtained from G by removing arc (a, b). We want to show that

$$h(G) \equiv h(G') \qquad (\text{mod. } 2) .$$

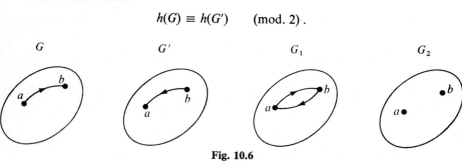

Fig. 10.6

Since graph G_2 is anti-symmetric, each hamiltonian path of G_2 (taken in its reverse direction) defines a hamiltonian path of $\overline{G}_1$, the complementary 1-graph of G_1, and vice versa. Therefore (from Theorem 1), we have

$$h(G_2) = h(\overline{G}_1) \equiv h(G_1) \qquad (\text{mod. } 2) .$$

Let $h_1(ab)$, $h_1(ba)$, $h_1(0)$ denote respectively the number of hamiltonian paths of G_1 that contain arc ab, arc ba, neither arc ab nor ba. Then,

$$h_1(0) = h(G_2) \equiv h(G_1) = h_1(0) + h_1(ab) + h_1(ba) .$$

Thus, $h_1(ab) \equiv h_1(ba)$ modulo 2, and consequently,

$$h(G) = h_1(0) + h_1(ab) \equiv h_1(0) + h_1(ba) = h(G') .$$

This shows that $h(G) \equiv h(G')$ modulo 2, and completes the proof.

Q.E.D.

3. Existence of a hamiltonian circuit

The first such existence theorem for a 1-graph G is a result due to Ghouila-Houri [1960] concerning the demi-degrees of G. This was generalized by Meyniel [1973]. A simpler proof was found by Bondy and Thomassen [1977], who used an idea due to Overbeck-Larisch [1977].

Lemma. *Let $\mu = [x_1, x_2, ..., x_k]$ be a path of a graph G, and let x be a vertex of G, $x \notin \mu$, such that $[x_1, x_2, ..., x_i, x, x_{i+1}, ..., x_k]$ is not a path, for any index i. Then the number $m(x, S)$ of arcs that join x and $S = \{ x_1, x_2, ..., x_k \}$ satisfies:*

$$m(x, S) \leqslant | S | + 1 .$$

Let
$$A = \{ x_i \, / \, x_i \in S, \, i \neq k, \, (x_i, x) \in U \} ,$$
$$B = \{ x_i \, / \, x_i \in S, \, i \neq k, \, (x, x_{i+1}) \in U \} .$$

By hypothesis, $A \cap B = \varnothing$, and since $(A \cup B) \subset (S - \{ x_k \})$, we have $| A \cup B | \leqslant | S | - 1$. Thus

$$m(x, S) \leqslant | A | + | B | + 2 \leqslant | A \cup B | + 2 \leqslant | S | + 1 .$$

Theorem 7 (Meyniel [1973]). *Let G be a strongly connected 1-graph, with no loops, of order n. If for every pair of non-adjacent vertices x, y,*

(1) $$d_G(x) + d_G(y) \geqslant 2n - 1 ,$$

then G contains a hamiltonian circuit.

Recall that $d_G(x) = d_G^+(x) + d_G^-(x)$ for G a 1-graph with no loops.

Suppose that G satisfies condition (1) but does not contain a hamiltonian circuit (proof by contradiction). Let $\mu = [x_1, x_2, \ldots, x_k, x_1]$ be a longest circuit of G. Let $S = \{ x_1, x_2, \ldots, x_k \}$.

1. There exists a path whose endpoints are two distinct vertices in S and whose internal vertices form a non-empty set of $X - S$.

Suppose not (proof by contradiction). Since G is strongly connected, there is a circuit $[s_0, a_1, a_2, \ldots, a_p, s_0]$ with $s_0 \in S$ and $A = \{ a_1, a_2, \ldots, a_p \} \subset X - S$. Since no vertex of A is adjacent to a vertex of $S - \{ s_0 \}$, we have for $a \in A$, $s \in S - \{ s_0 \}$

$$m(a, A) \leqslant 2|A| - 2,$$
$$m(s, A) = 0,$$
$$m(a, S) \leqslant 2,$$
$$m(s, S) \leqslant 2|S| - 2.$$

Also

$$m[a, X - (A \cup S)] + m[s, X - (A \cup S)] \leqslant 2[X - (A \cup S)],$$

for otherwise there would exist a path from a to s with an internal vertex in $X - (A \cup S)$, and this path, together with a part of μ, would yield a forbidden path from S to S.

Thus a and s satisfy:

$$d_G(a) + d_G(s) \leqslant 2[|A| + |S| + |X - (A \cup S)|] - 2 \leqslant 2n - 2.$$

But, since a and b are non-adjacent, this contradicts (1).

2. Let $v = [s_1, a_1, \ldots, a_p, s_2]$ be a path whose endpoints are two distinct vertices in S, with $A = \{ a_1, \ldots, a_p \} \subset X - S$, $A \neq \varnothing$, chosen so that the portion $\mu[s_1, s_2]$ is as short as possible.

By the choice of v, since no vertex of A is adjacent to $S_1 = \mu[s_1, s_2] - \{s_1, s_2\}$, we have, for all $a \in A$

(2) $$m(a, S_1) = 0.$$

Since μ is a longest circuit of G, then by the Lemma the vertex a also satisfies

(3) $$m(a, S - S_1) \leqslant |S - S_1| + 1.$$

Lastly,

(4) $$m(a, A) \leqslant 2|A| - 2.$$

Let $\bar{\mu}[s_2, s_1]$ be the longest path from s_2 to s_1 whose vertex set $\bar{S}$ satisfies $S - S_1 \subset \bar{S} \subset S$. Since μ is a longest circuit and $A \neq \varnothing$, $\bar{S}$ must be a proper subset of S. Therefore there exists a vertex $s \in S - \bar{S} \subset S_1$, and by the Lemma

(5) $m(s, \bar{S}) \leqslant |\bar{S}| + 1.$

Also

(6) $m(s, S - \bar{S}) \leqslant 2|S - \bar{S}| - 2.$

Since no point of S_1 is adjacent to A,

(7) $m(s, A) = 0.$

Also (as with Section 1 above), we have

(8) $m[a, X - (A \cup S)] + m[s, X - (A \cup S)] \leqslant 2|X - (A \cup S)|.$

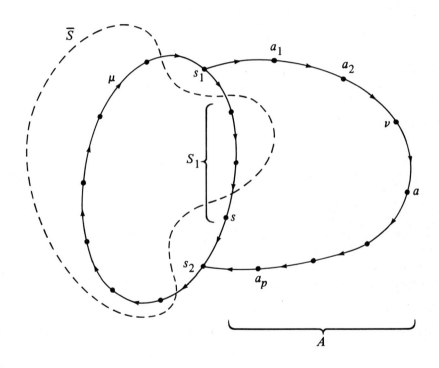

Fig. 10.7

Successively combining inequalities (2) through (8) gives

$$d_G(a) + d_G(s) \leqslant |S - S_1| + 1 + 2|A| - 2 + |\overline{S}| + 1 + 2|S - \overline{S}| - 2$$
$$+ 2|X - (A \cup S)| \leqslant 2|\overline{S}| + 2|S - \overline{S}| + 2|A| + 2|X - (A \cup S)| - 2 \leqslant 2n - 2.$$

Since a and s are non-adjacent, this contradicts (1).

Q.E.D.

Corollary 1 (Theorem 4, Section 2). *Every complete strongly connected 1-graph has a hamiltonian circuit.*

Corollary 2 (Ghouila-Houri's Theorem, 1960). *If a strongly connected 1-graph G of order n, with no loops, satisfies*

$$d_G(x) \geqslant n \qquad (x \in X),$$

then G has a hamiltonian circuit.

In fact, Theorem 7 guarantees the existence of a hamiltonian circuit if G has a vertex x_0, such that

$$d_G(x) \geqslant n \quad (x \neq x_0), \qquad d_G(x_0) \geqslant n - 1.$$

Corollary 3. *Let G be a 1-graph, with no loops, of order n. If for every pair of non-adjacent vertices x, y*

$$d_G(x) + d_G(y) \geqslant 2n - 3,$$

then G has a hamiltonian path.

Form the strongly connected graph G' by adding a vertex a to G, and by joining a with two parallel arcs to each vertex in G. Every pair of non-adjacent vertices x, y in G' are also non-adjacent in G, and thus

$$d_{G'}(x) + d_{G'}(y) = d_G(x) + d_G(y) + 4 \geqslant 2n - 3 + 4 = 2(n + 1) - 1.$$

By Theorem 7, G' has a hamiltonian circuit which implies that G has a hamiltonian path.

Q.E.D.

(Note that this corollary implies the existence of a hamiltonian path in a complete graph).

Corollary 4 (Bermond [1971]). *Let $G = (X, U)$ be a 1-graph of order n with no loops and let k be an integer, $0 \leqslant k \leqslant n - 1$. If*

(1) $\qquad d_G^+(x) + d_G^-(y) \geqslant n + k \qquad (x, y \in X, x \neq y, (x, y) \notin U)$,

then each family of vertex-disjoint elementary paths with total length k is contained in a hamiltonian circuit.

The result is immediate for $k = 0$, because for any pair of non-adjacent vertices x, y

$$d_G^+(x) + d_G^-(y) \geq n, \qquad d_G^+(y) + d_G^-(x) \geq n.$$

Therefore

$$d_G(x) + d_G(y) \geq 2n \geq 2n - 1$$

and then Theorem 7 guarantees the existence of a hamiltonian circuit.

Suppose now that $k > 0$, and that the result is true for all integers less than k. Let G be a graph satisfying (1). Consider a family $(\mu_1, \mu_2, \ldots)$ of pair-wise disjoint paths with total length k.

Associate with the path $\mu_1 = [a_0, a_1, \ldots, a_h]$ a graph G_1, obtained by removing the vertices of μ_1 from G and by adding a vertex z, adding the arc (x, z) if $x \in X - \mu_1$ and $(x, a_0) \in U$, or the arc (z, x) if $x \in X - \mu_1$ and $(a_h, x) \in U$. A vertex x, other than z, of G_1 satisfies

$$d_{G_1}^+(x) \geq d_G^+(x) - h, \qquad d_{G_1}^-(x) \geq d_G^-(x) - h.$$

Also,

$$d_{G_1}^+(z) \geq d_G^+(a_h) - h, \qquad d_{G_1}^-(z) \geq d_G^-(a_0) - h.$$

If $(x, y) \notin U_{G_1}$, $x, y \neq z$, then $(x, y) \notin U$; hence by (1)

$$d_{G_1}^+(x) + d_{G_1}^-(y) \geq d_G^+(x) + d_G^-(y) - 2h \geq n + k - 2h = (n - h) + (k - h).$$

If $(x, z) \notin U_{G_1}$, then $(x, a_0) \notin U$; therefore

$$d_{G_1}^+(x) + d_{G_1}^-(z) \geq d_G^+(x) - h + d_G^-(a_0) - h \geq (n - h) + (k - h).$$

If $(z, x) \notin U_{G_1}$, then $(z, x) \notin U$; therefore

$$d_{G_1}^+(z) + d_{G_1}^-(x) \geq d_G^+(a_h) - h + d_G^-(x) - h \geq (n - h) + (k - h).$$

Thus, in all these cases (1) is satisfied for $k - h$, and by induction G_1 has a hamiltonian circuit containing $\mu_2, \mu_3, \ldots$ (of total length $k - h$). This implies that G has a hamiltonian circuit containing $\mu_1, \mu_2, \mu_3, \ldots$ (of total length k).

Q.E.D.

Corollary 5 (Bermond [1971]). *If $G = (X, U)$ is a 1-graph such that*

$$d_G^+(x) + d_G^-(y) \geq n + 1 \qquad (x, y \in X, x \neq y, (x, y) \notin U),$$

then for every (ordered) pair of distinct vertices a, b there exists a hamiltonian path from a to b.

Consider the graph G' formed by adding to G (if necessary) the arc (b, a). The desired path is found in G by applying Corollary 4 to G', with $k = 1$ and $\mu_1 = [b, a]$.

Corollary 6 (Lewin [1975]). *Every strongly connected 1-graph with at least $(n - 1)(n - 2) + 3$ arcs has a hamiltonian circuit.*

This result follows easily from Corollary 2. Note that all strongly connected graphs with $(n - 1)(n - 2) + 2$ arcs and no hamiltonian circuits have been tabulated (Bermond, Germa, Sotteau, 1980).

Nash-Williams (1969) conjectured that if every vertex x satisfies $d_G^+(x) \geqslant \dfrac{n}{2}$ and $d_G^-(x) \geqslant \dfrac{n}{2}$, then there exist two arc-disjoint hamiltonian circuits. Several counter-examples to this conjecture have been found. However, Zhang Cun-Quan [1982] has shown: *If G is a strongly connected graph of order $n \geqslant 9$ with $d_G^+(x) \geqslant \dfrac{n}{2}$ and $d_G^-(x) \geqslant \dfrac{n}{2}$ for every vertex x, then G has a pair of arc-disjoint circuits, one of which is hamiltonian and the other of length $\geqslant n - 1$.*

Jackson [1981] showed: *If an anti-symmetric 1-graph G satisfies $d_G^+(x) \geqslant \dfrac{n}{2} - 1$, $d_G^-(x) \geqslant \dfrac{n}{2} - 1$ for each vertex x, then G has a hamiltonian circuit.*

He also proved: *If an anti-symmetric 1-graph G satisfies $d_G^+(x) \geqslant k$, $d_G^-(x) \geqslant k$ for all x, then G has a path of length $2k$, and this bound is the best possible.*

Thomassen (see Jackson [1981]) conjectured that if $d_G^+(x) \geqslant \dfrac{n}{3}$, $d_G^-(x) \geqslant \dfrac{n}{3}$, the 1-graph G has a hamiltonian circuit. As well, he proved: *If G is a 1-graph of order $n = 2k + 1$ with $d_G^+(x) = d_G^-(x) = k$ for each vertex x, then G has a hamiltonian circuit, or else is one of the two graphs in Fig. 10.8.*

In many cases, analogous conditions on the demi-degrees imply the existence of circuits of all lengths. For example, Moon [1968] has shown: *If a complete anti-symmetric graph ("tournament") G is strongly connected, then for each vertex x and for each integer k, $3 \leqslant k \leqslant n$, there exists a circuit of length k passing through x.*

Thomassen [1977] proved: *If G is a strongly connected 1-graph of order n, and if every pair of non-adjacent vertices x, y satisfies $d_G(x) + d_G(y) \geqslant 2n$, then G has a circuit of length k for all k from 2 to n, unless G is a tournament, or unless $n = 2q$ and G is the symmetric complete-bipartite graph $K_{q, q}^*$.*

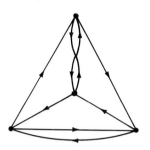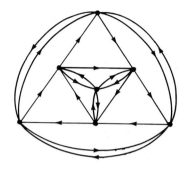

Fig. 10.8

Alspach [1967] showed: *If G is a regular tournament of order n, then for every arc u and for each k, $3 \leq k \leq n$, there exists a circuit of length k containing u.*

Generalizations have been obtained by Jakobson [1972], by Thomassen [1980], by Zhang [1982], by Zhu and Chen [1981].

Certain authors have worked on partitioning the set of arcs into hamiltonian circuits. Kelly conjectured that this partition is possible for a regular tournament. Tillson [1980] showed: *for K_n^*, $n \neq 4$, 6 this partition is possible.* Bermond and Faber [1976] showed: *this partition is possible for $K_{n,n}^*$.*

A *dissection* of graph G is defined to be a set of elementary paths of G such that each vertex of the graph is contained in exactly one path. Call a path *closed* if it also defines a circuit, otherwise call it *open*.

The *value* of a dissection is defined to be the number of open paths in the dissection. The value is always less than or equal to the number n of vertices since a single vertex is a path of length 0. A hamiltonian path is a dissection of value 1. A hamiltonian circuit is a dissection of value 0.

Dissection Theorem. *For a 1-graph $G = (X, \Gamma)$, the minimum value of a dissection equals*

$$\delta_0 = \max_{S \subset X} \left(|S| - |\Gamma(S)| \right).$$

Consider a dissection $a = (\alpha^1, \alpha^2, ..., \alpha^p, \beta^1, \beta^2, ..., \beta^q)$ of value q, where α^i is a closed path, and β^j is an open path. Let A^i denote the set of all vertices α^i, and let B^j denote the set of all vertices in β^j.

Make two copies X and $\bar{X}$ of X, and form the bipartite graph $H = (X, \bar{X}, \bar{E})$ where $(x_i, \bar{x}_j) \in \bar{E}$ if, and only if, $x_j \in \Gamma(x_i)$.

Each circuit α^i defines uniquely in H a matching of $|A^i|$ edges, and each open path defines in H a matching of $| B^j | - 1$ edges. Therefore the dissection defines a matching E_0 of cardinality

$$| E_0 | = \sum_i | A^i | + \sum_j (| B^j | - 1) .$$

Thus

$$n = \sum_i | A^i | + \sum_j | B^j | = \sum_i | A^i | + \sum_j (| B^j | - 1) + q = | E_0 | + q .$$

This correspondence between the matchings and the dissections is a bijection. Therefore, from the König theorem (Chapter 7), the minimum value of a dissection equals

$$n - \max | E_0 | = n - (n - \delta_0) = \delta_0 .$$

This gives the required formula.

Q.E.D.

Corollary 1. *If G has a hamiltonian circuit, then $\delta_0 = 0$.*

The proof is immediate.

4. Existence of a hamiltonian cycle

Although hamiltonian circuit existence theorems for 1-graphs have not been widely studied, there are numerous theorems in the literature on the existence of hamiltonian cycles in simple graphs, of which the most well-known are:

Dirac's Theorem [1952]. *A simple graph G of order n which satisfies*

$$d_G(x) \geqslant \frac{n}{2}$$

for every vertex x has a hamiltonian cycle.

Ore's Theorem [1961] *A simple graph G of order n such that every pair of non-adjacent vertices x, y satisfies*

$$d_G(x) + d_G(y) \geqslant n$$

has a hamiltonian cycle.

Ore's Theorem is a generalization of Dirac's Theorem, and follows directly from the results on 1-graphs of the preceding section. In fact, if G^* is the symmetric graph obtained by replacing each edge of G by two parallel arcs, then Ore's condition on G implies on G^* the relation $d_{G^*}(x) + d_{G^*}(y) \geqslant 2n$; thus, by Theorem 7, G^* has a hamiltonian circuit.

In this section we study different extensions of Ore's Theorem. The reader is also referred to an excellent comparative study of several different theorems due to Bondy [1978].

Theorem 8. *Let* $G = (X, E)$ *be a simple graph of order* n *with degrees* $d_1 \leqslant d_2 \leqslant \cdots \leqslant d_n$. *Let* q *be an integer,* $0 \leqslant q \leqslant n - 3$. *If, for every* k *with* $q < k < \frac{1}{2}(n + q)$, *the following condition holds*:

(A) $\qquad\qquad d_{k-q} \leqslant k \;\;\Rightarrow\;\; d_{n-k} \geqslant n - k + q$

then for each subset F *of edges with* $|F| = q$ *such that the connected components of* (X, F) *are elementary chains, there exists a hamiltonian cycle of* G *that contains* F.

Furthermore, this result is the best possible in the following sense: each sequence of degrees that does not satisfy condition (A) *is majorized by a sequence of degrees of a graph that does not have the desired property.*

1. We shall first show that condition (A) implies: $d_1 > q$.

Suppose that $d_1 \leqslant q$, and let $k = q + 1$. Then

$$q < k = \frac{q}{2} + \frac{q + 2}{2} < \frac{q + n}{2} \, .$$

Furthermore, since $d_1 \leqslant k$, it follows from condition (A) that

$$d_{n-q-1} \geqslant n - q - 1 + q = n - 1 \, .$$

This shows that there exist at least $q + 2$ vertices joined to all the other vertices of G; therefore $d_1 \geqslant q + 1$, which is a contradiction.

2. We shall show that if a simple graph G satisfies condition (A), the graph G' obtained from G by adding a new edge also satisfies condition (A). Let

$$S_k = \{ x \, / \, d_G(x) \leqslant k \} \, .$$

and

$$S'_k = \{ x \, / \, d_{G'}(x) \leqslant k \} \, .$$

Clearly, $|S'_k| \leqslant |S_k|$. Note that condition (A) is equivalent to

(A') $\qquad\qquad |S_k| \geqslant k - q \;\;\Rightarrow\;\; |S_{n-k+q-1}| < n - k \, .$

Since G satisfies (A'),

$$|S'_k| \geqslant k - q \;\;\Rightarrow\;\; |S_k| \geqslant k - q \;\;\Rightarrow\;\; |S_{n-k+q-1}| < n - k \;\;\Rightarrow$$
$$\Rightarrow\;\; |S'_{n-k+q-1}| < n - k \, .$$

This shows that G' also satisfies condition (A).

3. To prove the main part of the theorem, suppose that there exists a graph satisfying condition (A) that does not have the desired property. By adding new edges as many times as needed, we obtain a graph $G = (X, E)$ satisfying condition (A), that does not have the desired property, and such that the addition of any new edge gives a graph with the desired property.

Let $F \subset E$ be a set with $|F| = q$ that forms pairwise disjoint elementary chains, and that is not contained in any hamiltonian cycle of G.

Since G is not complete, there exist two non-adjacent vertices y_1 and y_n; choose y_1 and y_n with $d_G(y_1) + d_G(y_n)$ as large as possible and with $d_G(y_1) \leqslant d_G(y_n)$.

There is a hamiltonian cycle μ' containing the edges of F in graph $G' = G + [y_1, y_n]$, and, necessarily, $[y_1, y_n] \in \mu'$; therefore, there is a hamiltonian chain μ in G that contains F, of the form

$$\mu = [y_1, y_2, ..., y_n] .$$

Let

$$I = \{ i / 1 \leqslant i \leqslant n - 1, [y_1, y_{i+1}] \in E, [y_i, y_{i+1}] \notin F \} .$$

Set I is not empty, since

(B) $$|I| \geqslant d_G(y_1) - q \geqslant d_1 - q > 0 .$$

Moreover,

(C) $$i \in I \implies [y_i, y_n] \notin E$$

because, otherwise, the cycle

$$[y_1, y_2, ..., y_i, y_n, y_{n-1}, ..., y_{i+1}, y_1]$$

would be a hamiltonian cycle containing F.

4. Let $k = d_G(y_1)$. We shall show that

(D) $$q < k < \frac{n + q}{2} .$$

Since $d_1 > q$, the first inequality of (D) holds. From (C), it follows that

$$\Gamma_G(y_n) \subset (X - \{ y_n \}) - \{ y_i / i \in I \} .$$

Hence,

$$d_G(y_n) \leqslant (n - 1) - [d_G(y_1) - q] .$$

Thus

(E) $$d_G(y_1) + d_G(y_n) \leqslant n + q - 1 .$$

Finally,

$$k = d_G(y_1) < \frac{n + q}{2} \; .$$

5. From (C), we know that, for each $i \in I$,

$$[y_i, y_n] \notin E$$

and

$$d_G(y_i) + d_G(y_n) \leqslant d_G(y_1) + d_G(y_n) \, .$$

Thus,

$$d_G(y_i) \leqslant k \quad (i \in I) \, .$$

Since $|I| \geqslant k - q$, there are $k - q$ vertices of degree $\leqslant k$. Hence

$$d_{k-q} \leqslant k \, ,$$

which implies that

$$d_{n-k} \geqslant n - k + q \, .$$

This shows that there are at least $k + 1$ vertices of degree $\geqslant n - k + q$. Hence, for some $x \neq y_1$,

$$\begin{cases} [y_1, x] \notin E \\ d_G(x) \geqslant n - k + q \, . \end{cases}$$

Thus,

$$n + q \leqslant d_G(x) + d_G(y_1) \leqslant d_G(y_1) + d_G(y_n) \, .$$

This contradicts (E) and completes the proof of the theorem.

6. To show that the conditions of the theorem are the best possible, consider a non-decreasing sequence $d_1 \leqslant d_2 \leqslant \cdots \leqslant d_n$ and an integer k such that

$$\begin{cases} q < k < \dfrac{n + q}{2} \\ d_{k-q} \leqslant k \\ d_{n-k} \leqslant n - k + q - 1 \, . \end{cases}$$

Sequence (d_i) is majorized by sequence (d_i'), where

$$d_i' = \begin{cases} k & \text{if} & 0 < i \leqslant k - q \\ n - k + q - 1 & \text{if} & k - q < i \leqslant n - k \\ n - 1 & \text{if} & n - k < i \leqslant n . \end{cases}$$

We shall show that there exists a graph G' with the sequence (d_i') for its degrees. Let the vertex set of G' be the union of three disjoint sets A, B, C, where A is a set S_{k-q} of $k - q$ isolated vertices, B is a clique K_k of k vertices, and C is a clique K_{n-2k+q} of $n - 2k + q$ vertices. Join each vertex of B to each vertex of $A \cup C$. Graph G' has order n, and $d_{G'}(a) = k$ if $a \in A$, $d_{G'}(b) = n - 1$ if $b \in B$, and $d_{G'}(c) = n - k + q - 1$ if $c \in C$.

Consider a set F of edges of G' that forms an elementary chain of length q in B, and suppose that there exists a hamiltonian cycle μ that contains F. In sequence μ, at most $k - q$ elements of B are followed by an element of $A \cup C$. Since $|A| = k - q$, there are exactly $k - q$ elements of B followed by an element of A. Thus there is no element of B followed by an element of C, which is impossible, since $|C| = n - 2k + q > 0$.

$$\text{Q.E.D.}$$

Corollary 1. *Let G be a simple graph with degrees $d_1 \leqslant d_2 \leqslant \cdots \leqslant d_n$. Let q be an integer, where $0 \leqslant q \leqslant n - 3$. If the following condition holds*

$$\left. \begin{array}{l} 1 \leqslant i < j \leqslant n \\ d_i \leqslant i + q \\ d_j \leqslant j + q - 1 \end{array} \right\} \Rightarrow d_i + d_j \geqslant n + q ,$$

then, for each set F of edges with $|F| = q$ such that the components of (X, F) are vertex disjoint elementary chains, there exists a hamiltonian cycle that contains F.

Suppose that G is a simple graph satisfying this condition. We shall show that the graph G also satisfies the conditions of Theorem 8.

If the conditions of Theorem 8 were not satisfied, there would exist an integer k such that

$$\begin{cases} q < k < \dfrac{n + q}{2} \\[2mm] d_{k-q} \leqslant k \\[2mm] d_{n-k} \leqslant n - k + q - 1 . \end{cases}$$

Let $i = k - q$ and $j = n - k$; then $i < j$ since, otherwise, $k \geqslant \dfrac{n + q}{2}$. Thus, we have

$$\begin{cases} 1 \leqslant i < j \leqslant n \\ d_i \leqslant i + q \\ d_j \leqslant j + q - 1 \,, \end{cases}$$

and, consequently,

$$d_i + d_j \geqslant n + q \,.$$

On the other hand, we have also

$$d_i + d_j \leqslant (k - q) + q + (n - k) + q - 1 = n + q - 1 \,,$$

which yields a contradiction.

Q.E.D.

Corollary 1 was proved directly by Bondy [1970]. It has been generalized by Las Vergnas [1970], who proved the following stronger result:

Let $G = (X, E)$ be a simple graph of order $n \geqslant 3$. Let the vertices x_i of G be indexed arbitrarily, and let q be an integer, where $0 \leqslant q \leqslant n - 1$. If

$$\left. \begin{array}{l} 1 \leqslant i < j \leqslant n \\ i + j \geqslant n - q \\ d_G(x_i) \leqslant i + q \\ d_G(x_j) \leqslant j + q - 1 \\ [x_i, x_j] \notin E \end{array} \right\} \Rightarrow d_G(x_i) + d_G(x_j) \geqslant n + q$$

then, for each $F \subset E$ with $|F| = q$ such that the connected components of (X, F) are elementary cycles, there exists a hamiltonian cycle that contains F.

The following result is an easy consequence of Corollary 1.

Corollary 2 (Kronk [1969]). *Let G be a simple graph of order n, and let q be an integer, where $0 \leqslant q \leqslant n - 2$. Suppose that*
  (1) *for each k such that $q < k < \frac{1}{2}(n + q - 1)$, the set S_k of vertices of degree $\leqslant k$ has cardinality $< k - q$,*
  (2) *if $n + q$ is odd, and if $k = \frac{1}{2}(n + q - 1)$, then $|S_k| \leqslant k - q$.*

Then for each set F of edges with $|F| = q$ such that (X, F) consists of vertex disjoint elementary chains, there exists a hamiltonian cycle that contains F.

If the degrees of G are indexed in increasing order, conditions (1) and (2) are equivalent to

(1') $d_{k-q} > k$ if $q < k < \dfrac{n+q-1}{2}$

(2') $d_{k-q+1} > k$ if $k = \dfrac{1}{2}(n+q-1)$.

We have $d_{k-q} > k$ for all $k < \dfrac{n+q-1}{2}$ from (1'). For $k = \dfrac{n+q-1}{2}$, we have from (2'):

$$d_{n-k} = d_{k-q+1} \geqslant \frac{1}{2}(n+q+1) = n - k + q.$$

In either case, the conditions of Theorem 8 are satisfied, and there exists a hamiltonian cycle that contains F.

Q.E.D.

Corollary 3 (Chvátal [1971]). *Let G be a simple graph of order $n \geqslant 3$ with degrees $d_1 \leqslant d_2 \leqslant \cdots \leqslant d_n$. If*

$$d_k \leqslant k < \frac{n}{2} \;\Rightarrow\; d_{n-k} \geqslant n - k,$$

then there exists a hamiltonian cycle in G. Furthermore, if the sequence (d_i) does not satisfy the condition, there exists a graph G', with a sequence of degrees majorizing (d_i) that has no hamiltonian cycles.

This Corollary is established by setting $q = 0$ in Theorem 8.

Note that Corollary 3 does not characterize all the sequences that imply the existence of a hamiltonian cycle. For example, Nash-Williams has shown that each regular graph of degree h with $2h + 1$ vertices has a hamiltonian cycle, and the corresponding sequence does not satisfy the above condition.

Corollary 4 (Bondy [1969]). *Let G be a simple graph of order $n \geqslant 3$ with degrees $d_1 \leqslant d_2 \leqslant \cdots \leqslant d_n$. If*

$$\left. \begin{array}{l} i \neq j \\ d_i \leqslant i \\ d_j \leqslant j \end{array} \right\} \;\Rightarrow\; d_i + d_j \geqslant n,$$

then G has a hamiltonian cycle.

This corollary is established by setting $q = 0$ in Corollary 1.

Corollary 5 (Chvátal [1972]). *Let* $G = (X, Y, E)$ *be a simple bipartite graph such that* $|X| = |Y| = n \geq 2$, *and*

$$d_G(x_1) \leq d_G(x_2) \leq \cdots \leq d_G(x_n)$$
$$d_G(y_1) \leq d_G(y_2) \leq \cdots \leq d_G(y_n) .$$

If

$$d_G(x_k) \leq k < n \quad \Rightarrow \quad d_G(y_{n-k}) \geq n - k + 1 ,$$

then G has a hamiltonian cycle.

Let G' be the graph obtained from G by joining each pair of vertices in Y. Since a hamiltonian cycle of G' cannot contain an edge of $G' - G$, it is sufficient to show that graph G' has a hamiltonian cycle.

The degree sequence of G' is $d_k = d_G(x_k)$ for $k \leq n$, and $d'_k = d_G(y_{k-n}) + (n - 1)$ for $k > n$. Clearly, this degree sequence satisfies the conditions of Corollary 3 and, consequently, G has a hamiltonian cycle.

Q.E.D.

Corollary 6 (Chvátal [1972]). *If* $d_1 \leq d_2 \leq \cdots \leq d_n$ *are the degrees of a simple graph G of order* $n \geq 2$, *and if*

$$d_k \leq k - 1 \leq \frac{n}{2} - 1 \quad \Rightarrow \quad d_{n+1-k} \geq n - k$$

then G has a hamiltonian chain.

Let G' be the graph obtained from G by adding a vertex x_{n+1} and by joining this vertex to all the vertices of G.

Let (d'_i) be the degree sequence of G'; for $k \leq \dfrac{(n + 1) - 1}{2}$, we have

$$d'_k = d_k + 1 \leq k \quad \Rightarrow \quad d'_{n+1-k} = d_{n+1-k} + 1 \geq n + 1 - k .$$

From Corollary 3, graph G' has a hamiltonian cycle, and therefore, graph G has a hamiltonian chain.

Q.E.D.

It is easy to verify that this result is the best possible in the same sense as Theorem 8. Besides, Las Vergnas [1971] has extended this corollary to prove a sufficient condition for the existence of a spanning tree of G with at most h pendant vertices.

The following corollaries are classical results:

Corollary 7 (Pósa [1962]). *Let G be a simple graph of order* $n \geq 3$, *such that*

(1) *for each* k *such that* $1 \leqslant k < \dfrac{n-1}{2}$ *, the number of vertices of degree*
$\leqslant k$ *is* $< k$,

(2) *(if n is odd) the number of vertices of degree* $\leqslant \dfrac{n-1}{2}$ *is* $\leqslant \dfrac{n-1}{2}$.

Then G has a hamiltonian cycle.
This corollary is established by setting $q = 0$ in Corollary 2.

Corollary 8. *Let G be a simple graph of order $n \geqslant 3$ with degrees $d_1 \leqslant$*
$d_2 \leqslant \cdots \leqslant d_n$. *If $k < \dfrac{n}{2}$ implies $d_k > k$, then G has a hamiltonian cycle.*

Since $k \leqslant \dfrac{n-1}{2}$ implies $d_k > k$, the number of vertices of degree $\leqslant k$ is
$< k$. From Corollary 7, G contains a hamiltonian cycle.

$$\text{Q.E.D.}$$

Corollary 9 (Erdös, Gallai [1959]). *Let G be a simple graph of order $n \geqslant 3$.*
Let x_1 be the vertex of minimum degree. If $d_G(x_1) \geqslant 2$, and if $d_G(x) \geqslant \dfrac{n}{2}$ for all
$x \neq x_1$, *then G has a hamiltonian cycle.*

If the vertices are indexed as in Corollary 1, we have $d_G(x_1) > 1$. For
$1 < k < \dfrac{n}{2}$, we have $d_G(x_k) \geqslant \dfrac{n}{2} > k$. From Corollary 8, there exists a hamiltonian cycle.

$$\text{Q.E.D.}$$

Corollary 10 (Ore [1961]). *Let $G = (X, E)$ be a simple graph of order $n \geqslant 3$,*
such that
$$d_G(x) + d_G(y) < n \quad \Rightarrow \quad [x, y] \in E.$$

Then G has a hamiltonian cycle.

Let $k < \dfrac{n}{2}$. From Corollary 2, it suffices to show that the set $S_k =$
$\{ x \mid x \in X, d_G(x) \leqslant k \}$ has cardinality $< k$.

From the hypothesis, S_k is a clique because $k < \dfrac{n}{2}$. Also $|S_k| \leqslant k + 1$,
since the degrees of its vertices are $\leqslant k$.

$|S_k| \neq k + 1$ because, otherwise, S_k would be a connected component of
G, and two vertices $x \in S_k$ and $y \in X - S_k$ would not be adjacent, implying
$$d_G(x) + d_G(y) \leqslant k + (n - k - 2) = n - 2 < n.$$

This contradicts the hypothesis of the corollary.

$| S_k | \neq k$ because, otherwise, each vertex of S_k is adjacent to at most one vertex of $X - S_k$, and the number of edges leaving S_k is

$$m_G(S_k, X - S_k) \leqslant | S_k | = k < \frac{n}{2}.$$

Since $| X - S_k | > \frac{n}{2}$, this implies that there exists a vertex $y \in X - S_k$ non-adjacent to S_k. Let $x \in S_k$. Vertices x and y are non-adjacent, and

$$d_G(x) + d_G(y) \leqslant k + (n - k - 1) = n - 1 < n.$$

This contradicts the statement of the corollary.

Thus $| S_k | < k$, and G has a hamiltonian cycle.

Q.E.D.

Theorem 9 (Erdös, Chvátal [1972]). *Let $G = (X, E)$ be a simple graph, and let h and q be two integers such that $h \geqslant 2$ and $0 \leqslant q \leqslant 2$. If graph G is h-connected, and if G contains no stable sets of $h - q + 1$ vertices, then, for each $F \subset E$ with $| F | \leqslant q$, G has a hamiltonian cycle that contains F.*

Recall that a set of vertices is called *stable* if no two of its vertices are adjacent. Let F be a set of edges with $| F | \leqslant q$. From Corollary 4 to Theorem (3, Ch. 9), there exists a cycle μ that contains F. Assume that μ is the longest cycle with this property. Then, $| \mu | \geqslant h + 1$ because Corollary 4 to Theorem (3, Ch. 5) shows the existence of a cycle that contains any $h - 2$ vertices and any two edges not incident to these $h - 2$ vertices.

If μ is not a hamiltonian cycle, there exists a vertex $a \notin \mu$. Since $| \mu | \geqslant h$, Corollary 3 to Theorem (3, Ch. 9), shows that a can be joined to the vertices of μ by h elementary chains $\mu_1[a, x_1], \mu_2[a, x_2], ..., \mu_h[a, x_h]$, that are pairwise vertex-disjoint (except at a). Let x_i denote the unique vertex of μ_i in μ, and let y_i denote the vertex of μ that immediately follows x_i in μ.

Let $I = \{ i / [x_i, y_i] \notin F \}$. No y_i, $i \in I$, is adjacent to a, because, otherwise, there exists a cycle longer than μ. Since $| I | \geqslant h - q$, and since the set $\{ y_i / i \in I \} \cup \{ a \}$ cannot be stable (its cardinality is $\geqslant h - q + 1$), there is an edge $[y_i, y_j]$ in G with $i, j \in I$ and $i < j$.

Consider the cycle

$$\mu[y_i, x_j] - \mu_j[a, x_j] + \mu_i[a, x_i] - \mu[x_i, y_i] + [y_j, y_i]$$

that contains F. The length of this cycle is greater than the length of μ. This contradicts the maximality of μ.

Q.E.D.

Corollary 1 (Erdos–Chvátal Theorem). *For a simple graph G, let $\varkappa(G)$ be the vertex-connectivity number and let $\alpha(G)$ be the stability number (i.e., the cardinality of the maximum stable set). If $\alpha(G) \leqslant \varkappa(G)$, then G has a hamiltonian cycle.*

Bondy [1978] showed that Ore's Theorem follows easily from the Erdös–Chvátal Theorem; the latter also implies the following:

Theorem 10 (Nash-Williams [1970]). *A simple regular graph of degree k and order $n = 2k + 1$ has a hamiltonian cycle.*

Clearly, such a graph $G = (X, E)$ is connected but not complete. Thus, let A be any articulating set and S a stable set; by the Erdös–Chvátal Theorem, it suffices to show that $|S| \leqslant |A|$. Let (A, C, D) be a partition of X, where C and D consist of collections of connected components of G_{X-A}.

CASE 1: $S \cap C = \emptyset$, $S \cap D = \emptyset$. In this case, $S \subset A$, thus $|S| \leqslant |A|$ and the Theorem is proved.

CASE 2: $S \cap C \neq \emptyset$, $S \cap D = \emptyset$. Let $c \in S \cap C$ and $d \in D$. Then

$$0 = |\Gamma_G c| + |\Gamma_G d| - 2k$$
$$\leqslant (|C| + |A - S|) + (|A| + |D| - 1) - |A| - |C| - |D| + 1$$
$$= |A| - |S|.$$

Again, $|S| \leqslant |A|$ and the Theorem is proved.

CASE 3: $S \cap C \neq \emptyset$, $S \cap D \neq \emptyset$. Let $c \in S \cap C$ and $d \in S \cap C$. Then

$$0 = |\Gamma_G c| + |\Gamma_G d| - 2k$$
$$= |\Gamma_G c \cap C| + |\Gamma_G c \cap A| + |\Gamma_G d \cap D| + |\Gamma_G d \cap A| - k$$
$$\leqslant |C - S| + |D - S| + |A - S| + |\Gamma_G c \cap \Gamma_G d \cap (A - S)| - 2k$$
$$\leqslant 2k + 1 - |S| + |A| - |A \cap S| - 2k = 1 - |S| + |A| - |A \cap S|.$$

Suppose that $|S| > |A|$ (proof by contradiction). Then $|S| = |A| + 1$ and $|A \cap S| = 0$. Since all the inequalities must be equalities, it follows that

$$|\Gamma_G c \cap C| = |C - S|, \qquad |\Gamma_G d \cap D| = |D - S|$$
$$\text{and} \quad \Gamma_G c \cap \Gamma_G d \cap (A - S) = A - S = A.$$

Thus

$$|C - S| = |\Gamma_G c \cap C| = k - |A|,$$
$$|D - S| = k - |A|.$$

The number of edges $m(C, A)$ between C and A satisfies

$$m(C, A) \geq |C \cap S|(k - |X - S|) + |C - S|(k - |C| + 1)$$

$$= |A| \cdot |C \cap S| + (k - |A|)(k - |C| + 1).$$

The number $m(D, A)$ satisfies a similar inequality, and combining the two inequalities term by term gives

$$m(A, X - A) \geq |A|(|C \cap S| + |D \cap S|) + (k - |A|)(n + 1 - |C| - |D|)$$

$$\geq |A|(|A| + 1) + (k - |A|)(|A| + 1) = k(|A| + 1) > k|A|.$$

Since G is of degree k, we have the desired contradiction.

Remark. Regular graphs have been the object of more elaborate existence theorems, of which certain ones generalize Theorem 10. Lovasz [1970] conjectured that connected vertex-transitive graphs (a special class of regular graphs) each contain a hamiltonian chain. This conjecture has been proved only for n prime (Turner [1967]). Erdös and Hobbs [1977] showed that if G is a 2-connected regular graph of order n even and with degree $n - q$, with $q \geq 3$ and $\dfrac{n}{2} \geq q^2 + q + 1$, then G has a hamiltonian cycle. The result also holds if n is odd and if $\dfrac{n + 1}{2} \geq 2q^3 - 3q + 3$. Improving on a result of Jackson [1980], Chu, Liu and Yu [1982] showed: every regular graph of order n and degree $k \geq \dfrac{n}{3} - 1$ is hamiltonian, with the exception of Petersen's graph (Fig. 10.11). This result is the best possible, for there exist 2-connected k-regular graphs with $3k + 4$ vertices (for even $k \geq 4$) or with $3k + 5$ vertices (for any k). Finally, Woodall [1978] proved that if every non-empty set of vertices A satisfies $|\Gamma_G A| \geq \dfrac{1}{3}(n + |A| + 3)$, then G has a hamiltonian cycle.

Theorem 11 (Las Vergnas [1970]). *Let $G = (X, Y, E)$ be a simple bipartite graph, where $X = \{x_1, x_2, \ldots, x_n\}$, $Y = \{y_1, y_2, \ldots, y_n\}$, $n \geq 2$; let q be an integer, $0 \leq q \leq n - 1$. If*

(A) $\left.\begin{array}{l} [x_j, y_k] \notin E \\ d_G(x_j) \leq j + q \\ d_G(y_k) \leq k + q \end{array}\right\} \Rightarrow d_G(x_j) + d_G(y_k) \geq n + q + 1$

then each set F of q edges that forms vertex-disjoint chains is contained in a hamiltonian cycle of G.

Suppose that there exists a bipartite graph that satisfies condition (A) but does not have the required property. By adding as many new edges to this graph as needed, we obtain a bipartite graph $G = (X, Y, E)$ satisfying condition (A), but not the required property, and such that the addition of any new edge gives a graph with the required property.

Let $F \subset E$ be a set of q edges that is not contained in any hamiltonian cycle of G. Let $j(x)$ denote the index of vertex $x \in X$, and let $k(y)$ denote the index of vertex $y \in Y$.

1. Let $a_1 \in X$ and $b_1 \in Y$ be two nonadjacent vertices that maximize $j(a_1) + k(b_1)$. The graph obtained from G by adding edge $[a_1, b_1]$ has a hamiltonian cycle that contains F and, therefore, graph G has a hamiltonian chain μ that contains F. Let

$$\mu = [a_1, b_2, a_2, b_3, ..., a_{i-1}, b_i, ..., a_n, b_1] .$$

2. Let

$$I = \{ i / \quad 2 \leqslant i \leqslant n, [a_1, b_i] \in E, \quad [a_{i-1}, b_i] \notin F \} .$$

Then, $|I| \geqslant d_G(a_1) - q$. If $i \in I$, then $[a_{i-1}, b_1] \notin E$ because, otherwise,

$$[a_1, b_i, a_i, b_{i+1}, ..., a_n, b_1, a_{i-1}, b_{i-1}, ..., b_2, a_1]$$

would be a hamiltonian cycle containing F. Thus

$$\Gamma_G(b_1) \subset X - \{ a_{i-1} / i \in I \}$$

and so

$$d_G(b_1) \leqslant n - |I| \leqslant n - d_G(a_1) + q ,$$

or

(B) $$d_G(a_1) + d_G(b_1) \leqslant n + q .$$

3. Let $j(a_1) = s$, $k(b_1) = t$, so that $a_1 = x_s$, $b_1 = y_t$.
If $i \in I$, it follows from Part 2 that $[a_{i-1}, b_1] \notin E$, and, therefore,

$$j(a_{i-1}) + k(b_1) \leqslant s + t .$$

Hence

$$j(a_{i-1}) \leqslant s .$$

Thus there are at least $|I|$ vertices x with $j(x) \leqslant s$, and, therefore,

$$s \geqslant |I| \geqslant d_G(a_1) - q .$$

Hence

$$d_G(x_s) \overset{\leq}{} s + q .$$

By the same argument,

$$d_G(y_t) \leqslant t + q .$$

Since G satisfies condition (A), and since $[x_s, y_t] \notin E$, it follows that

$$d_G(x_s) + d_G(y_t) \geqslant n + q + 1 .$$

This contradicts (B).

Q.E.D.

Corollary 2 (Bondy [1969]). *Let $G = (X, Y, E)$ be a simple bipartite graph with $|X| = |Y| = n \geqslant 2$. If the vertices of X and of Y are indexed with increasing degrees, and if*

$$d_G(x_j) \leqslant j, \; d_G(y_k) \leqslant k \quad \Rightarrow \quad d_G(x_j) + d_G(y_k) \geqslant n + 1 ,$$

then G has a hamiltonian cycle.

The corollary is established by setting $q = 0$ in Theorem 11.

Corollary 3. *Let $G = (X, Y, E)$ be a simple bipartite graph with $|X| = |Y| = n \geqslant 2$, and let q be an integer, where $0 \leqslant q < n$, such that*

(1) *for each integer $j \leqslant \dfrac{n + q}{2}$, the set $S_j = \{ x / x \in X, d_G(x) \leqslant j \}$ has cardinality $< j - q$,*

(2) *for each integer $k \leqslant \dfrac{n + q}{2}$, the set $T_k = \{ y / y \in Y, d_G(y) \leqslant k \}$ has cardinality $< k - q$.*

Then for each set F of q edges that forms vertex disjoint elementary chains, there is a hamiltonian cycle that contains F.

This follows easily from Theorem 11.

Corollary 4 (Moon, Moser [1963]). *Let $G = (X, Y, E)$ be a simple bipartite graph with $|X| = |Y| = n \geqslant 2$. If $|S_k| < k$ and $|T_k| < k$ for each integer $k \leqslant \dfrac{n}{2}$. Then G has a hamiltonian cycle.*

The corollary is established by letting $q = 0$ in Corollary 2.

Corollary 5. *Let $G = (X, Y, E)$ be a simple bipartite graph with $|X| = |Y| = n \geqslant 2$. If each vertex $z \in X \cup Y$ satisfies $d_G(z) > \dfrac{n}{2}$, except for two vertices, $x_1 \in X$ with $d_G(x_1) \geqslant 2$, and $y_1 \in Y$ with $d_G(y_1) \geqslant 2$, then G has a hamiltonian cycle.*

If $k = 1$, then

$$|S_k| = 0 < k ; \quad |T_k| = 0 < k .$$

If $k > 1$ and $k \leqslant \dfrac{n}{2}$, then

$$| S_k | \leqslant 1 < k ; \qquad | T_k | \leqslant 1 < k .$$

Corollary 3 can now be used to show the existence of a hamiltonian cycle in G.

Q.E.D.

Corollary 6. *If* $G = (X, Y, E)$ *is a simple bipartite graph with* $| X | = | Y | = n \geqslant 2$ *such that*

$$d_G(x) + d_G(y) \leqslant n \quad \Rightarrow \quad [x, y] \in E ,$$

then G *has a hamiltonian cycle.*

The corollary is established by letting $q = 0$ in Theorem 11.

Corollary 7. *Let* $G = (X, Y, E)$ *be a simple bipartite graph with*

$$| X | = | Y | = n, \qquad | E | = m.$$

If

$$m \geqslant n^2 - n + 2 ,$$

then G *has a hamiltonian cycle.*

It suffices to show that there do not exist in G two nonadjacent vertices x and y with $d_G(x) + d_G(y) \leqslant n$. If two such vertices exist, then graph G can be obtained from the complete bipartite graph $K_{n, n}$ by removing a set of edges of cardinality at least

$$\big(n - d_G(x)\big) + \big(n - d_G(y)\big) - 1 = 2 n - \big(d_G(x) + d_G(y)\big) - 1 \geqslant n - 1 .$$

On the other hand, the number of edges eliminated from $K_{n, n}$ is

$$n^2 - m \leqslant n^2 - (n^2 - n + 2) = n - 2 ,$$

which is a contradiction.

Q.E.D.

5. Hamilton-connected graphs

Consider a simple graph $G = (X, E)$. Graph G is defined to be *Hamilton-connected* if for each pair x, y of distinct vertices, there is a hamiltonian chain with endpoints x and y. In this section we shall study sufficient conditions for a graph to be Hamilton-connected.

The above definition can be generalized. Let F be a set of q edges that form vertex-disjoint elementary chains. Let x, y be a pair of distinct vertices which are the endpoints of two distinct chains of (X, F). Graph G is defined to be *q-Hamilton-connected* if for each such F and for each such pair x, y, there is a hamiltonian chain with endpoints x and y that contains F.

For $q = 0$, a q-Hamilton-connected graph is a Hamilton-connected graph.

Lemma. *For two integers q and n with $0 \leqslant q \leqslant n - 2$, let $\mathcal{H}(n, q)$ be a class of simple graphs of order n satisfying the two following conditions:*
(1) *If $G \in \mathcal{H}(n, q)$, each set F of q edges forming a system of vertex-disjoint elementary chains is contained in a hamiltonian cycle.*
(2) *If $G \in \mathcal{H}(n, q)$, the graph G' obtained from G by adding any new edge also belongs to $\mathcal{H}(n, q)$.*
Then each graph of $\mathcal{H}(n, q)$ is $(q - 1)$-Hamilton-connected.

Let $G = (X, E)$ be a graph of $\mathcal{H}(n, q)$, and let F be a set of $q - 1$ edges of G such that the connected components of (X, F) are elementary chains. Let x, y be two vertices that are the endpoints of two distinct chains of (X, F). If $[x, y] \in E$, let $G' = G$; otherwise, let $G' = G + [x, y]$.

From condition (2), graph G' belongs to $\mathcal{H}(n, q)$; since $F \cup \{ [x, y] \}$ constitutes a system of vertex-disjoint elementary chains of total length q, graph G' has a hamiltonian cycle that contains $F \cup \{ [x, y] \}$. Therefore, F is contained in a hamiltonian chain of G with endpoints x and y.

Q.E.D.

Theorem 12. *Let G be a simple graph of order $n \geqslant 3$ with degrees $d_1 \leqslant d_2 \leqslant \cdots \leqslant d_n$ that satisfies the following condition:*

$$\left. \begin{array}{c} d_{k-1} \leqslant k \\ 2 \leqslant k \leqslant \dfrac{n}{2} \end{array} \right\} \;\Rightarrow\; d_{n-k} \geqslant n - k + 1 \,.$$

Then graph G is Hamilton-connected.

The graphs that satisfy this condition form a class $\mathcal{H}(n, 1)$ by Theorem 8.

Q.E.D.

Theorem 13. *Let G be a simple graph of order $n \geqslant 3$ such that*
(1) *for each integer k, where $1 < k < \dfrac{n}{2}$, the set S_k of vertices of degree $\leqslant k$ has cardinality $| S_k | < k - 1$,*
(2) *if n is even, $| S_{n/2} | \leqslant \dfrac{n}{2} - 1 \,.$*
Then, graph G is Hamilton-connected.

The graphs that satisfy these conditions form a class $\mathscr{H}(n, 1)$, by Corollary 2 to Theorem 8.

Q.E.D.

Theorem 14. *Let $G = (X, Y, E)$ be a bipartite graph, with $|X| = |Y| = n \geqslant 2$, in which the vertices $x_i \in X$ and $y_i \in Y$ are indexed such that:*

$$d_G(x_1) \leqslant d_G(x_2) \leqslant \cdots \leqslant d_G(x_n),$$
$$d_G(y_1) \leqslant d_G(y_2) \leqslant \cdots \leqslant d_G(y_n).$$

Suppose that, if j and k are the smallest two indices such that

$$d_G(x_j) \leqslant j + 1, \qquad d_G(y_k) \leqslant k + 1,$$

(if they exist), we have $\quad d_G(x_j) + d_G(y_k) \geqslant n + 2$.

Then, for every pair x, y, there exists a hamiltonian chain with endpoints x and y.

From Theorem 11, the graphs that satisfy these conditions form a class $\mathscr{H}(2n, 1)$.

Q.E.D.

Theorem 15. *Let $G = (X, Y, E)$ be a bipartite graph with $|X| = |Y| = n \geqslant 2$ satisfying the two following conditions:*

(1) *for each $j \leqslant \dfrac{n + 1}{2}$, the set of vertices x_i of degree $\leqslant j$ has cardinality $< j - 1$,*

(2) *for each $k \leqslant \dfrac{n + 1}{2}$, the set of vertices y_i of degree $\leqslant k$ has cardinality $< k - 1$.*

Then, graph G is Hamilton-connected.

From Corollary 2 to Theorem 11, the graphs that satisfy these conditions form a class $\mathscr{H}(2n, 1)$.

Q.E.D.

Theorem 16 (Erdös, Gallai [1959]). *Let G be a simple graph of order $n \geqslant 3$ with $d_G(x) + d_G(y) > n$ for each pair x, y of distinct, non-adjacent vertices. Then, graph G is Hamilton-connected.*

From the Lemma, it suffices to show that

(1) $\qquad x \neq y, \quad d_G(x) + d_G(y) \leqslant n \quad \Rightarrow \quad [x, y] \in E$

implies

(2) *each edge of G is contained in some hamiltonian cycle.*

The proof is similar to the proof for the first part of Theorem 8. Let G be a simple graph that satisfies (1), but not (2). We may assume as before that the addition of any new edge causes the graph to satisfy property (2).

Let e_0 be an edge of G that is not contained in any hamiltonian cycle. There exist two non-adjacent vertices a and b, since $G \neq K_n$. Since $G + [a, b]$ has a hamiltonian cycle containing e_0, there exists in G a hamiltonian chain containing e_0 that is of the form

$$\mu[a, b] = [a, y_2, y_3, ..., y_{n-1}, b].$$

Let $y_1 = a$, and let

$$I = \{ i / i \geq 2, [a, y_i] \in E, [y_{i-1}, y_i] \neq e_0 \}.$$

Then,

$$|I| \geq d_G(a) - 1.$$

For $i \in I$, we have $y_{i-1} \notin \Gamma_G(b)$ because, otherwise,

$$[a, y_i, y_{i+1}, y_{i+2}, ..., y_n, y_{i-1}, y_{i-2}, ..., a]$$

would be a hamiltonian cycle containing e_0. Thus,

$$n - 1 - d_G(b) \geq |I| \geq d_G(a) - 1$$

and

$$d_G(a) + d_G(b) \leq n.$$

But, from (1), this implies that $[a, b] \in E$ which is a contradiction.

Q.E.D.

Theorem 17 (Ore [1963]). *If G is a simple graph of order n with m edges such that*

$$m \geq \frac{(n-1)(n-2)}{2} + 3,$$

then G is Hamilton-connected.

Let $G = (X, E)$ be a graph that satisfies the inequality. Consider two non-adjacent vertices a and b. Let E_0 denote the set of edges adjacent to neither a nor b. Then,

$$|E| = d_G(a) + d_G(b) + |E_0|,$$

and

$$|E_0| \leq \frac{1}{2}(n-2)(n-3),$$

$$|E| \geq \frac{1}{2}(n-1)(n-2) + 3.$$

Thus,

$$d_G(a) + d_G(b) = |E| - |E_0| \geqslant$$

$$\geqslant \frac{n-2}{2}[(n-1) - (n-3)] + 3 = n+1.$$

Therefore, from Theorem 16, G is Hamilton-connected.

Q.E.D.

The inequality presented by this theorem is the best possible: Consider the complete graph K_n with n vertices and remove all the edges incident to vertex a except for two edges.

If $n \geqslant 4$, this graph is not Hamilton-connected because the two vertices adjacent to a cannot be the endpoints of a hamiltonian chain. The number of edges of this graph equals

$$m = \frac{1}{2}n(n-1) - (n-3) = \frac{1}{2}(n-1)(n-2) + 2.$$

Theorem 18 (Moon [1965]). *The minimum number of edges in a simple Hamilton-connected graph of order $n \geqslant 4$ is $[\frac{1}{2}(3n+1)]$.*

1. If $m < [\frac{1}{2}(3n+1)]$, then a graph G with m edges contains at least one vertex a with $d_G(a) \leqslant 2$ because, otherwise,

$$\frac{3n}{2} \leqslant m < \left[\frac{1}{2}(3n+1)\right],$$

which is impossible.

Therefore, this graph G is not Hamilton-connected (because $n \geqslant 4$ and the vertices adjacent to a cannot be the endpoints of a hamiltonian chain).

2. For $m = [\frac{1}{2}(3n+1)]$, we shall show that there exists a Hamilton-connected graph G_n with n vertices and m edges.

For $n = 2k$, graph G_n is shown in Fig. 10.9. For $n = 2k - 1$, graph G_n is shown in Fig. 10.10.

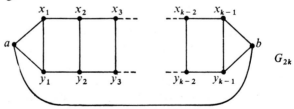

Fig. 10.9

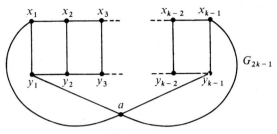

Fig. 10.10

It is left to the reader to verify the existence of hamiltonian chains of the forms

$$\mu[a, b], \quad \mu[a, x_i], \quad \mu[a, y_j], \quad \mu[x_i, y_j]$$

in graph G_{2k}, and hamiltonian chains of the forms

$$\mu[a, x_i], \quad \mu[x_i, x_j], \quad \mu[x_i, y_j]$$

in graph G_{2k-1}.

The number of edges in G_{2k} equals

$$m = \frac{3n}{2} = 3k = \left[3k + \frac{1}{2}\right] = \left[\frac{1}{2}(3n + 1)\right].$$

The number of edges in G_{2k-1} equals

$$m = \frac{3.2(k - 1) + 4}{2} = 3k - 3 + 2 =$$

$$= 3k - 1 = \left[\frac{1}{2}(6k - 2)\right] = \left[\frac{1}{2}(3n + 1)\right].$$

Q.E.D.

Theorem 19 (Karaganis [1968]). *If G is a connected graph of order $n \geqslant 2$, then its cube G^3 (the composition product $G \cdot G \cdot G$) is Hamilton-connected.*

Graph G^3 has the same vertex set as G. Two vertices are adjacent in G^3 if, and only if, their distance in G is $\leqslant 3$ (see Chapter 4).

Consider a spanning tree H of G. We shall show by induction that H^3 is Hamilton-connected. Clearly, this is true for $n = 2$. Suppose that that is true for all trees of order $< n$ and consider a tree H of order $n > 2$. Let a and b be two distinct vertices of H. We shall show that H^3 has a hamiltonian chain between a and b.

Since H is a tree, it possesses a unique chain $\mu[a, b] = [a, x_1, ..., b]$ between a and b. If edge $[a, x_1]$ is removed from H, two trees H_a and H_b are formed that respectively contain a and b.

Let $\mu[a, a']$ be a hamiltonian chain in $(H_a)^3$ between a and a neighbour of a in H_a (where a' is distinct from a if H_a is not a single vertex). Let $\mu[b', b]$ be a hamiltonian chain of $(H_b)^3$ between b and vertex b', the neighbour of x_1 in H_b (where b' is distinct from b if H_b is not a single vertex).

Vertices a' and b' are separated in H by at most three edges. Thus, they are adjacent in H^3. It follows that $\mu[a, a'] + [a', b'] + \mu[b', b]$ is the required hamiltonian chain between a and b in H^3.

                                                                        Q.E.D.

Theorem 19 shows that the vertices of a tree of order n can be numbered from 1 to n with numbers 1 and n assigned arbitrarily, such that any two vertices with consecutive numbers are separated by at most 3 edges.[1]

6. Hamiltonian cycles in planar graphs (abstract)

It has often been conjectured that each regular 3-connected graph of degree 3 possesses a hamiltonian cycle.

The first known counter-example was the Petersen graph (see Fig. 10.11).

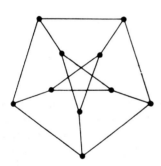
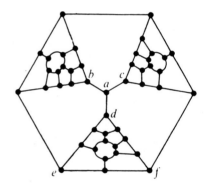

Fig. 10.11. Petersen graph   **Fig. 10.12.** Tutte graph

---

[1] A proof for Theorem 19 already appeared in 1960 when Sekanina showed that the cube of a connected graph has a hamiltonian chain. (M. Sekanina, "On an Ordering of the Set of Vertices of a Connected Graph," *Publ. Fac. Sc. Brno*, **412**, 1960, pp. 137–141.) Sekanina also asked for graphs whose square contained a hamiltonian cycle. It has recently been proved that the square of a 2-connected graph has a hamiltonian cycle (Fleischner [1974]), that it is Hamilton-connected (Hobbs, Nash-Williams), and that it is also panconnected, i.e., if the distance between two vertices x and y is k, then there exists an elementary chain joining x and y with length p ($p = k, k + 1, ..., n - 1$) (Faudree, Schelp). If G is connected of order at least 4, then G^3 is panconnected (Alavi, Williams). If G is connected of order at least 5, then the representative graph of G or $\bar{G}$ has a hamiltonian cycle (Nebesky). The square $(L(G))^2$ of the representative graph of G is vertex-pancyclic, i.e., for every vertex e of $(L(G))^2$ and for every integer $k(3 \leqslant k \leqslant m)$, there exists a cycle of length k containing e (Bermond, Rosenstiehl).

In fact, the Petersen graph is the only graph of order $\leqslant 10$ without hamiltonian cycles, such that the removal of any vertex creates a graph with a hamiltonian cycle (R. Sousselier, in Berge [1963]).

The conjecture is false even for planar graphs. Tutte [1946] constructed the first 3-connected planar graph regular of degree 3 that has no hamiltonian cycles (see Fig. 10.12)

Barnette and Jucovič [1970] have shown that for a 3-connected planar graph without hamiltonian cycles, the smallest number of vertices is 11; an example with 11 vertices is given in Fig. 10.13. Clearly, since this graph is bipartite with an odd number of vertices, no hamiltonian cycle exists.

M. Balinski [1961] conjectured the existence of a hamiltonian chain in all planar cubic 3-connected graphs. B. Grünbaum and T. S. Motzkin [1962]

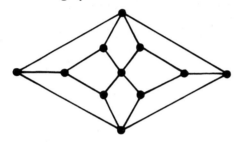

Fig. 10.13. Herschel graph

presented a counter-example by constructing a cubic, planar 3-connected graph with 944 vertices in which no elementary chain could contain more than 939 vertices (see also Brown [1961], Jucovič [1968]). The simplest example, given in Fig. 10.14, contains only 88 vertices (Zamfirescu, in Grünbaum [1975]).

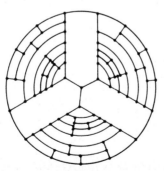

Fig. 10.14. Zamfirescu graph

Tutte [1956] has shown that each 4-connected planar graph has a hamiltonian cycle (see Ore [1968], for a proof).

A graph is defined to be (k)-*edge-connected* if it cannot be disconnected into two connected components, each containing a cycle, by the removal of less than k edges. In 1884, Tait conjectured that each simple (3)-edge-connected planar graph regular of degree 3 has a hamiltonian cycle. The Tutte graph (Fig. 10.12) provides a counter-example to this conjecture. Moreover, (4)-edge-connected planar graphs without hamiltonian cycles have been constructed by different authors (Hunter [1962], Ledeberg [1966], Tutte [1960]).

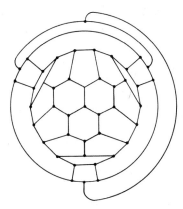

Fig. 10.15. Another Tutte graph

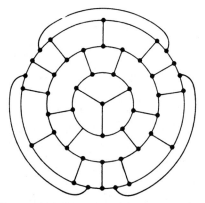

Fig. 10.16. Graph of Kozyrev and Grinberg

The problem is more difficult for planar (5)-edge-connected graphs. The first example (Walther [1965] of a planar (5)-edge-connected graph without hamiltonian cycles had 114 vertices. Figures 10.15 and 10.16 present the two simplest known examples with 46 vertices (Kozyrev and Grinberg, in Sachs [1968]) and 44 vertices (Tutte [1972]), respectively.

Planar graphs with or without hamiltonian cycles have been constructed by Bosák [1967], Čulik [1964], Kotzig [1962] and Sachs [1967] (cf. Grünbaum [1975]).

EXERCISES

1. Show that the complete bipartite graph $K_{p,q}$ has no hamiltonian cycles if $p \neq q$, and it has $[p / 2] - 1$ disjoint hamiltonian cycles if $p = q$.

2. Show using the theorem due to Pósa (Corollary 7 to Theorem 8) that if G is a graph with n vertices, m edges and of minimum degree k, and if

$$m \geq 1 + \max \left\{ \binom{n-k}{2} + k^2, \binom{n - \left[\frac{n-1}{2}\right]}{2} + \left[\frac{n-1}{2}\right]^2 \right\},$$

then G has a hamiltonian cycle. Show that this is the best possible result.

(Erdös [1962])

3. Show that if $G = (X, Y, E)$ is a bipartite graph with $|X| = |Y| = n, |E| = m$, with minimum degree $k \leq \frac{n}{2}$, and satisfies

$$m \geq 1 + \max \left\{ n(n-k) + k^2, n\left(n - \left[\frac{n}{2}\right]\right) + \left[\frac{n}{2}\right]^2 \right\},$$

then G has a hamiltonian cycle.

4. Show that the Tutte graph (Fig. 10.12) has no hamiltonian cycles.

Hint: Use reductions to show that if the graph consisting of the triangle edf and its interior had a hamiltonian chain, then this chain cannot have both e and f for endpoints.

5. Show that the graph in Fig. 10.15 has no hamiltonian cycles.

6. Show that the graph of Kozyrev and Grinberg (Fig. 10.16) has no hamiltonian cycles.

Hint: Suppose the existence of a hamiltonian cycle μ. Denote by f_i' the number of faces with a contour of i edges that lie in the interior region of μ. Denote by f_i'' the number of faces with a contour of i edges that lie in the exterior region of μ. Show that

$$\Sigma (i - 2) f_i' = n - 2 = \Sigma (i - 2) f_i''.$$

Since $f_i' = f_i'' = 0$ for $i \neq 5, 8, 9$, then

$$3 f_5' + 6 f_8' + 7 f_9' = 3 f_5'' + 6 f_8'' + 7 f_9''.$$

Hence $f_9' \equiv f_9'' \mod 3$, which contradicts $f_9' + f_9'' = 1$.

7. If $G = (X, U)$ is a complete symmetric graph whose arcs are divided into two classes U' and U'', show that there exist two vertices a and b and a hamiltonian circuit μ such that the portion of μ from a to b contains only arcs of U', and the portion of μ from b to a contains only arcs of U''.

(H. Raynaud [1970])

8. In a 1-graph $G = (X, \Gamma)$, a *hamiltonian bi-path* is defined to be two elementary paths such that each vertex belongs to exactly one of these two paths. Let $\bar{G}$ denote the complementary 1-graph of G. Let $h_0(G)$ denote the number of hamiltonian circuits in G, $h_1(G)$ denote the number of hamiltonian paths in G, and $h_2(G)$ denote the number of hamiltonian bi-paths in G. Show that

$$h_0(G) \equiv h_0(\bar{G}) + h_1(G) + h_2(G) \quad \text{(mod. 2)} .$$

Show that the number of hamiltonian bi-paths in a complete anti-symmetric 1-graph is odd.

(Berge, [1967]).

Hint: Use the proof of Theorem 1, replacing the word "arrangement" by "circular permutation of degree n".

9. A graph G of order n is defined to be *hypohamiltonian* if it has no hamiltonian cycles, but the subgraph G_x obtained by removing any vertex x has a hamiltonian cycle. Prove the following:
(1) If G is hypohamiltonian, then $n > 3$.
(2) If G is hypohamiltonian, then $d_G(x) \geqslant 3$ for each vertex x.
(3) If G is hypohamiltonian, and if y and z are consecutive vertices of an elementary cycle of length $n - 1$ in G_x, then x is not adjacent to both y and z.

(4) If G is hypohamiltonian, then $d_G(x) \leqslant \dfrac{n-1}{2}$ for each vertex x.

(5) If G is regular of degree h, then $hn = 2m$.
(6) Let x, a, b, a', b' be distinct vertices of G such that x is adjacent to a and to a', b is adjacent to b', and arcs (a, b) and (a', b') are both contained in some hamiltonian cycle of G_x. Then, G has a hamiltonian cycle.
(7) If G is hypohamiltonian, then $n \geqslant 7$.
(8) If G is hypohamiltonian, then $n \neq 7$.
(9) If G is hypohamiltonian, then $n \neq 8$.
(10) If G is hypohamiltonian, then $n \neq 9$.
(11) If G is hypohamiltonian of order 10, then G is regular of degree 3.
(12) The Petersen graph is hypohamiltonian.
(13) Each hypohamiltonian graph of order 10 is isomorphic to the Petersen graph.
(For a complete proof, see Herz, Duby, Vigué [1967].)

10. In a simple, regular graph of degree 3, consider two edges incident to a vertex a. Show that the parity of the number of hamiltonian cycles that contain these two edges does not depend upon which two edges incident to a have been chosen.

11. Let $G = (X, U)$ be a complete strongly connected graph of order $n \geqslant 3$; show that each vertex lies on some circuit of length k, for $k = 3, 4, ..., n$.

12. Let G be a complete strongly connected graph of order $n \geqslant 4$; show that there exist two distinct vertices a and b such that the subgraphs $G_{x-\{a\}}$ and $G_{x-\{b\}}$ are both strongly connected.

13. Show that the incidence graph faces versus vertices of a convex polyhedron other than a cube is hamilton-connected.

(Brualdi [1980])

CHAPTER 11

Covering Edges with Chains

1. Eulerian cycles

One of the oldest combinatorial problems, due to Euler, can be stated as follows: An *eulerian chain* (respectively, *eulerian cycle*) is defined to be a chain (respectively, cycle) that uses each edge exactly once. *When does a multigraph have an eulerian chain or an eulerian cycle?*

EXAMPLE 1. Is is possible to trace out the graph in Fig. 11.1 without lifting your pencil from the paper and without repeating any edge? After several attempts, the reader will find that this is impossible. However, this is possible for the graph in Fig. 11.2, which has an eulerian chain.

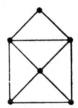

Fig. 11.1   Fig. 11.2

EXAMPLE 2 (Euler). The city of Königsberg (today known as Kaliningrad) is divided by the Pregel River that surrounds the Island of Kneiphof. There are seven bridges in the city as shown in Fig. 11.3. Can a pedestrian traverse

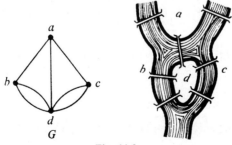

Fig. 11.3
225

each bridge exactly once? This problem puzzled the residents of Königsberg until Euler showed in 1736 that no solution exists.

Consider the multigraph G in Fig. 11.3 whose vertices represent the districts a, b, c, d of Königsberg and whose edges represent the bridges of Königsberg. Clearly, the bridge problem is solved by finding an eulerian chain in this graph.

Theorem 1 (Euler [1766]). *A multigraph G has an eulerian chain if, and only if, it is connected (except for isolated vertices) and the number of vertices of odd degree is 0 or 2.*

1. *Necessity.* If there is an eulerian chain μ in G, then G is clearly connected. Furthermore, only the two endpoints of μ (if they are distinct) can be of odd degree. Thus, there can only be 0 or 2 vertices of odd degree.

2. *Sufficiency.* We shall prove by induction: *If there are only two vertices a and b whose degree is odd, then there exists an eulerian chain that starts at a and finishes at b; if there are no points whose degree is odd, then the graph possesses an eulerian cycle.*

We shall assume this statement to be true for graphs with fewer than m edges, and prove that it also holds for a graph G with m edges. To be more definite, assume that G has two vertices a and b of odd degree.

The required chain μ will be defined by a traveller who traverses the graph from a and moves in such a way that he never uses the same edge twice. If he reaches a vertex $x \neq b$, he will have been along an odd number of the edges incident to x, and therefore he will be able to leave x by an edge which has not yet been used; when this is no longer possible, he must necessarily have arrived at b. However, it is possible that not all edges have been used.

After removing the used edges, we obtain a partial graph G' whose vertices are all of even degree. Let $C_1, C_2, ..., C_k$ be the connected components of G' that have at least one edge. By the induction hypothesis, these components have eulerian cycles, $\mu_1, \mu_2, ..., \mu_k$. Since G is connected, chain μ encounters each of the C_i. Without loss of generality, suppose that μ first encounters component C_1 at vertex x_1, then encounters component C_2 at vertex x_2, etc. Consider the chain

$$\mu[a, x_1] + \mu_1 + \mu[x_1, x_2] + \mu_2 + \cdots + \mu[x_k, b] .$$

Clearly, this chain is an eulerian chain from a to b.

Q.E.D.

The reader can now see that no solution exists for the Königsberg Bridge problem (Example 2).

Similarly, it is possible to show that the edges of a connected multigraph with exactly $2\,q$ vertices of odd degree can be covered with only q chains.

LOCAL ALGORITHM TO CONSTRUCT AN EULERIAN CYCLE. Consider a connected multigraph G in which all vertices have even degree. The following rules construct an eulerian cycle in G:

RULE 1. *Starting from any vertex a, follow a chain without using the same edge twice.*

RULE 2. *If we arrive at a vertex x different from a after the kth step, never depart from vertex x along an edge that is an isthmus of the partial Graph G_k generated by the unused edges, unless x is a pendant vertex of G_k.*

RULE 3. *If vertex a is revisited, depart from vertex a along any unused edge if it exists. If no unused edge exists, stop.*

We shall show that the chain generated by the algorithm is eulerian.

1. *It is always possible to follow the rules.*

Upon arrival at a vertex $x \neq a$, there is always in G_k an edge incident to x, because $d_G(x)$ is even, and therefore $d_{G_k}(x)$ is odd. If this edge is unique, it is a pendant edge in G_k and can be used. If this edge is not unique, there is at least one other edge that is not an isthmus, since, otherwise, there are two isthmi in G_k joining x to two distinct connected components C and D. Note that C contains at least one vertex of odd degree, because, otherwise,

$$0 \equiv \sum_{x \in C} d_{G_k}(x) = 1 + 2\,m_{G_k}(C, C) \bmod 2,$$

which is impossible. Similarly, D contains at least one vertex of odd degree. Since G_k contains exactly two vertices of odd degree, and since one of them is x, the contradiction follows.

2. *If the rules are followed, the chain determined during the first k steps is the beginning of an eulerian cycle.*

Graph G_k is connected except for isolated vertices. Upon arrival at a vertex $x \neq a$, there are only two vertices x and a of odd degree in G_k. Therefore, from Theorem 1, G_k has an eulerian chain from x to a.

Q.E.D.

We shall now apply the Euler theorem to the study of the *factors* of a graph. A *factor* is defined to be a system of vertex-disjoint elementary cycles such that each vertex is contained in exactly one cycle.

Theorem 2 (Petersen [1891]). *If $G = (X, E)$ is a regular multigraph of even degree $h = 2\,k$, then G has k edge-disjoint factors.*

1. First we shall show that G has a factor. From Theorem 1, each connected component of G has an eulerian cycle. By directing each edge along the direction of travel in this eulerian cycle, we obtain a graph $H = (X, U)$ with

$$d_H^+(x) = d_H^-(x) = k \qquad (x \in X) \, .$$

The bipartite multigraph $\overline{H} = (X, \overline{X}, E)$ obtained by making two copies X and $\overline{X}$ of X, and by taking

$$m_{\overline{H}}(x_i, \overline{x}_j) = m_H^+(x_i, x_j),$$

is regular of degree k. Thus, from Corollary 4 to Theorem (5, Ch. 7), $\overline{H}$ contains a perfect matching E_0, that corresponds to a factor in G.

2. The theorem is obviously valid for $k = 1$. Assume it is valid for every regular multigraph of degree $2\,k' < 2\,k$; we shall show that the theorem is valid for a regular multigraph $G = (X, E)$ of degree $2\,k$.

From part (1), we know that G has a factor $E_0 \subset E$. The partial multigraph $(X, E - E_0)$ of degree $2(k - 1)$ has $k - 1$ disjoint factors $E_1, E_2,...,$ E_{k-1} by the induction hypothesis. Thus, G has k disjoint factors $E_0, E_1,...,$ E_{k-1}.

$$\text{Q.E.D.}$$

Corollary. *If G is a regular multigraph of odd degree $2\,k + 1$ such that*

$$\min_{\substack{S \neq X \\ S \neq \varnothing}} m_G(S, X - S) \geqslant 2\,k \, ,$$

then G has k edge-disjoint factors.

Without loss of generality, we may assume that G is connected. Clearly, G is a graph with an even number of vertices. Then, from Theorem (13, Ch. 8), G possesses a perfect matching. The edges not contained in the matching form a regular graph of degree $2\,k$, which from Theorem 2 can be decomposed into k disjoint factors.

$$\text{Q.E.D.}$$

Remark. Bäbler [1938] extended this corollary as follows: *A regular graph of odd degree $h = 2\,k + 1$ contains q edge-disjoint factors, $q \leqslant k$, if*

$$\min_{\substack{S \neq X \\ S \neq \varnothing}} m_G(S, X - S) \geqslant 2\,q \, .$$

2. Covering edges with disjoint chains

In this section we shall study the following problem: *Given a simple graph $G = (X, E)$, cover its edges with as few as possible edge-disjoint chains of a certain type. What is the smallest number of such chains needed to cover E?*

EXAMPLE 1 (Kirkman). Each day, n knights meet at a round table. No knight wants to sit next to the same neighbour twice. How many days can they meet? The answer is the maximum number of disjoint hamiltonian cycles in a complete graph with n vertices. Later we shall show that this number is $\left[\dfrac{n-1}{2}\right]$.

EXAMPLE 2 (Lucas). Six boys a, b, c, d, e, f and six girls $\bar{a}$, $\bar{b}$, $\bar{c}$, $\bar{d}$, $\bar{e}$, $\bar{f}$ join hands and dance in a circle. How many dances can they dance so that no boy ever joins hands with the same girl twice and no boy ever joins hands with another boy? The problem is solved by finding the maximum number of disjoint hamiltonian cycles in the complete bipartite graph $K_{6,6}$. There are three disjoint hamiltonian cycles in $K_{6,6}$:

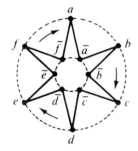

$$a\,\bar{a}\,b\,\bar{b}\,c\,\bar{c}\,d\,\bar{d}\,e\,\bar{e}\,f\,\bar{f}\,a$$

$$a\,\bar{c}\,b\,\bar{d}\,c\,\bar{e}\,d\,\bar{f}\,e\,\bar{a}\,f\,\bar{b}\,a$$

$$a\,\bar{e}\,b\,\bar{f}\,c\,\bar{a}\,d\,\bar{b}\,e\,\bar{c}\,f\,\bar{d}\,a\,.$$

Fig. 11.4

These cycles are found by rotating the labels of the tips of the star in Fig. 11.4. Since each vertex has degree 6, clearly, these cycles cover all the edges.

Theorem 3. *The maximum number of disjoint hamiltonian cycles in the complete graph K_n is $\left[\dfrac{n-1}{2}\right]$.*

1. First, suppose that $n = 2k + 1$ is odd. We shall show that $k = \dfrac{n-1}{2}$ disjoint hamiltonian cycles can be found.

Number the vertices $0, 1, 2, ..., 2k$ and place them on a circle as in Fig. 11.5. Then, the sequence

$$[0, 1, 2, 2k, 3, 2k - 1, 4, ..., k + 3, k, k + 2, k + 1, 0]$$

is a hamiltonian cycle. If 1 is added (modulo 2 k) to the index of each vertex $\neq 0$, i.e. if the system of solid lines in Fig. 11.5 is rotated around point 0, then a new hamiltonian cycle is obtained. This cycle is disjoint from the preceding one because the sum of two successive vertices is 2 or 3 in the first cycle and 4 or 5 in the second cycle. Thus $k - 1$ rotations around point 0 can be made, and these rotations produce k disjoint hamiltonian cycles.

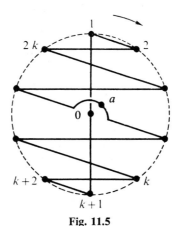

Fig. 11.5

2. Suppose that $n = 2 k + 2$ is even. Number the vertices of the graph 0, 1, 2, ..., $2 k, a$, and place them on a circle as shown in Fig. 11.5, with vertex a placed on another plane above the centre of the circle. Similarly, k disjoint hamiltonian cycles can be produced by rotating the solid lines in Fig. 11.5 around point 0. Thus K_n contains $k = \left[\dfrac{n-1}{2} \right]$ disjoint hamiltonian cycles.

$$\text{Q.E.D.}$$

Corollary 1. *If n is odd, the edges of graph K_n can be covered by $\dfrac{n-1}{2}$ disjoint hamiltonian cycles.*

Since the degree of each vertex is $n - 1 = 2 k$, the k disjoint hamiltonian cycles cover all the edges.

$$\text{Q.E.D.}$$

Corollary 2. *If n is even, the edges of graph K_n can be covered by a perfect matching and $\left[\dfrac{n-1}{2} \right]$ disjoint hamiltonian cycles.*

If the k hamiltonian cycles are removed, the remaining partial graph is regular of degree $(n - 1) - 2k = 1$.

Q.E.D.

Corollary 3. *If n is even, the edges of graph K_n can be covered by $\frac{n}{2}$ disjoint hamiltonian chains.*

The edges of K_{n+1} can be covered by disjoint Hamiltonian cycles as shown in Corollary 1. If we remove one vertex, these cycles correspond to $\frac{n}{2}$ disjoint hamiltonian chains of K_n.

Q.E.D.

Corollaries 1 and 3 are also consequences of the following general theorem:

Theorem 4 (Lovász [1968]). *The edges of a simple graph G with n vertices can always be covered by $\left[\frac{n}{2}\right]$ elementary chains and cycles that are pairwise edge-disjoint, or less.*

1. We may assume that the graph G has no isolated vertices and no isolated edges. Otherwise, we could prove the theorem for the graph obtained by removing isolated vertices and edges.

2. Let m denote the number of edges in graph G. If $2m - n \leqslant 0$, then

$$\sum_{i=1}^{n} d_G(x_i) \leqslant n \ .$$

Since $d_G(x_i) \geqslant 1$ for all i, we have $d_G(x_i) = 1$ for all i. This contradicts the hypothesis that G has no isolated edges. Thus $2m - n > 0$.

3. We shall assume that the theorem is valid for each graph G' of order n' with

$$\left[\frac{n'}{2}\right] = \left[\frac{n}{2}\right] \qquad \text{and} \qquad 2m' - n' < 2m - n$$

and we shall prove the theorem by induction on $2m - n$.

4. First, suppose that there is a vertex of even degree. Let x be such that:

$$\Gamma_G(x) = \{ a_1, a_2, ..., a_k, b_1, b_2, ..., b_l \} \ ,$$
$$d_G(a_i) \text{ even} \ , \quad d_G(b_j) \text{ odd} \ , \quad k \geqslant 1 \ .$$

Let G' be the graph obtained from G by removing edges $[a_i, x]$ for $i = 1, 2, ..., k$. Graph G' satisfies $2m' - n' < 2m - n$, and contains a minimum covering $M' = \{\mu'_1, \mu'_2, ...\}$ of less than $\left[\frac{n}{2}\right]$ disjoint elementary chains and cycles (by the induction hypothesis).

Since vertices a_i and b_j have odd degree in G', each of these vertices is the endpoint of an elementary path of M' of length > 0. Since the covering M' is minimum, each of these vertices is the endpoint of exactly one path of M'.

5. For each $i = 1, 2, \ldots, k$, form a sequence $(y_i^0, y_i^1, y_i^2, \ldots)$ of elements of $\Gamma_G(x)$, defined in the following manner:

– $y_i^0 = a_i$; thus $y_i^0 \in \Gamma_G(x)$.

– If y_i^p is defined and in $\Gamma_G(x)$, then it has odd degree in G', thus it is the endpoint of a path $\mu'(y_i^p, z) \in M'$. If μ' does not include x, then the sequence stops; otherwise, define y_i^{p+1} as the vertex preceding x on μ'. Thus $y_i^{p+1} \in \{ b_1, b_2, \ldots \}$.

Note that $y_i^p = y_j^q$ implies $i = j$ and $p = q$. If $p \leqslant q$, this is obvious for $p = 0$ (since $y_i^0 = a_i$). If $p \geqslant 1$, then $y_i^0 \in \mu'[y_i^{p-1}]$, and this path contains the edge $[y_i^q, x]$. Since the paths of M' have no common edges, these two are the same, and $y_i^{p-1} = y_j^{q-1}$. Thus $y_i^0 = y_j^{q-p}$, thus $q - p = 0$ and $i = j$.

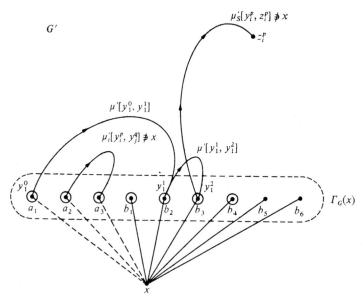

Fig. 11.6

6. Now, we shall show that a covering (μ_i) of G is obtained from covering M' of G' in the following way:

If $\mu_r' \in M'$ has an endpoint in Y, let

$$\mu_r = \mu_r'.$$

Thus, μ_r' is an elementary chain or an elementary cycle of G.

If $\mu_s'[y_i^p, z_j^q] \in M'$ has in Y only one endpoint, say y_i^p, let

$$\mu_s = \mu_s' + [y_i^p, x] \qquad\qquad\qquad \text{if} \quad x \notin \mu_s'[y_i^p, z_i^p]$$

$$= \{ \mu_s' + [y_i^p, x] \} - \{ [y_i^{p+1}, x] \} \qquad \text{otherwise.}$$

Thus, μ_s is an elementary chain of G.

If $\mu_t'[y_i^p, y_j^q] \in M'$ has two endpoints y_i^p and y_j^q in Y, let

$$\mu_t = \mu_t' + [y_i^p, x] + [y_j^q, x] \qquad\qquad \text{if} \quad x \notin \mu_t'[y_i^p, y_j^q]$$

$$= \{ \mu_t' + [y_i^p, x] + [y_j^q, x] \} - \{ [y_i^{p+1}, x], [y_j^{q+1}, x] \} \qquad \text{otherwise.}$$

Thus, in the first case, μ_t is an elementary cycle of G, and in the second case, μ_t is an elementary chain of G.

To show that the set M of all the chains μ_j forms a covering of G, note first that

— an edge of G that is not of the form $[y_i^p, x]$ is also an edge of G' and is covered by a single chain of M,

— an edge of the form $[y_i^p, x]$ is covered by the only chain of M that comes from the chain μ_s' (or μ_t') that has y_i^p as an endpoint.

7. Now, suppose that each vertex of G is of odd degree; consequently, n is even.

Replace G by a graph G_1 obtained by taking a vertex x of odd degree $\geqslant 3$ and adding a vertex a_1 in the middle of some edge incident to x, say $[x, z]$.

Since graph G_1' (obtained from graph G_1 as in Part 4) satisfies

$$2\, m_1' - n_1' = 2\, m - (n + 1) < 2\, m - n \ ,$$

G_1 can be covered by $\left[\dfrac{n_1}{2}\right] = \dfrac{n}{2}$ disjoint chains. Each vertex of G_1 that is distinct from a_1 is the endpoint of a covering chain. Hence, a_1 is not the endpoint of a covering chain. Thus, the $\dfrac{n}{2}$ covering chains of G_1 also cover G.

Q.E.D.

Corollary. *Let* $G = (X, E)$ *be a simple graph of order n with only vertices of odd degree; then its edges can be covered by* $\dfrac{n}{2}$*-edge disjoint elementary chains so that each vertex of G is the endpoint of exactly one of these chains.*

Since $\sum d_G(x) = 2\, m$ is even, then n is even. Therefore, from Theorem 3, the edges of G can be covered by $\dfrac{n}{2}$ disjoint chains and cycles. Furthermore, each vertex is necessarily the endpoint of at least one covering chain, because its degree is odd.

Q.E.D.

We shall now state three conjectures related to Theorem 4.

Conjecture 1 (Hajós). *If all the n vertices have even degree, the edges can be covered by $\left[\dfrac{n}{2}\right]$ edge-disjoint elementary cycles, or less.*

Conjecture 2 (Gallai). *The edges of a connected graph of order n can be covered by $\left[\dfrac{n+1}{2}\right]$ edge-disjoint elementary chains.*

Conjecture 3 (Nash-Williams). *For $n \geqslant 15$, if the number of edges is divisible by 3 and if each vertex has even degree $\geqslant \dfrac{3\,n}{4}$, then the edges can be covered by edge-disjoint triangles.*

Theorem 5 (Erdös, Goodman, Pósa [1966]). *Let G be a simple graph of order n; then the edges of G can be covered by $[n^2/4]$ edge-disjoint cliques, each of cardinality $\leqslant 3$.*

1. The theorem is true for $n = 2$. Assume that the theorem is valid for all graphs with $n - 1$ vertices, we shall show that it is also valid for a graph G of order $n > 2$.

First, we shall show that

$$\left[\frac{n^2}{4}\right] = \left[\frac{(n-1)^2}{4}\right] + \left[\frac{n}{2}\right] .$$

If $n = 2\,k$, then

$$\left[\frac{n^2}{4}\right] = k^2 = \left[\frac{(2\,k-1)^2}{4} - \frac{1}{4}\right] + k = \left[\frac{(n-1)^2}{4}\right] + \left[\frac{n}{2}\right] .$$

If $n = 2\,k + 1$, then

$$\left[\frac{n^2}{4}\right] = \left[\frac{(2\,k+1)^2}{4}\right] = k^2 + k = \left[\frac{(n-1)^2}{4}\right] + \left[\frac{n}{2}\right] .$$

2. If G contains a vertex x_0 with degree $d_G(x_0) \leqslant \left[\dfrac{n}{2}\right]$, the subgraph G_0 obtained by removing x_0 can be covered by $\left[\dfrac{(n-1)^2}{4}\right]$ cliques of cardinality $\leqslant 3$ that are pairwise disjoint. Thus, the minimum number of disjoint cliques that cover the edges of G is less than or equal to

$$\left[\frac{(n-1)^2}{4}\right] + \left[\frac{n}{2}\right] = \left[\frac{n^2}{4}\right].$$

3. If each vertex x satisfies $d_G(x) > \left[\frac{n}{2}\right]$, the number $k = \min_{x \in X} d_G(x)$ can be written as

$$k = \left[\frac{n}{2}\right] + r \quad , \quad r > 0 \quad ,$$

where

$$r = k - \left[\frac{n}{2}\right] \leqslant k - \frac{n-1}{2} \leqslant k - \frac{k}{2} = \frac{k}{2} \quad .$$

Let x_1 be a vertex of degree $k \geqslant 2r$. We shall show that *the subgraph generated by $\Gamma_G(x_1)$ contains a matching of cardinality r.*

In fact, if this subgraph contains a matching of only $r - 1$ edges $[y_1, y_2]$, $[y_3, y_4]$, ..., $[y_{2r-3}, y_{2r-2}]$, vertex y_{2r-1} is not adjacent to $\Gamma_G(x_1) - \{y_1, ..., y_{2r-2}\}$. Thus,

$$d_G(y_{2r-1}) = m_G(y_{2r-1}, \Gamma_G(x_1)) + m_G(y_{2r-1}, X - \Gamma_G(x_1)) \leqslant$$

$$\leqslant (2r - 2) + n - k =$$

$$= 2r - 2 + n - \left[\frac{n}{2}\right] - r = n - \left[\frac{n}{2}\right] + r - 2 \leqslant$$

$$\leqslant \left[\frac{n}{2}\right] + r - 1 = k - 1,$$

which contradicts that k equals the minimum degree.

4. Let $\Gamma_G(x_1) = \{y_1, y_2, ..., y_k\}$, and let G_1 be the partial subgraph obtained by removing x_1 and the r edges of the matching $[y_1, y_2]$, ..., $[y_{2r-1}, y_{2r}]$ defined above. The edges of graph G can be covered by:

— r triangles $[x_1, y_1, y_2, x_1]$, ..., $[x_1, y_{2r-1}, y_{2r}, x_1]$,
— $k - 2r$ isolated edges $[x_1, y_{2r-1}]$, ..., $[x_1, y_k]$,
— the $\left[\frac{(n-1)^2}{4}\right]$ disjoint cliques that cover G_1.

Thus the minimum number of disjoint cliques that cover G is less than or equal to

$$\left[\frac{(n-1)^2}{4}\right] + r + (k - 2r) = \left[\frac{(n-1)^2}{4}\right] + \left[\frac{n}{2}\right] = \left[\frac{n^2}{4}\right].$$

Q.E.D.

Remark. This theorem is the best possible, i.e., the edges of some graph

G_n of order n cannot be covered by less than $\left[\dfrac{n^2}{4}\right]$ cliques of cardinality $\leqslant 3$.

For an even $n = 2\,k$, let G_n be the complete bipartite graph $K_{k,k}$. The minimum number of cliques needed to cover all the edges is

$$k^2 = \frac{n^2}{4} = \left[\frac{n^2}{4}\right].$$

For an odd $n = 2\,k + 1$, let G_n be the complete bipartite graph $K_{k,k+1}$. The minimum number of cliques needed to cover all the edges is

$$k(k+1) = \frac{(2\,k+1)^2}{4} - \frac{1}{4} = \left[\frac{n^2}{4}\right].$$

3. Counting eulerian circuits

Let $G = (X, U)$ be a (directed) graph. A circuit of G is defined to be *eulerian* if it uses each arc exactly once. In this section, we shall study the number of distinct eulerian circuits in G.

Graph G is defined to be *pseudo-symmetric* if the number of arcs entering vertex x equals the number of arcs leaving vertex x, for all vertices x, i.e.

$$d_G^+(x) = d_G^-(x) \qquad (x \in X).$$

This terminology is consistent because each symmetric graph is also pseudo-symmetric.

Theorem 6. *A graph possesses an eulerian circuit if, and only if, it is connected (except for isolated vertices) and is pseudo-symmetric.*

The proof is exactly the same as the proof of Theorem 1.

EXAMPLE. What is the longest circular sequence that can be formed from the digits 0 and 1 so that no k-tuple of k consecutive digits occurs twice? Since 2^k distinct k-tuples can be formed from the numbers 0 and 1, the sequence cannot have more than 2^k entries. Using Theorem 6, we shall show that there does exist such a sequence with 2^k entries.[1]

Consider a graph G whose vertices represent the different $(k-1)$-tuples of 0 and 1 with arcs from vertex $(\alpha_1, \alpha_2, ..., \alpha_{k-1})$ to vertex $(\alpha_2, \alpha_3, ..., \alpha_{k-1}, 0)$ and to vertex $(\alpha_2, \alpha_3, ..., \alpha_{k-1}, 1)$. Since graph G is pseudo-symmetric, it possesses an eulerian circuit. If $(\alpha_1, \alpha_2, ..., \alpha_{k-1})$ is the first vertex of this circuit, $(\alpha_2, \alpha_3, ..., \alpha_k)$ the second vertex, $(\alpha_3, \alpha_4, ..., \alpha_{k+1})$ the third vertex, etc., the required sequence will be $\alpha_1, \alpha_2, \alpha_3, \alpha_4, \ldots$.

For $k = 4$, the graph in Fig. 11.7 provides several circular sequences of $2^4 = 16$ entries. For example,

[1] This problem occurs in telecommunications when it is necessary to determine the current position of a labelled cylinder by reading from the cylinder a sequence of only k of its labels (see Posthumus [1894]).

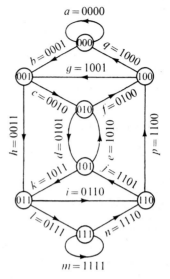

abcdefgh*ij*k*lmn*pq = 0000101001101111
abcdk*ij*efgh*lmn*pq = 0000101101001111
abcdk*i*pghl*mn*jefq = 0000101100111101
abcfgh*ij*edk*lmn*pq = 0000100110101111
abh*ij*k*lmn*pgcdefq = 0000110111100101
abh*ij*edk*lmn*pgcfq = 0000110101111001
abh*ij*efgcdk*lmn*pq = 0000110100101111
abh*i*pgcdk*lmn*jefq = 0000110010111101

Fig. 11.7

(Circular sequences obtained by permuting *i* and *lmn* have been omitted.) There are 16 different solutions.

The problem of counting the eulerian circuits is closely related to the problem of counting the spanning arborescences, see Theorem (21, Ch. 3).

Theorem 7. *Let G be a connected pseudo-symmetric graph, and let x_1 be a vertex of G. Then G has a partial subgraph which is an arborescence rooted at x_1. Moreover, this arborescence H can be constructed by travelling through an eulerian circuit starting at x_1 and placing in H the first arc used to enter each vertex.*

From Theorem (13, Ch. 3), *H* is an arborescence rooted at x_1 since:
1. each vertex distinct from x_1 is the terminal endpoint of exactly one arc of *H*,
2. x_1 is not the terminal endpoint of any arc of *H*,
3. *H* does not contain any circuits, because if a path formed from the arcs of *H* goes from *x* to *y*, then vertex *x* has been visited before vertex *y*, and therefore no paths exist in *H* from *y* to *x*.

Q.E.D.

Theorem 8 (Aardenne-Ehrenfest, de Bruijn [1951]). *In a connected pseudo-symmetric graph G, let Δ_1 denote the number of distinct arborescences rooted* *at x_1 that are partial graphs, and let r_k denote the outer demi-degree (or inner demi-degree) of vertex x_k. There are exactly*

$$\Delta_1 \prod_{k=1}^{n} (r_k - 1)!$$

distinct eulerian circuits.

Note that two eulerian circuits are not considered as distinct if one can be obtained from the other by a circular permutation of the arcs.

Consider an arborescence H rooted at x_1 that is a partial graph of G. We shall show that there are exactly

$$\prod_{k=1}^{n} (r_k - 1)!$$

distinct eulerian circuits that produce H as in Theorem 7. Number the arcs entering x_k from 1 to r_k and let

$$\omega^-(x_k) = \{ u_k(1), u_k(2), ..., u_k(r_k) \} .$$

Suppose that $u_k(r_k)$ is the arc of H in $\omega^-(x_k)$, for each $k \neq 1$. Suppose that arc $u_1(1)$ is fixed. Then there are exactly

$$\prod_{k=1}^{n} (r_k - 1)!$$

possible numberings. Each numbering corresponds to exactly one circuit as follows: Starting from x_1, travel in reverse through all the arcs of the graph by choosing at vertex x_k the unused arc of $\omega^-(x_k)$ with the smallest possible number.

This procedure defines a circuit because the route can terminate only at x_1 (when any other vertex is encountered there is always an exit arc because the graph is pseudo-symmetric). We shall show that this circuit is eulerian.

If it were not eulerian, then there is an arc $u_k(j)$ that is not used by the circuit. Since $r_k \geqslant j$, arc $u_k(r_k)$ has not been used. Hence, arc $u_p(r_p)$, incident into the initial endpoint x_p of arc $u_k(r_k)$, has not been used, etc. Since each of these unused arcs belongs to H, the process eventually arrives at x_1. But this is a contradiction, since the procedure that generated the circuit stopped at x_1, and consequently, all arcs incident to x_1 have been used.

Thus each numbering corresponds uniquely to an eulerian circuit such that the first arc of the circuit to enter a vertex is present in the arborescence H. Since the number of distinct arborescences H is Δ_1, and is known by Theorem (21, Ch. 3), the proof is complete.

Q.E.D.

Corollary 1. *In a pseudo-symmetric graph, the number of arborescences rooted at x_k that are partial graphs is independent of which vertex x_k is chosen.*

Theorem 8 can also be stated for vertex x_k instead of for vertex x_1. Since the number of eulerian circuits does not change, $\Delta_1 = \Delta_k$.

Q.E.D.

Corollary 2. *In a graph G with m arcs and with order n that satisfies $m > 2n$, the number of eulerian circuits is even.*

If G is not connected and pseudo-symmetric, the result is obvious. If $d_G^+(x) \leqslant 2$ for each vertex x, then

$$m = \sum_x d_G^+(x) \leqslant 2n,$$

which is impossible. If $d_G^+(x) \geqslant 3$ for $x = x_k$, then $(r_k - 1)!$ is even and the formula in Theorem 8 shows that the number of distinct eulerian circuits is even.

Q.E.D.

Note that this result is the best possible: each graph shown in Fig. 11.8 satisfies $m = 2n$ and contains exactly one eulerian circuit.

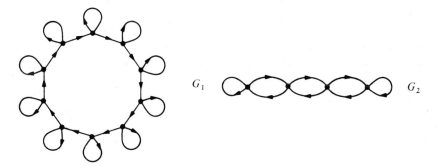

Fig. 11.8

Corollary 2 can be extended by using a fundamental theorem of algebra, due to Amitsur and Levitzki [1950], that was reformulated in more combinatorial terms by M. P. Schützenberger [1958] as follows.

Let G be a graph of order n whose arcs are numbered 1, 2, ..., m. To each eulerian circuit $(1, j_2, j_3, ..., j_m)$ beginning with arc 1, there corresponds a permutation

$$\pi = \begin{pmatrix} 1 & 2 & \cdots & m \\ 1 & j_2 & \cdots & j_m \end{pmatrix}.$$

An eulerian circuit μ is defined to be *odd* (respectively, *even*) if the corresponding permutation is odd (respectively, even). We shall show that if $m > 2n$, the number of even eulerian circuits equals the number of odd eulerian circuits. (Note that if the arcs are numbered differently, the equality is preserved.)

Lemma. *Let* $\pi = (j_1, j_2, ..., j_m)$ *be a permutation of* 1, 2, ..., *m and let*

$$\pi_k = (j_1, j_2, ..., j_{k-1}, j_{k+2}, j_{k+3}, ..., j_m)$$

be a permutation of degree $m - 2$ *obtained from* π *by the elimination of two consecutive indices* j_k, j_{k+1}; *then* π *and* π_k *have the same parity if and only if one of the following conditions holds:*

(1) $j_k < j_{k+1}$ *and* $j_k + j_{k+1}$ *odd.*

(2) $j_k > j_{k+1}$ *and* $j_k + j_{k+1}$ *even.*

Permutation $\pi = (j_1, j_2, ..., j_m)$ becomes $(j_1, j_2, ..., j_{k-1}, j_{k+1}, ..., j_m, j_k)$ by a product of $m - k$ transpositions of two consecutive terms. Similarly, this new permutation becomes $(j_1, j_2, ..., j_{k-1}, j_{k+2}, j_{k+3}, ..., j_m, j_k, j_{k+1})$ by a product of $m - k$ transpositions. Thus π becomes $(j_1, j_2, ..., j_m, j_k, j_{k+1})$ by a product of $2(m - k)$ transpositions.

Two permutations are of the same parity if one can be obtained from the other by an even number of transpositions; thus, permutations π and $(j_1, j_2, ..., j_m, j_k, j_{k+1})$ have the same parity. By removing j_k and j_{k+1} from this latter sequence, the number of inversions is decreased by

$$| \{ i \, / \, i \neq j_k, j_{k+1}, i > j_k \} | + | \{ i \, / \, i \neq j_{k+1}, i > j_{k+1} \} |.$$

If $j_k < j_{k+1}$, this number equals

$$(m - j_k - 1) + (m - j_{k+1}) = 2m - 1 - (j_k + j_{k+1}).$$

If $j_k > j_{k+1}$, this number equals

$$(m - j_k) + (m - j_{k+1}) = 2m - (j_k + j_{k+1}).$$

Since permutations π and π_k have the same parity if, and only if, this number is even, the lemma follows.

Q.E.D.

Theorem 9 (Schützenberger [1958]). *In a graph* $G = (X, U)$ *of order n with m arcs that satisfies* $m > 2n$, *the number of even eulerian circuits equals the number of odd eulerian circuits.*

1. We may assume that $m = 2n + 1$. Otherwise we have $m - 2n = k > 1$.

Then, construct a graph G' from G by adding $k - 1$ new vertices and by replacing one arc $(x, y) \in U$ by an elementary path from x to y that passes through each of these $k - 1$ new vertices. Graph G' satisfies $m' - 2n' = (m + k - 1) - 2(n + k - 1) = m - 2n - k + 1 = 1$; hence, G' has as many even eulerian circuits as it has odd eulerian cycles, and the same result follows for G.

2. Note that the theorem is true for a graph of order 1 or 2. We shall assume that it is valid for all graphs of order $< n$, and we shall show that it is also valid for a graph G of order $n > 2$, with $m = 2n + 1$.

3. We may assume that G is connected and satisfies

$$d_G^+(x) = d_G^-(x) \qquad (x \in X).$$

Otherwise, from Theorem 6, the number of eulerian circuits equals 0, and the theorem is proved.

4. We may assume that G is a 1-graph. Otherwise, there are two arcs i_1 and i_2 with the same initial endpoint and the same terminal endpoint. With the eulerian circuit corresponding to a permutation π, we can associate the eulerian circuit μ' corresponding to the permutation π' obtained from π by transposing i_1 and i_2. Since μ and μ' have different parities, the theorem follows.

5. If G contains a vertex x with $d_G^+(x) = d_G^-(x) = 2$ and with a loop at x, the theorem can be proved as follows: The three arcs incident to x are of the form
$$(x_0, x) = i_0, \quad (x, x) = i_1, \quad (x, x_2) = i_2.$$

Consider the graph G' obtained from G by removing x and all arcs incident to x, and by adding an arc $i_0 = (x_0, x_2)$. Graph G' has order $n' = n - 1$ and satisfies $m' - 2n' = (m - 2) - 2(n - 1) = m - 2n = 1$. Thus, from part 2, G' has as many even eulerian circuits as it has odd eulerian circuits. To each of the corresponding permutations, add the entries i_1, i_2 after the entry i_0. In this way, each eulerian circuit of G is obtained. If, for example, $i_1 < i_2$ and $i_1 + i_2$ is even, two circuits of the same parity in G' correspond to two circuits with the same parity in G. Thus, the theorem is valid for G.

6. If G contains a vertex x with $d_G^+(x) = d_G^-(x) = 1$, the theorem can be proved as follows: By making the transformation shown in Fig. 11.9, a graph G_i is obtained for each arc $i \in \omega^-(x_0)$.

From Part 5, each graph G_i has the required property.

To each eulerian circuit in G_i, there corresponds an eulerian circuit in G.

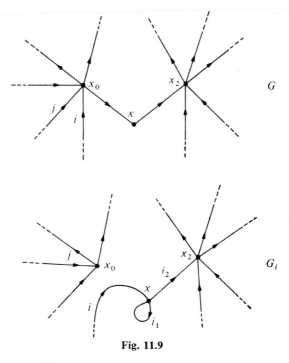

Fig. 11.9

To each eulerian in G there corresponds an eulerian circuit in only one of the G_i. Hence G has the required property.

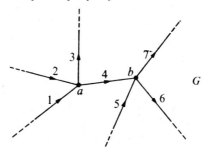

Fig. 11.10.

7. In all the other cases, we shall show that the outer demi-degree of each vertex in G equals 2, except for one vertex x_1 with $d_G^+ (x_1) = 3$. From Part 6, we may assume that $d_G^+ (x) \geqslant 2$ for all x, and consequently

$$2n + 1 = m = \sum_{x \in X} d_G^+(x) \geqslant 2n .$$

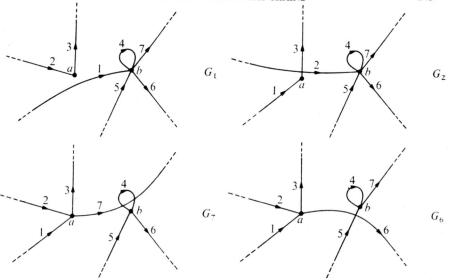

Fig. 11.11.

Thus, $d_G^+(x) = 2$ for all $x \neq x_1$. Since G has order $n > 2$, it contains the configuration shown in Fig. 11.10.

8. The configuration in Fig. 11.10 can be altered in four ways as shown in Fig. 11.!1. These alterations yield the graphs G_1, G_2, G_7, G_6, each of order n and with $m - 2n = 1$.

Let $n(G)$, $n_0(G)$ and $n_1(G)$ respectively denote the number of eulerian circuits, the number of even eulerian circuits and the number of odd eulerian circuits in graph G. Then

$$n(G) = n(G_1) + n(G_2) - n(G_6) - n(G_7).$$

From Part 6,

$$n_0(G_1) = n_1(G_1), \quad n_0(G_2) = n_1(G_2).$$

Furthermore, from Part 5,

$$n_0(G_6) = n_1(G_6), \quad n_0(G_7) = n_1(G_7)$$

Thus,

$$n_0(G) = n_0(G_1) + n_0(G_2) - n_0(G_6) - n_0(G_7) =$$
$$= n_1(G_1) + n_1(G_2) - n_1(G_6) - n_1(G_7) =$$
$$= n_1(G).$$

Thus, G has the required property.

$$Q.E.D.$$

EXERCISES

1. Let $G = (X, E)$ be a simple connected graph regular of degree 4, and if (E', F') is a partition of the edges into 2 factors; show that there exists an eulerian cycle in G that alternately passes through edges of E' and F'.   (Kotzig [1956])

2. Use Theorem (15, Ch. 3), and Theorem (6, Ch. 11), to give a necessary and sufficient condition that a semi-functional 1-graph possesses a hamiltonian circuit.

3. A simple graph $L(G)$ is defined to be the *line-graph* of $G = (X, E)$ if each vertex of $L(G)$ represents an edge of G and if two vertices are adjacent in $L(G)$ if they represent adjacent edges of G. (These graphs are studied in Chapter 17, Section 4.) Show that if G possesses an eulerian cycle, then $L(G)$ possesses also an eulerian cycle. Show that the converse is not true.

Hint: Consider a graph G with 4 vertices a, b, c, d and 3 edges ab, bc, bd.
                                                                (Chartrand [1964])

4. Show that if G possesses an eulerian cycle, then $L(G)$ possesses a hamiltonian cycle. Show that the converse is not true.

Hint: Consider the graph G with 4 vertices a, b, c, d and 6 edges ab, ac, ad, bc, bd, cd.
                                                                (Chartrand [1964])

5. Show that if G possesses a hamiltonian cycle, then $L(G)$ possesses a hamiltonian cycle. Show that the converse is not true.

Hint: Consider the graph G with 5 vertices, a, b, c, d, e and 6 edges ab, ac, ad, eb, ec, ed.
                                                                (Chartrand [1964])

6. Show that if G is a regular multigraph of order n and odd degree $h = 2k + 1$ with $h \geqslant n + 1$, then G has a factor.

Hint: Use the proof of Theorem 2, and Corollary 1 to Theorem (5, Ch. 7).

6. For $n(r - 1) \equiv 0 \pmod 2$, show that the complete r-partite graph $K_{r \times n}$ is decomposable into hamiltonian cycles.

                                                                (Auerbach, Laskar [1976])

7. For n even $\geqslant 8$, the complete symmetric graph K_n^* is decomposable into hamiltonian circuits (Tillson; conjecture due to Bermond–Faber). Show that if K_n^* is thus decomposable, then so is K_{2n-2}^*.

                                                                (Bouchet)

8. Show that the edges of a simple graph G can be covered by a family of $\leqslant \frac{3}{4} n$ edge-disjoint elementary chains.

                                                                (Donald [1980])

CHAPTER 12

Chromatic Index

1. Edge colourings

The *chromatic index $q(G)$* of a graph G is defined to be the smallest number of colours needed to colour the edges of G so that no two adjacent edges have the same colour. A *q-colouring of the edges* is defined to be a partition of the edge set into q subsets that are matchings.

Clearly,

$$q(G) \geqslant \max_{x \in X} d_G(x) .$$

In this chapter, we may assume without loss of generality that G is a multi-graph without loops.

EXAMPLE 1. *Scheduling an oral examination.* At the end of the academic year each student must be examined orally by each of his professors. How can the examinations be scheduled so that they end as soon as possible?

Let X be the set of students, and let Y be the set of professors. Form a bipartite multigraph $G = (X, Y, E)$ in which $a \in X$ is joined to $b \in Y$ by k edges if, and only if, student a must be examined by professor b exactly k times.

If the edges of this graph are coloured so that no two adjacent edges have the same colour, each colour can correspond with one examination period. Hence all examinations can be completed in $q(G)$ time periods, and not less.

EXAMPLE 2. *The Lucas schoolgirls problem.* Each day the $2p$ schoolgirls of a boarding house take a walk in p rows of two; can they take $2p - 1$ walks consecutively without any two girls walking together more than once? This is possible, if and only if the chromatic index of the complete graph K_{2p} with $2p$ vertices is equal to $2p - 1$, each colour representing a walk.

EXAMPLE 3. *Latin square.* A latin square is an $n \times n$ matrix with entries $1, 2, ..., n$ such that no entry appears twice in the same row and no entry appears twice in the same column.

Let $k < n$; in an $n \times n$ tableau, place only entries $1, 2, ..., k$, so that no entry appears twice in the same row or in the same column. Is it possible to place the entries $k + 1, k + 2, ..., n$ in the empty positions so that a latin square is formed? Represent the rows of the matrix by vertices $a_1, a_2, ..., a_n$, and the columns by vertices $b_1, b_2, ..., b_n$; join a_i and b_j if and only if the position in row i, column j is empty. The latin square can be completed if, and only if, the edges of this bipartite graph can be coloured with $n - k$ colours $k + 1, k + 2, ..., n$, each edge $[a_i, b_j]$ coloured with α corresponding to an entry α at the intersection of row i and column j.

Theorem 1. *The chromatic index of a simple complete graph G of order n (and maximum degree $h = n - 1$) is*

$$q(G) \begin{cases} = h & \text{if n is even} \\ = h + 1 & \text{if n is odd.} \end{cases}$$

CASE 1: *n is even.* Number the vertices $0, 1, 2, ..., n - 1$ and place the vertices as shown in Fig. 12.1.

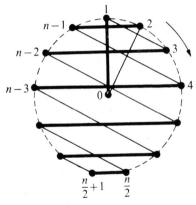

Fig. 12.1

Let the first perfect matching (i.e., the edges of the first colour) be

$$[0, 1], [2, n - 1], [3, n - 2], ..., \left[\frac{n}{2}, \frac{n}{2} + 1\right].$$

These edges are the dark edges in Fig. 12.1.

By adding 1 modulo n to the index of each vertex except vertex 0 in the above matching, we obtain another perfect matching, which corresponds to a rotation around 0 of the dark edges in the figure. This operation can be performed $n - 1$ times, and each time a new colour is assigned to the resulting matching. No edge is coloured twice, because each time the coloured edges form a different angle with the horizontal. However, each edge of G has been coloured, and therefore $q(G) = n - 1$.

CASE 2: *n is odd.* Consider the graph G' formed from G by adding a vertex x_0 that is joined to each vertex of G. From Case 1 above, we have

$$q(G') = (n + 1) - 1 = n.$$

Thus the edge of G can be coloured with n colours. The edges of G cannot be coloured with $n - 1$ colours, because then each colour would correspond to a perfect matching, and a graph with odd order has no perfect matching. Thus

$$q(G) = n = h + 1.$$

Q.E.D.

Theorem 2. *The chromatic index of a bipartite multigraph $G = (X, Y, E)$ with maximum degree h is*

$$q(G) = h.$$

From Corollary 4 to Theorem (5, Ch. 7), graph G contains a matching E_1 that saturates every vertex of degree h. Colour E_1 with the first colour.

Consider the bipartite multigraph $(X, Y, E - E_1)$, whose maximum degree is $h - 1$. This graph contains a matching E_2 that saturates every vertex of degree $h - 1$. Colour E_2 with the second colour. By repeating this operation, the edges of G will be coloured with h colours.

Q.E.D.

APPLICATION TO LATIN SQUARES. The theory of latin squares, which was founded by Euler, can use Theorem 2 in a very interesting way: a latin square of order n is formed when the entries 1, 2, ..., n are distributed into the n^2 positions of an $n \times n$ square so that no entry is repeated in any row and no entry is repeated in any column. We shall show:

Let T be a p × q rectangle whose positions contain the numbers 1, 2, ..., n such that no number is repeated in any row and no number is repeated in any column. Let m(k) denote the number of times that k appears in T. A necessary and sufficient condition that n − p rows and n − q columns can be added to form a latin square is that

$$m(k) \geqslant p + q - n \qquad (k = 1, 2, ..., n).$$

Necessity. Let $m'(k)$ denote the number of rows of T in which k does not appear. If a latin square can be formed, then $m'(k) \leqslant n - q$. Hence

$$m(k) = p - m'(k) \geqslant p - n + q.$$

Sufficiency. Let

$$P = \{ 1, 2, ..., p \}, \qquad N = \{ 1, 2, ..., n \},$$

form the bipartite graph $G = (P, N, E)$ in which $i \in P$ and $k \in N$ are joined by an edge if number k does not appear in the i-th row of T. Then,

$$d_G(i) = n - q \qquad (i \in P)$$
$$d_G(k) = m'(k) \leqslant n - q \qquad (k \in N).$$

From Theorem 2, the edges of G can be coloured with $n - q$ colours. Complete T by adding $n - q$ columns; the first additional column will be defined by the endpoints in N of the edges of the first colour, the second column by the endpoints in N of the edges of the second colour, etc. In this way, we obtain a $p \times n$ rectangle $\overline{T}$, that satisfies also the hypothesis of the theorem because $\overline{m}(y) = p = (p + n) - n$. Then $n - p$ rows can now be added to complete $\overline{T}$, and this yields a latin square.

$$\text{Q.E.D.}$$

Consider the problem of colouring the edges of a bipartite graph with colours $\alpha_1, \alpha_2, ..., \alpha_q$ such that there are exactly m_i edges with colour α_i. The following theorem gives conditions for the existence of such a colouring.

Theorem 3 (Folkman, Fulkerson [1966]). *Let $G = (X, Y, E)$ be a bipartite multigraph with maximum degree $\leqslant q$. Consider a sequence*

$$m_1 \geqslant m_2 \geqslant ... \geqslant m_q,$$

with

$$\sum_{i=1}^{q} m_i = m = |E|.$$

Form the conjugate sequence $(m_1^*, m_2^*, ...)$ where m_j^* denotes the number of m_i that are $\geqslant j$ (see Chapter 6). If there exists a colouring of the edges in q colours $\alpha_1, \alpha_2, ..., \alpha_q$ with exactly m_i edges with colour α_i, for all i, then

$$m_G(A, B) \geqslant \sum_{j > |X - A| + |Y - B|} m_j^* \qquad (A \subset X, B \subset Y).$$

This necessary condition is also sufficient for the case:

$$m_1 = m_2 = \cdots = m_k \geqslant m_{k+1} = m_{k+2} = \cdots = m_q.$$

Necessity. If there are m_i edges with colour α_i, then the number of edges with colour α_i that join A and B is

$$m_i(A, B) = m_i - m_i(X - A, Y) - m_i(X, Y - B) + m_i(X - A, Y - B) \geqslant$$
$$\geqslant m_i - |X - A| - |Y - B|.$$

Hence

$$m_G(A, B) = \sum_{i=1}^{q} m_i(A, B) \geqslant$$

$$\geqslant \sum_{i=1}^{q} \max \left\{ 0, m_i - |X - A| - |Y - B| \right\} = \sum_{j > |X-A| + |Y-B|} m_j^* .$$

(The last equality can be verified from the Ferrers diagram, see Chapter 6, Section 1.)

Sufficiency (*for the special case*). Consider such a sequence (m_i). See Fig. 12.2. Let

$$m' = \sum_{i=1}^{k} m_i ; \qquad m'' = \sum_{i=k+1}^{q} m_i ; \qquad m' + m'' = m ;$$

$$h' = k ; \qquad h'' = q - k ; \qquad h' + h'' = q .$$

We shall show first that G is the union of two disjoint partial multigraphs $G' = (X, Y, E')$ and $G'' = (X, Y, E'')$ with

$$|E'| = m' , \qquad |E''| = m'' ,$$

$$\max d_{G'}(z) \leqslant h' , \qquad \max d_{G''}(z) \leqslant h'' .$$

From the lemma to Theorem (7, Ch. 7), it suffices to show that for all A, B,

(1) $m_G(A, B) \geqslant m' - h' |X - A| - h' |Y - B| ,$

(2) $m_G(A, B) \geqslant m'' - h'' |X - A| - h'' |Y - B| .$

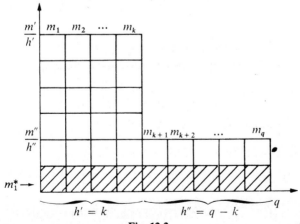

Fig. 12.2

Inequality (1) can be easily verified from the Ferrers diagram (Fig. 12.2). If

$$\frac{m''}{h''} \leqslant |X - A| + |Y - B|,$$

then

$$\sum_{j > |X-A| + |Y-B|} m_j^* = h' \left(\frac{m'}{h'} - |X - A| - |Y - B| \right) =$$

$$= m' - h'|X - A| - h'|Y - B|.$$

If

$$\frac{m''}{h''} > |X - A| + |Y - B|,$$

then

$$\sum_{j > |X-A| - |Y-B|} m_j^* \geqslant m' - h'|X - A| - h'|Y - B|.$$

In both cases the inequality of the hypothesis yields inequality (1). To verify inequality (2), note that if

$$\frac{m''}{h''} > |X - A| + |Y - B|,$$

then

$$\sum_{j > |X-A| + |Y-B|} m_j^* \geqslant h'' \left(\frac{m''}{h''} - |X - A| - |Y - B| \right) =$$

$$= m'' - h''|X - A| - h''|Y - B|,$$

and if

$$\frac{m''}{h''} \leqslant |X - A| + |Y - B|,$$

then

$$\sum m_j^* \geqslant 0 \geqslant h'' \left(\frac{m''}{h''} - |X - A| - |Y - B| \right).$$

Thus inequality (2) is satisfied in both cases.

From Theorem (7, Ch. 7), multigraph G' (with m' edges and maximum degree $\leqslant h'$) contains a matching E_1' with $\frac{m'}{h'} = m_1$ edges that saturates each vertex of maximum degree. Colour the edges of E_1' with the first colour.

The partial multigraph generated by $E' - E_1'$ has maximum degree $\leqslant h - 1$. From Theorem (7, Ch. 7), this graph contains a matching E_2' with $\dfrac{m' - m_1}{h' - 1} = m_1$ edges. Colour the edges of E_2' with the second colour, etc.

Thus G' can be decomposed into $h' = k$ matchings of m_1 edges. Similarly, G'' can be decomposed into $h'' = q - k$ matchings of m_{k+1} edges. Thus G can be coloured with q colours so that m_i edges are of colour α_i for all i.

Q.E.D.

2. The Vizing theorem and related results

This section presents some bounds for the chromatic index of a multigraph without loops. The following very simple theorem gives a lower bound.

Theorem 4. *Let G be a multigraph without loops, with m edges, with maximum degree h, and let t be the cardinality of a maximum matching. Then,*

$$q(G) \geqslant \max \left\{ h, \left[\frac{m}{t} \right]^* \right\}.$$

Consider a colouring of the edges with $q = q(G)$ colours $\alpha_1, \alpha_2, ..., \alpha_q$, and let E_i denote the set of edges with colour α_i. We have

$$m = |E_1| + |E_2| + \cdots + |E_q| \leqslant qt.$$

Hence, $q \geqslant \dfrac{m}{t}$, and $q \geqslant \left[\dfrac{m}{t} \right]^*$. The theorem follows.

Q.E.D.

Uncoloured edge lemma. *Let G be a multigraph without loops and let $q(G) = q + 1$. Suppose that a set $C = \{ \alpha_1, \alpha_2, ..., \alpha_q \}$ of q colours has been used to colour all the edges of G except one edge $[a, b]_0$, and let C_x denote the set of colours used for the edges incident to vertex x. Then*

$$\begin{cases} |C_a \cup C_b| = q \\ |C_a \cap C_b| = d_G(a) + d_G(b) - q - 2 \\ |C_a - C_b| = q - d_G(b) + 1 \\ |C_b - C_a| = q - d_G(a) + 1 . \end{cases}$$

No colour of C can be missing from both C_a and C_b, since then $[a, b]_0$ could be coloured with this missing colour. Thus $C = C_a \cup C_b$, and

$$q = |C| = |C_a \cup C_b| = |C_a \cap C_b| + |C_a - C_b| + |C_b - C_a| .$$

Furthermore,

$$d_G(a) - 1 = |C_a| = |C_a \cap C_b| + |C_a - C_b|,$$
$$d_G(b) - 1 = |C_b| = |C_a \cap C_b| + |C_b - C_a|.$$

By elimination, these three equalities yield:

$$|C_a \cap C_b| = d_G(a) - 1 + d_G(b) - 1 - q,$$
$$|C_a - C_b| = q - (d_G(a) - 1),$$
$$|C_b - C_a| = q - (d_G(b) - 1).$$

Q.E.D.

Theorem 5. *If G consists of a cycle $[x_1, x_2, ..., x_n, x_1]$, with possibly more than one edge between two consecutive vertices, and if G has m edges and maximum degree h, then*

$$q(G) = \begin{cases} h & \text{if } n \text{ is even} \\ \max\left\{ h, \left[\dfrac{2m}{n-1}\right]^* \right\} & \text{if } n \text{ is odd.} \end{cases}$$

If n is even, the proof is evident because then G is bipartite and Theorem 2 applies. Thus we may assume that $n = 2k + 1$ odd.
Note from Theorem 4 that

$$q(G) \geqslant \max\left\{ h, \left[\frac{m}{t}\right]^* \right\} = \max\left\{ h, \left[\frac{m}{k}\right]^* \right\}.$$

It remains to show that

$$q(G) \leqslant \max\left\{ h, \left[\frac{m}{k}\right]^* \right\}.$$

This is true for a multigraph with $2k + 1$ edges, because $q(G) = 3$ for a cycle G without multiple edges. Therefore, we may assume that $m > 2k + 1$. Suppose that the theorem is valid for all multigraphs with less than m edges, and consider a multigraph G with $m > 2k + 1$ edges.

1. If we remove from G an edge $[a, b]_0$ so that a and b remain adjacent, the resulting graph G' is of the same type but with $m' = m - 1$ edges. Thus, from the induction hypothesis, the edges of G' can be coloured with a set C of q colours, where

$$q = \max\left\{ h, \left[\frac{m}{k}\right]^* \right\} \geqslant \max\left\{ h', \left[\frac{m'}{k}\right]^* \right\}.$$

For $\alpha \in C$, let E_α denote the set of edges with colour α.

2. We shall assume that $q(G) = q + 1$ and we shall show that this leads to a contradiction.

There is in C a colour γ with $|E_\gamma| < k$ since, otherwise,

$$m - 1 = \sum_{\alpha \in C} |E_\alpha| = kq,$$

and

$$q = \frac{m - 1}{k} < \frac{m}{k},$$

which contradicts $q \geqslant \left[\frac{m}{k}\right]^*$.

Two cases must be considered: $\gamma \in C_b - C_a$ (or $\gamma \in C_a - C_b$) and $\gamma \in C_a \cap C_b$. From the Uncoloured Edge Lemma, $C = C_a \cup C_b$, and no other cases are possible.

CASE 1: $\gamma \in C_b - C_a$. From the Uncoloured Edge Lemma,

$$|C_a - C_b| = q - d_G(b) + 1 \geqslant h - d_G(b) + 1 \geqslant 1.$$

Thus there exists a colour $\alpha \in C_a - C_b$, and clearly $\alpha \neq \gamma$. Let $G(\alpha, \gamma)$ denote the partial graph of G generated by the edges with colour α or γ. The connected component of $G(\alpha, \gamma)$ that contains b is an elementary chain with endpoint b (because $\alpha \notin C_b$) and does not contain vertex a (because $|E_\gamma| < k$). By interchanging the colours γ and α along this chain, colour γ is no longer incident to vertices a and b and, therefore, $[a, b]_0$ can be coloured with γ, which is a contradiction.

CASE 2: $\gamma \in C_a \cap C_b$. From the Uncoloured Edge Lemma, there is a colour $\alpha \in C_a - C_b$ and a colour $\beta \in C_b - C_a$. Clearly, $\gamma \neq \alpha, \beta$.

Consider the connected component of graph $G(\beta, \gamma)$ that contains a. This component is an open chain $\mu_{\alpha\beta}[a, x]$ that contains b if γ joins vertices a and b, and does not contain b if γ does not join vertices a and b.

Since this chain begins with an edge with colour γ, by interchanging γ and β along this chain a new colouring with $|E_\gamma| < k$ is obtained.

Furthermore, in this new colouring, $\gamma \in C_b - C_a$, which is Case 1.

Q.E.D.

Another very general tool, first introduced for hypergraphs, has been shown to be very fruitful in dealing with edge colouring problems (Fournier [1973, 1977]). Let G be a multigraph with no loops, and let k be a positive integer. By an *edge k-colouring (in the general sense)* of G, we will mean a partition π of the set of edges into k classes; by 'an edge of colour i' we will mean an edge of the ith class. For each vertex x, the set of colours that appear in the neighbourhood of x will be denoted by $C_x(\pi)$, or simply C_x when there is no ambiguity. Obviously, $\mid C_x \mid \leq d_G(x)$ and $\mid C_x \mid \leq k$. We will be interested in finding conditions in which $\mid C_x \mid = \min \{ d_G(x), k \}$ for all x, that is in which there exists a 'proper' colouring of the neighbourhood of each vertex.

The integer $\delta_\pi(x) = \min \{ d_G(x), k \} - \mid C_x \mid$ is called the *deficiency* of the colouring π at the point x: thus, we wish to know for what values of k there exists a k-colouring with total deficiency $\Sigma_x \delta_\pi(x) = 0$.

If $k \geq \max d_G(x)$, then a k-colouring π with deficiency zero is a *k-colouring (in the usual sense)* as defined previously.

Lemma (Fournier [1973]). *Let G be a multigraph with no loops. If, in a minimum deficiency edge k-colouring π, there exist a vertex x_0 and colours α and β such that β does not appear in the neighbourhood of x_0 and α appears more than once, then the partial graph of G induced by the colours α and β has, in the connected component $H_{\alpha, \beta}(x_0)$ containing x_0, an elementary odd cycle whose edges are coloured α, β, α, ..., α.*

(1) We prove the lemma for $k = 2$; more precisely, we show: if G is connected and has no odd cycle, then G has a zero deficiency 2-colouring.
– If all vertices of G have degree ≤ 2, the result is trivial. Suppose then that there exists a vertex y with degree > 2.
– If all vertices of G have even degree, then G has an eulerian cycle (cf. Ch. 11). Alternately colouring the edges of this cycle with the colours α and β gives a zero deficiency 2-colouring.
– If G has a vertex of odd degree, then by adding a vertex to G and joining it to all vertices of odd degree, one obtains a multigraph G' with all vertices of even degree which can be 2-coloured as above. This colouring induces a zero-deficiency 2-colouring in G.
This completes the proof of the lemma for $k = 2$.
(2) We now prove the lemma for $k > 2$. If $H_{\alpha, \beta}(x_0)$ is not an odd cycle of the desired form, then the edges of H can be recoloured giving a zero deficiency 2-colouring (as per the previous section 1). This yields a k-colouring π' with:

$$\mid C_{x_0}(\pi') \mid = \mid C_{x_0}(\pi) \mid + 1,$$

$$\mid C_x(\pi') \mid \geq \mid C_x(\pi) \mid \qquad \text{for } x \neq x_0.$$

Thus π' has a smaller deficiency than does π, contradicting the minimality of π.

Q.E.D.

Theorem 6 (Gupta [1978]). *Let G be a bipartite multigraph. For each positive integer k there exists a zero deficiency k-colouring.*

In fact, if a *k*-colouring is of minimum but not zero deficiency, then there exist a vertex x_0 and colours α and β such that β does not appear in the neighbourhood of x_0, and α appears more than once. By Fournier's Theorem, there exists an odd cycle, contradicting the fact that G is bipartite.

Corollary 1. *In a bipartite multigraph, the chromatic index is equal to the maximum degree.*

It suffices to set $k = \max d_G(x)$. This yields Theorem 2.

Corollary 2. (Gupta [1974]). *In a bipartite multigraph with minimum degree k, it is possible to k-colour the edges such that every colour occurs in the neighbourhood of each vertex.*

It suffices to set $k = \min d_G(x)$.

Note that Theorem 6 has been extended to larger classes of multigraphs by De Werra [1976]. For more general multigraphs, this method has also been used by Hilton [1975].

In the following section, $p = \max_{x,\,y} m_G(x, y)$ denotes the maximum multiplicity of the graph G, and $p(x) = \max_y m_G(x, y)$ denotes the maximum multiplicity at x.

The following essential theorem is due to Vizing, and was proved by Fournier [1973, 1977] as part of a stronger result announced by Gupta without proof [1966].

Theorem 7 (Fournier [1977]). *Let $G = (X, E)$ be a multigraph, with no loops, of maximum degree h, and let k be an integer $\geq h$. Then there exists an edge k-colouring of G such that:*

$$| C_x(\pi) | = d_G(x) \qquad if \quad d_G(x) \leqslant k - p(x),$$

$$| C_x(\pi) | \geqslant k - p(x) \qquad if \quad d_G(x) > k - p(x).$$

Let

$$\varrho(x) = \begin{cases} 0 & if \quad d_G(x) \leqslant k - p(x), \\ d_G(x) - k + p(x) & if \quad d_G(x) > k - p(k). \end{cases}$$

Given a k-colouring π, let:

$$\varepsilon(\pi) = \sum_{x \in X} \max\{0, \delta_\pi(x) - \varrho(x)\}$$

We must show the existence of a k-colouring such that $\varepsilon(\pi) = 0$. Let π_0 be a k-colouring that minimizes $\varepsilon(\pi)$. Suppose $\varepsilon(\pi_0) \neq 0$ (proof by contradiction). There exists a vertex x_0 such that $\delta_{\pi_0}(x_0) > \varrho(x_0) \geq 0$. Since $|C_{x_0}(\pi_0)| < k$, there is a colour β that does not appear in the neighbourhood of x_0; and since $|C_{x_0}(\pi_0)| < d_G(x_0)$, there is a colour α_0 that appears more than once. Let e_0 be any edge incident to x_0 with colour α_0. We will define a succession of colours $\alpha_0, \alpha_1, \ldots$, of k-colourings $\pi_0, \pi_1, \ldots$, and of edges $e_0, e_1, \ldots$ incident to x_0, such that for $i = 0, 1, \ldots$

(1) in π_i, the colour α_i appears more than once in the neigbourhood of x_0, and the colour β does not appear,

(2) in π_i, the edge e_i has colour α_i,

(3) $\varepsilon(\pi_i)$ is minimum,

(4) $\alpha_i \neq \alpha_{j-1}$ for all j such that $j < i$ and $y_j = y_i$.

Clearly, α_0, π_0 and e_0 satisfy these conditions. Assume that α_{i-1}, π_{i-1} and e_{i-1} are already defined for some $i \geq 1$. In π_{i-1}, the component $H_{\alpha_{i-1}, \beta}(x_0)$ is an odd cycle whose edges are coloured alternately α_{i-1} and β, except that both edges incident to x_0 are coloured α_{i-1}. Otherwise, it would be possible to recolour with colours α_{i-1} and β the edges of $H_{\alpha_{i-1}, \beta}(x_0)$ with zero deficiency (Fournier's lemma), and thus obtain a k-colouring π'_{i-1} of G for which

$$\delta_{\pi'_{i-1}}(x_0) < \delta_{\pi_{i-1}}(x_0) \quad \text{and} \quad \delta_{\pi'_{i-1}}(x) \leq \delta_{\pi_{i-1}}(x) \quad \text{for all } x \neq x_0.$$

where $\varepsilon(\pi'_{i-1}) < \varepsilon(\pi_{i-1})$, contradicting (3).

Let e_i be the edge of $H_{\alpha_{i-1}, \beta}(x_0)$ incident to x_0, other than e_{i-1}. Let y_i be the endpoint of e_i other than x_0. Then

$$k \geq |C_{y_i}(\pi_{i-1})| + p(y_i).$$

Otherwise, we would have

$$\delta_{\pi_{i-1}}(y_i) = d_G(y_i) - |C_{y_i}(\pi_{i-1})| < d_G(y_i) - k + p(y_i) = \varrho(y_i)$$

(the vertex y_i is 'over-coloured'), and exchanging α_{i-1} and β on e_i would yield a k-colouring π'_{i-1} for which $\varepsilon(\pi'_{i-1}) < \varepsilon(\pi_{i-1})$ (since $\delta_{\pi'_{i-1}}(x_0) < \delta_{\pi_{i-1}}(x_0)$ and $\delta_{\pi'_{i-1}}(y_i) = \delta_{\pi_{i-1}}(y_i) + 1 \leq \varrho(y_i)$).

Thus is implied the existence of a colour α_i which does not appear in the neighbourhood of y_i, and which also satisfies (4). In fact, this colour α_i must on one hand be different from $|C_{y_i}(\pi_{i-1})|$ colours, and on the other hand, to satisfy (4), be selected from at most $m_G(y_i, x_0) - 1$ colours (the number of edges parallel to the edge e_i). Since

$$k \geqslant |\, C_{y_i}(\pi_{i-1}) \,| + p(y_i) > |\, C_{y_i}(\pi_{i-1}) \,| + m_G(y_i, x_0) - 1$$

the colour α_i definitely exists. Let π_i be the k-colouring obtained from π_{i-1} by recolouring edge e_i with α_i. Obviously, $\varepsilon(\pi_i) \leqslant \varepsilon(\pi_{i-1})$. As $\varepsilon(\pi_{i-1})$ is minimum by hypothesis, $\varepsilon(\pi_i)$ is also minimum. Finally, the colour β does not appear in the neighbourhood of x_0 in π_i, and α_i appears more than once. Else, we would have $\delta_{\pi_i}(x_0) < \delta_{\pi_{i-1}}(x_0)$ and $\varepsilon(\pi_i) < \varepsilon(\pi_{i-1})$. Thus, we have found α_i, π_i and e_i which satisfy conditions (1) to (4).

The procedure stops as soon as a colour α_t is selected that is the same as a colour α_s with $0 \leqslant s < t - 1$ ($s \neq t - 1$, since $\alpha_t \neq \alpha_{t-1}$ by construction).

In π_{t-1}, $H_{\alpha_s, \beta}(x_0)$ is an elementary chain with endpoints x_0 and y_{s+1}. In fact, it is an elementary cycle in π_s, π_{s+1}, and thus in all the following k-colourings (since the colour α_s is not further altered) this cycle does not include edge e_{s+1}. Also, $y_t \neq y_{s+1}$, for otherwise condition (4) would be contradicted for $i = t$ and $j = s + 1$. The result of all this is that in π_t, $H_{\alpha_s, \beta}(x_0)$ is not an odd cycle, but a chain with endpoints y_{s+1} and y_t. Reasoning as before (applying Fournier's Lemma), we deduce the existence of a k-colouring π_t' such that $\varepsilon(\pi_t') < \varepsilon(\pi_t)$, which contradicts the fact that $\varepsilon(\pi_t)$ is a minimum.

Q.E.D.

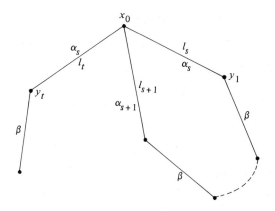

Fig. 12.3

Corollary 1 (Ore [1967]). *If G is a multigraph with no loops, then*

$$q(G) \leqslant \max_{x} \{\, d_G(x) + p(x) \,\}$$

In fact, for $k = \max\, d_G(x) + p(x)$, we have $d_G(x) \leqslant k - p(x)$, and the

k-colouring π defined by Theorem 7 satisfies $|C_x(\pi)| = d_G(x)$, and thus C' is an edge colouring in the usual sense.

Corollary 2 (Vizing's Theorem [1964]). *Let G be a multigraph, with no loops, with maximum degree h and maximum multiplicity p. Then $h \leqslant q(G) \leqslant h + p$.*

This follows easily from Corollary 1.

Corollary 3. *Let G be a simple graph. For all $k \geqslant 1$, there is a k-colouring for which the deficiency of each vertex is at most 1, and is 0 if $d_G(x) \leqslant k - 1$.*

Set $p(x) = 1$ in Theorem 7.

Remark. Let G be a multigraph with no loops. Let S be a set such that the subgraph induced by

$$S \cap \{x \mid k - p(x) < d_G(x) < k + p(x)\}$$

has no cycles. Fourier [1977] proved the existence of a k-colouring π such that

$$\delta_\pi(x) = 0 \qquad\qquad \text{if } x \in S$$

$$\delta_\pi(x) \leqslant \max\{0, d_G(x) - k + p(x)\} \qquad \text{if } d_G(x) \leqslant k$$

$$\delta_\pi(x) \leqslant \max\{0, k - d_G(x) + p(x)\} \qquad \text{if } d_G(x) \geqslant k$$

In particular, this stronger result implies that if G is a simple graph in which every cycle has a vertex of degree $\neq k$, then G has a zero deficiency k-colouring.

Theorem 8 (Shannon [1949]). *Let G be a multigraph, with no loops, with maximum degree h. Then $q(G) \leqslant \left\lceil \dfrac{3h}{2} \right\rceil$.*

Let $h > 0$, and suppose there exists a multigraph G with maximum degree $\leqslant h$ such that $q(G) = \left\lceil \dfrac{3h}{2} \right\rceil + 1$. By removing some edges if necessary, we may assume that G is critical with respect to this property, i.e. if we remove from G any edge, the resulting graph has chromatic index $\left\lceil \dfrac{3h}{2} \right\rceil$.

G contains two distinct vertices a and b with

$$m_G(a, b) \geqslant \left\lceil \frac{h}{2} \right\rceil + 1$$

because, otherwise, $m_G(x, y) \leqslant \left[\dfrac{h}{2}\right]$ for each pair x, y, and the Vizing theorem would imply

$$q(G) \leqslant h + \left[\frac{h}{2}\right] = \left[\frac{3\,h}{2}\right].$$

Let $[a, b]_0$ be an edge joining a and b. The multigraph $G' = G - [a, b]_0$ can be coloured with $q = \left[\dfrac{3\,h}{2}\right]$ colours (because G is critical); for one such colouring,

$$| C_a \cup C_b | \leqslant \big(d_G(a) - m_G(a, b)\big) + \big(d_G(b) - m_G(a, b)\big) + \big(m_G(a, b) - 1\big)$$

$$\leqslant d_G(a) + d_G(b) - m_G(a, b) - 1$$

$$\leqslant 2\,h - \left[\frac{h}{2}\right] - 1 - 1$$

$$\leqslant \left[\frac{3\,h}{2}\right] - 1 .$$

Thus edge $[a, b]_0$ can be coloured with one of the q colours, which contradicts $q(G) = q + 1$.

Remark. It is easy to show that the bound $\left[\dfrac{3\,h}{2}\right]$ given by the corollary can be attained for each value of h. Consider the multigraph G_h consisting of three vertices a, b, c, $\left[\dfrac{h}{2}\right]^*$ edges between a and b and $\left[\dfrac{h}{2}\right]$ edges between a and c and between b and c.

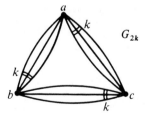 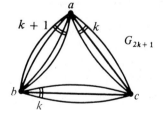

Fig. 12.4

(See Fig. 12.4). Each edge of G_h requires a different colour, and the total number of edges is

$$3 k = \left[\frac{3 h}{2}\right] \quad \text{if} \quad h = 2 k$$

$$3 k + 1 = \left[\frac{3 h}{2}\right] \quad \text{if} \quad h = 2 k + 1 .$$

Thus G_h is a multigraph with maximum degree h and chromatic index $\left[\frac{3 h}{2}\right]$.

Vizing [1965] has also shown that if a multigraph G with maximum degree h does not contain a subgraph G_h as defined above, then

$$q(G) < \left[\frac{3 h}{2}\right] .$$

If h is even, this result can be improved by the following theorem:

Theorem 9. *Let G be a multigraph without loops and with even maximum degree $h = 2 k$. Let $2 \leqslant p \leqslant k$. If G does not contain G_{2p} as a subgraph, then*

$$q(G) \leqslant \frac{3 h}{2} - \left[\frac{h}{2 p}\right] .$$

Two cases must be considered:

CASE 1: $p = k$.

Let G be a multigraph with maximum degree $\leqslant 2 k$ such that $q(G) = 3 k$. We may assume that G is critical with respect to this property. It remains to show that G is identical to the graph G_{2k} shown in Fig. 12.4. G contains two distinct vertices a and b with

$$m_G(a, b) \geqslant k ,$$

because, otherwise, the Vizing theorem would imply

$$q(G) \leqslant \max_{x} d_G(x) + \max_{x,y} m_G(x, y) \leqslant 2 k + (k - 1) = 3 k - 1 ,$$

which contradicts $q(G) = 3 k$.

Let $[a, b]_0$ be an edge joining these two vertices. Let

$$G' = G - [a, b]_0 .$$

Multigraph G' can be coloured with $3 k - 1$ colours, and for such a colouring,

$3 k - 1 = | C_a \cup C_b | \leqslant$

$\leqslant (d_G(a) - m_G(a, b)) + (d_G(b) - m_G(a, b)) + (m_G(a, b) - 1) =$

$= d_G(a) + d_G(b) - m_G(a, b) - 1 \leqslant 2 k + 2 k - k - 1 = 3 k - 1 .$

Thus

$$m_{G'}(a, b) = k .$$

Similarly, graph G' contains a vertex b_1 adjacent to b with

$$m_{G'}(b, b_1) \geqslant k$$

because, otherwise,

$$\max_{b_1} (d_{G'}(b_1) + m_{G'}(b, b_1)) \leqslant 2 k + (k - 1) = 3 k - 1 ,$$

which by Theorem 7 implies that $q(G) \leqslant 3 k - 1$, which contradicts $q(G) = 3 k$.

Since $m_{G'}(b, a) = m_G(a, b) - 1 = k - 1$, then $b_1 \neq a$. Furthermore, since $d_G(b) \leqslant 2 k$ and $m_G(b, a) = k$, we have

$$m_G(b, b_1) = k .$$

By applying the same argument to the pair b, b_1, we see that G contains a vertex $b_2 \neq b_1$ with $m_G(b_1, b_2) = k$, etc. Vertices $a, b, b_1, b_2, \ldots$ form a chain of k-tuple edges, that can terminate only at vertex a. Therefore, G has a cycle μ such that each pair of two consecutive vertices is joined by k edges. Since G is critical, G is connected and, therefore, G can have no other edges.

If cycle μ has even length, then $q(G) = 2 k$ (from Theorem 5), which contradicts $q(G) = 3 k$.

If cycle μ has odd length $n_0 \geqslant 5$, then, from Theorem 5,

$$q(G) = \left[\frac{2 m}{n_0 - 1} \right]^* = \left[\frac{2 n_0 k}{n_0 - 1} \right]^* =$$

$$= 2 k + \left[\frac{2 k}{n_0 - 1} \right]^* \leqslant$$

$$\leqslant 2 k + \left[\frac{2 k}{4} \right]^* < 3 k ,$$

which contradicts $q(G) = 3 k$.

Thus, cycle μ has length 3, and $G = G_{2k}$.

CASE 2: $p < k$.

Let G be a multigraph with maximum degree $2k$ that does not contain a subgraph G_{2p}. We shall show that G can be coloured with $3k - \left[\dfrac{k}{p}\right]$ colours.

By adding vertices, G can be imbedded in a regular multigraph of degree $2k$ that contains no subgraph G_{2p}; thus, we may assume that G is regular.

From the Petersen theorem (Theorem 2, Ch. 11), the edge set E of G can be decomposed into k factors $E_1, E_2, ..., E_k$. The graphs $H_i = (X, E_i)$ are regular of degree 2; thus $q(H_i) \leqslant 3$, and, consequently,

$$q(G) = \sum_i q(H_i) \leqslant 3k .$$

Note that, if h is even, this is Shannon's result.

Let $F = E_1 \cup E_2 \cup \cdots \cup E_p$. Multigraph (X, F) has maximum degree $2p$ and contains no subgraph G_{2p}. From Case 1, (X, F) has chromatic index $\leqslant 3p - 1$. Thus

$$q(G) \leqslant (3p - 1)\left[\frac{k}{p}\right] + 3\left(k - \left[\frac{k}{p}\right]p\right) \leqslant 3k - \left[\frac{k}{p}\right] .$$

Q.E.D.

Corollary (Fiamčik, Jucovič [1970]). *Let G be a multigraph without loops and with maximum degree h. If G does not contain a partial subgraph G_4 (see Fig. 12.4), then*

$$q(G) \leqslant 3\left[\frac{h + 1}{2}\right] - \left[\frac{h + 1}{4}\right] .$$

If $h = 2k$ is even, then, from Theorem 9 with $p = 2$,

$$q(G) \leqslant 3k - \left[\frac{k}{2}\right] = 3\left[\frac{h + 1}{2}\right] - \left[\frac{h + 1}{4}\right] .$$

If $h = 2k + 1$ is odd, then G can be embedded in a graph G' with maximum degree $h' = 2k + 2$ that contains no subgraph G_4. Thus

$$q(G) \leqslant q(G') \leqslant 3(k + 1) - \left[\frac{k + 1}{2}\right] = 3\left[\frac{h + 1}{2}\right] - \left[\frac{h + 1}{4}\right] .$$

Q.E.D.

The table below gives upper bounds for the chromatic index of a multigraph G of maximum degree h. These bounds are calculated from the corollary to Theorem 8 and from Theorem 9.

$h = 3$ $\quad q(G) \leqslant 4$

$h = 4$ $\quad q(G) \leqslant 6$; $\quad$ if $\quad G \not\supset G_4$, $\quad q(G) \leqslant 5$

$h = 5$ $\quad q(G) \leqslant 7$; $\quad$ if $\quad G \not\supset G_5$, $\quad q(G) \leqslant 6$

$h = 6$ $\quad q(G) \leqslant 9$; $\quad$ if $\quad G \not\supset G_6$, $\quad q(G) \leqslant 8$

$h = 7$ $\quad q(G) \leqslant 10$; $\quad$ if $\quad G \not\supset G_7$, $\quad q(G) \leqslant 9$

$h = 8$ $\quad q(G) \leqslant 12$; $\quad$ if $\quad G \not\supset G_8$, $\quad q(G) \leqslant 11$

$\qquad\qquad\qquad\qquad$ and if $\quad G \not\supset G_4$, $\quad q(G) \leqslant 10$.

Extension (Andersen [1977]). *Let G be a multigraph without loops and with maximum degree h. If $G \not\supset G_s$, where $4 \leqslant s \leqslant h$, then*
If $G \not\supset G_s$, where $4 \leqslant s \leqslant h$, then

$$q(G) \leqslant \left[\frac{3h}{2} \right] - \left[\frac{h}{s} \right].$$

If both h and s are even, the conjecture becomes Theorem 9. If $h = s$ is odd, the conjecture can be proved as in Case 1 of Theorem 9.

Another conjecture is the following:

Conjecture (Vizing [1965]). *Let G be a simple graph with maximum degree h. Then the vertices and edges of G can be coloured with $h + 2$ colours so that no two adjacent vertices have the same colour, no two adjacent edges have the same colour, and no vertex has the same colour as an edge incident to it.*

3. Edge colourings of planar graphs

From the Theorem of Vizing, we know that if G is regular of degree 3, then $q(G) \leqslant 4$. It has been conjectured that *the chromatic index of a regular planar graph of degree 3 without isthmi is exactly 3.*

The proof of this conjecture follows from the Four Colour Theorem (Appel-Hakan [1977]).

Theorem 10. *For a regular planar multigraph G of degree 3 without isthmi, the following conditions are equivalent:*

(1) *the faces of G can be coloured with 4 colours so that no two adjacent faces have the same colour,*

(2) *the edges of G can be coloured with 3 colours so that no two adjacent edges have the same colour,*

(3) *each vertex can be assigned a coefficient $p(x)$, where $p(x)$ equals $+ 1$ or $- 1$, such that each face μ satisfies*

$$\sum_{x \in \mu} p(x) \equiv 0 \pmod{3} .$$

(1) ⇒ (2) If the faces of G are coloured with 4 colours α, β, γ, δ, label with a "0" each edge separating an α-face from a β-face or separating a γ-face from a δ-face. Label with a "1" each edge separating an α-face from a γ-face or separating a β-face from a δ-face. Finally, label with a "2" each edge separating an α-face from a δ-face or separating a β-face from a γ-face.

Two adjacent edges cannot be labelled with the same symbol (since this would imply that two adjacent faces have the same colour). Hence condition (2) is satisfied.

(2) ⇒ (1) Suppose that the edges of G can be coloured with three colours 0, 1, 2. The edges with colours 0 and 1 form a regular graph of degree 2, and, therefore, its faces can be coloured with only two colours p and q. The edges with colours 0 and 2 also form a regular graph whose faces can be coloured with two colours r and s. Thus, each face of G can be coloured with one of the four combinations pr, ps, qr, qs. If two faces are separated by an edge of colour 0, they are coloured with distinct combinations. The same is true for two faces separated by an edge of colour 1 or of colour 2. Thus, the faces can be coloured with 4 colours so that no two adjacent faces have the same colour.

(2) ⇒ (3) Consider a 3-colouring of the edges with colours 0, 1, 2. If the three edges incident to vertex x are 0, 1, 2 in clockwise order, let $p(x) = +1$; otherwise, let $p(x) = -1$. Now follow the contour of a face μ in the clockwise direction. Each time a vertex x with $p(x) = +1$ is crossed, the index of the colour decreases by one, and each time a vertex x with $p(x) = -1$ is crossed, the index increases by one (modulo 3). Therefore, the algebraic sum of all the coefficients of the vertices on contour μ is necessarily a multiple of 3.

(3) ⇒ (2) Suppose that the coefficients $p(x)$ satisfy condition (3).

Starting from some arbitrary edge to which we assign the label 0, we shall now give each edge a label, either 0 or 1 or 2; this is done step by step with the help of the coefficients $p(x)$ in such a way that the three edges incident to a vertex x are 0, 1, 2 counter-clockwise if $p(x) = +1$, and in the reverse order if $p(x) = -1$.

If G is connected, each edge $[a, b]$ will be given a label $g(a, b)$. It remains to show that this labelling is consistent. Let v be an elementary cycle which contains a set F of faces. Let $x \in X_i$ if vertex x is incident to exactly i faces in F. Let $\varepsilon(x, v) = -1$ for $x \in X_1$,

+ 1 for $x \in X_2$, 0 for $x \in X_3$. If $[a, b]$ and $[b, c]$ are two consecutive edges of v in a counter-clockwise tour, then

$$g(b, c) \equiv g(a, b) + \varepsilon(x, v)p(x) \quad (\text{mod. } 3) .$$

Moreover, we have (modulo 3),

$$\sum_{x \in v} \varepsilon(x, v)p(x) \equiv \sum_{i=0}^{3} \left(i \sum_{x \in X_i} p(x) \right) = \sum_{\mu \in F} \sum_{x \in v} p(x) \equiv 0 .$$

The consistency of the labelling on v follows.

Q.E.D.

Corollary 1. *If the number of edges bounding each face is a multiple of 3, then the edges of the graph can be coloured with only three colours.*

Let $p(x) = + 1$ for each vertex x: condition (3) of the theorem is satisfied.

Q.E.D.

As 4-chromatic cubic graphs with no isthmus are rare, they have been baptised "snarks" by M. Gardner [1976]. A "snark" is now considered to be any connected 3-regular graph with no isthmus, from which it is necessary to remove at least four vertices in order to have a disconnected graph whose components have cycles. Isaacs [1975] has shown that every 3-regular 3-chromatic graph with no isthmus can be obtained from snarks using simple constructions. The only snark of order 10 is the Petersen graph. Fiorini and Wilson showed that there are no snarks of order 12 or 14, Fouquet showed that there are none of order 16, and Preissman [1982] showed that there are none of order 18. The reader should also consult: S. Fiorini and R. Wilson [1977].

Conjecture. *For a simple planar graph with maximum degree $h \geqslant 6$, the chromatic index equals h.*

For $h = 2, 3, 4, 5$, it is easy to construct a simple planar graph G with $q(G) = h + 1$. For $h \geqslant 8$, the conjecture was proved by Vizing [1965].

EXERCISES

1. Consider the complete graph K_n, where n is even. Denote its vertices by $0, 1, ..., n - 1$, and consider the function

$$
\begin{aligned}
f(a, b) &= a + b \quad &\text{mod } (n - 1) \quad &\text{if} \quad a, b \neq n - 1 , \\
&= 2a \quad &\text{mod } (n - 1) \quad &\text{if} \quad a \neq n - 1, b = n - 1 , \\
&= 2b \quad &\text{mod } (n - 1) \quad &\text{if} \quad a = n - 1, b \neq n - 1 .
\end{aligned}
$$

Show that f defines a $(n-1)$-colouring of the edges, i.e.,

(1) $0 \leqslant f(a, b) \leqslant n-2$,
(2) $f(a, b)$ is an integer for $a \neq b$,
(3) $f(a, b) = f(b, a)$,
(4) $f(a, b) \neq f(a, c)$ if $b \neq c$.

2. In a bridge tournament, $4p$ players simultaneously play on p tables, and no player wants to have the same partner twice. Show that it is possible to organize $4p - 1$ games so that, for a permutation f on the set of chairs, each player seated in chair x will play his next game in chair $f(x)$.

3. Let G be a simple connected graph that decomposes into two connected components C_1 and C_2 when vertex a is removed. If G_1 is the subgraph of G generated by $C_1 \cup \{a\}$, and if G_2 is the subgraph of G generated by $C_2 \cup \{a\}$, show that

$$q(G) = \max \left\{ q(G_1), q(G_2), d_G(a) \right\}.$$

4. Let $m_1 \geqslant m_2 \geqslant \cdots \geqslant m_q$ be a non-increasing q-tuple of integers. Consider the relation $(m_i') \prec (m_i)$ defined by

$$\sum_{i=1}^{k} m_i' \leqslant \sum_{i=1}^{k} m_i \qquad (k = 1, 2, ..., q-1)$$

$$\sum_{i=1}^{q} m_i' = \sum_{i=1}^{q} m_i.$$

If $m_i \geqslant m_j + 2$, the q-tuple (m_i') defined by

$$m_i' = m_i - 1$$
$$m_j' = m_j + 1$$
$$m_k' = m_k \qquad (k \neq i, j)$$

is called the *transfer of* (m_i) *by* (i, j). Show that a transfer of m defines a q-tuple $m' \prec m$, and that each q-tuple m'' with $m'' \prec m$ can be obtained from m by a finite number of transfers.

5. Let $m_1 \geqslant m_2 \geqslant \cdots \geqslant m_q$ be a non-increasing sequence of integers, and let G be a multigraph with

$$m = \Sigma \, m_i$$

edges coloured $\alpha_1, \alpha_2, ..., \alpha_q$, where the number of edges coloured α_i equals m_i, for all i. Using Exercise 4, show that if (m_i') is a sequence with $(m_i') \prec (m_i)$, then G can be coloured in q colours with m_i' edges coloured α_i, for all i.

Hint: Note that in the partial graph $G(\alpha_i, \alpha_j)$ generated by the edges with colour α_i and α_j, a connected component is an alternating open chain whose extremities have different colours. The interchange of colours α_i and α_j along this chain defines a transfer by (i, j).

(Folkman, Fulkerson [1969])

6. Show that a simple regular graph of degree h with order has chromatic index equal to $h + 1$.

7. Let G be a simple graph with maximum degree h such that $q(G) = h + 1$, and such that $q(G - e) = h$ for every edge e. Show that:
  1. Each vertex adjacent to a vertex of degree k is adjacent to at least $h - k + 1$ vertices of degree h.
  2. G possesses an elementary cycle of length $\geqslant h + 1$.

(Vizing [1965])

8. Let $K_{n, n, ..., n}$ be the complete r-partite graph formed by joining pairwise the r sets of n elements $X_1, X_2, ... X_r$ in all possible ways. Show that the chromatic number is $n(r - 1)$ if nr is even, or $n(r - 1) + 1$ if nr is odd.

(Laskar and Hare [1972])

9. Let G be a graph consisting of disjoint odd cycles, with certain pairs of vertices connected by additional chains of length 2. Show that $q(G) = 3$.

(Mycielski [1978])

10. Let $G = (X, E)$ be a multigraph, and consider a k-colouring of G as an edge partition $(E_1, ..., E_k)$. Let $G(i)$ be the partial graph whose edges are of colour i. For positive integer k, show that there is a k-colouring $(E_1, ..., E_k)$ such that for each vertex x:

$$| d_{G(i)}(x) - d_{G(j)}(x) | \leq 2$$

for at most one pair of colours $\{ i, j \}$,

$$| d_{G(p)}(x) - d_{G(q)}(x) | \leq 1$$

for every pair of colours $\{ p, q \} \neq \{ i, j \}$.

(A.J.W. Hilton and D. de Werra [1982])

11. Let G be a simple graph. Given a positive integer k such that k does not divide the degree of any vertex of G, show that there exists an edge k-colouring such that for each vertex x and for every pair of colours $\{ i, j \}$, $| d_{G(i)}(x) - d_{G(j)}(x) | \leq 1$.

(A.J.W. Hilton, D. de Werra [1982])

12. Given a multigraph $G = (X, E)$ and a positive integer k, associate with each vertex x an interval $I(x) = [\alpha(x), \beta(x)]$ such that $\alpha(x) \leq (1/k)d_G(x) \leq \beta(x)$, with $\alpha(x), \beta(x)$ integers. An edge k-colouring $(E_1, ..., E_k)$ is *regular* if $d_{G(i)}(x) \in I(x)$, for every vertex x and for every colour i.

(1) Show that if G contains no connected partial subgraph H with m_H odd and $d_H(x) = 2 \alpha(x)$ or $2\beta(x)$ for every vertex x in H, then G has a regular edge k-colouring.

(D. de Werra [1982])

(2) Let $m_1 \geq m_2 \geq \cdots \geq m_k$ be a non-increasing k-tuple of integers. Consider the relation $(m'_1, ..., m'_k) < (m_1, ..., m_k)$ defined by

$$\sum_{i=1}^{q} m'_i \leq \sum_{i=1}^{q} m_i \qquad (q = 1, ..., k - 1)$$

$$\sum_{i=1}^{k} m'_i = \sum_{i=1}^{k} m_i.$$

Show that if G has a regular k-colouring $(E_1, ..., E_k)$ with $| E_i | = m_i$ for $i = 1, ..., k$ and $m_1 \geq \cdots \geq m_k$, then for each k-tuple $(m'_1, ..., m'_k) < (m_1, ..., m_k)$ there exists a regular k-colouring $(E'_1, ..., E'_k)$ of G with $| E'_i | = m'_i$ ($i = 1, ..., k$).

(D. de Werra [1979])

(3) Show that if $G = (X, Y, E)$ is a bipartite multigraph, then for every positive integer k there exists a k-colouring $(E_1, ..., E_k)$ such that

(a) $| d_{G(i)}(x) - d_{G(j)}(x) | \leq 1$ $(x \in X; i, j \leq k)$

(b) $| m_{G(i)}(x, y) - m_{G(j)}(x, y) | \leq 1$ $(x \in X, y \in Y; i, j \leq k)$

(c) $| | E_i | - | E_j | | \leq 1$ $(i, j \leq k)$.

(D. de Werra [1975])

13. Let $G = (X, E)$ be a multigraph that contains no odd elementary cycle with length greater than 3. Show that

$$q(G) = \max(\max_{x} d_G(x), \max_{x, y, z} \Delta(x, y, z))$$

where $\Delta(x, y, z)$ is the number of edges in the subgraph induced by the vertices x, y, z.

(D. de Werra [1978])

14. Let $G = (X, U)$ be a pseudo-symmetric 1-graph, and let k be a positive integer. Then there exists a partition of the arcs of G such that
 (1) each $G_i = (X, U_i)$ is pseudo-symmetric,
 (2) if $c_i(x)$ is the number of elementary circuits of G_i containing the vertex x, then

$$| c_i(x) - c_j(x) | \leq 1 \qquad (i, j \leq k; x \in X)$$

 (3) for all G_i, the following property holds: if K is an elementary circuit of G_i, then K contains a vertex z such that

$$c_j(z) \geq c_i(z) \qquad \text{for} \quad j = 1, \ldots, i - 1.$$

(D. de Werra [1978])

CHAPTER 13

Stability Number

1. Maximum stable sets

Consider a simple graph $G = (X, E)$. A set $S \subset X$ is defined to be *stable* if no two distinct vertices of S are adjacent. In other words, S is stable if, and only if,

$$\Gamma_G(S) \cap S = \varnothing.$$

Let $\mathscr{S}$ denote the family of all stable sets of G. Then,

$$\varnothing \in \mathscr{S}$$
$$S \in \mathscr{S}, A \subset S \quad \Rightarrow \quad A \in \mathscr{S}.$$

The *stability number* $\alpha(G)$ of G is defined to be the maximum cardinality of a stable set,

$$\alpha(G) = \max_{S \in \mathscr{S}} |S|.$$

EXAMPLE 1 (Gauss). *Problem of the eight queens.* Can eight queens be placed on a chessboard so that no queen can capture another queen? This famous problem is equivalent to finding a maximum stable set of a simple graph G with 64 vertices and with $y \in \Gamma_G(x)$ if squares x and y are in the same row or in the same column or in the same diagonal.

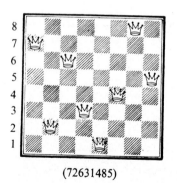

(72631485)   (61528374)

Fig. 13.1

269

This problem is more difficult that it would appear at first sight, and Gauss initially believed that it had only 76 solutions; in 1854, *Schachzeitung*, a Berlin chess journal, gave only 40 solutions. In fact, there are exactly 92 solutions, which can be obtained from the following permutations:

(72631485)	(61528374)	(58417263)
(35841726)	(46152837)	(57263148)
(16837425)	(57263184)	(48157263)
(51468273)	(42751863)	(35281746)

Each of these permutations gives a possible diagram (see Fig. 13.1) and from each of the diagrams we can obtain eight solutions by rotating the chessboard and by reflecting the chessboard with respect to the principal diagonal. The last permutation (35281746) gives only four distinct solutions because it yields the same diagram after rotation of the chessboard.

This problem has been generalized by Bruen and Dixon [1975], Kløve [1977] and Gabow [1979].

EXAMPLE 2. *Covering a chessboard with tetraminos.* Consider the problem of covering the 64 squares of a chessboard with 16 tetraminos, of various shapes, each covering exactly 4 squares. For example, if there are 15 L-shaped tetraminos and 1 square tetramino (Fig. 13.2), we can see that no covering is possible by the following argument: if the squares of the chessboard are coloured black and white as in Fig. 13.2, then the number n_1 of white squares and the number n_2 of black squares covered by an L-shaped tetramino always satisfies $n_1 - n_2 \equiv 2$ modulo 4.

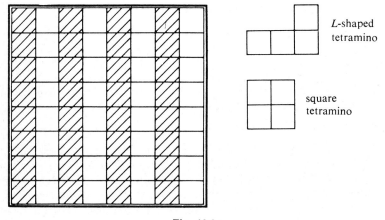

L-shaped
tetramino

square
tetramino

Fig. 13.2

In general, the problem reduces to verifying that the stability number of a certain graph G equals 16. To each tetramino i together with the position λ occupied by this tetramino, there corresponds a vertex $x(i, \lambda)$. Join vertices $x(i, \lambda)$ and $x(k, v)$ either if $i = k$, or if $i \neq k$ and the set of squares occupied by tetramino i placed in position λ overlaps with the squares occupied by tetramino k placed in position v. Each stable set with 16 vertices of the resulting graph corresponds to a covering of the chessboard with 16 tetraminos, and vice versa.

Numerous analogous problems have been proposed by Golomb [1965].

The construction of a maximum stable set is a particular case of the problem of finding a minimum transversal set of a hypergraph.

Consider a simple graph $G = (X, E)$. A *clique* is defined to be a set $C \subset X$ such that each pair of distinct vertices in C are adjacent. Let $\mathscr{C} = (C_1, C_2, ..., C_k)$ be a partition of X into cliques. Let

$$\theta(G) = \min_{\mathscr{C}} | \mathscr{C} |$$

denote the smallest possible number of cliques that partition X.

Theorem 1. *Let G be a simple graph then $\alpha(G) \leqslant \theta(G)$. Furthermore, if S is a stable set and if $\mathscr{C}$ is a partition into cliques such that $| S | = | \mathscr{C} |$, then S is a maximum stable set and $\mathscr{C}$ is a minimum partition.*

For each stable set S and each partition $\mathscr{C} = (C_1, C_2, ..., C_k)$, we have

$$| S \cap C_i | \leqslant 1 \qquad (i = 1, 2, ... , k) .$$

Hence, $| S | \leqslant | \mathscr{C} |$. Therefore

$$\alpha(G) = \max | S | \leqslant \min | \mathscr{C} | = \theta(G) .$$

Furthermore, if $| S_0 | = | \mathscr{C}_0 |$, the above inequality implies that

$$| S_0 | = \alpha(G), \qquad | \mathscr{C}_0 | = \theta(G) .$$

Q.E.D.

There are many classes of graph G with $\alpha(G) = \theta(G)$. (See for instance the graphs considered in Examples 1 and 2.) Other classes of graphs with this property are studied in Chapter 16.

Remark. Given a minimum partition $\mathscr{C} = (C_1, C_2, ..., C_k)$ for a graph G with $\alpha(G) = \theta(G)$, it is easy to construct a maximum stable set by a backtracing procedure: Select any vertex $x_1 \in C_1$. Choose a vertex $x_2 \in C_2$ that is non-adjacent to x_1. Choose a vertex $x_3 \in C_3$ that is non-adjacent to x_1, x_2, etc. When this procedure can no longer continue, backtrack (as in the exploration of an arborescence described in Chapter 3). Proceed until a maximum stable set has been found.

Given a simple graph $G = (X, E)$ and a stable set $S \subset X$, an *alternating sequence relative to S* is defined to be a sequence

$$\sigma = (a_1, b_1, a_2, b_2, a_3, \ldots)$$

of distinct vertices alternately belonging to $A = X - S$ and to $B = S$ that satisfies the following conditions:

(1) $a_1 \in A$,

(2) $b_i \in B - \{b_1, b_2, \ldots, b_{i-1}\}$ and

$$\Gamma_G(b_i) \cap \{a_1, a_2, \ldots, a_i\} \neq \varnothing,$$

(3) $a_{i+1} \in A - \{a_1, a_2, \ldots, a_i\}$ and

$$\Gamma_G(a_{i+1}) \cap \{b_1, b_2, \ldots, b_i\} \neq \varnothing,$$
$$\Gamma_G(a_{i+1}) \cap \{a_1, a_2, \ldots, a_i\} = \varnothing.$$

An alternating sequence is said to be *maximal* if no more vertices can be added to it without violating (2) or (3). We shall see that for the maximum stable sets, alternating sequences play a rôle similar to that of alternating chains for the maximum matchings.

The maximum stable set lemma. *B is a maximum stable set if and only if every stable set S disjoint from B can be matched into B.*

1. Let B be a maximum stable set, and let S be a stable set disjoint from B. By Theorem 5 (Ch. 7), S can be matched into B if and only if every set $A \subset S$ satisfies

$$|\Gamma_G(A) \cap B| \geqslant |A|.$$

But this condition is satisfied, for otherwise, there would exist $A \subset S$ with $|\Gamma_G(A) \cap B| < |A|$, and the set $[B - \Gamma_G(A) \cap B] \cup A$ would be stable, with cardinality $> |B|$, a contradiction.

2. Conversely, let B be a stable set satisfying the property in the lemma. If B is not maximum, then there exists a stable set T with

$$|T - B| > |B - T|.$$

Thus, the set $S = T - B$ cannot be matched into B, a contradiction.

Theorem 2. *A stable set B of a graph G is maximum if and only if there exists no maximal alternating sequence of odd length.*

1. Suppose that there does exist a maximal alternating sequence (relative to B), of the form

$$\sigma = (a_1, b_1, a_2, b_2, \ldots, a_{k-1}, b_{k-1}, a_k)$$

with $a_i \in X - B$, $b_j \in B$.

Let $\overline{A} = \{a_1, a_2, \ldots, a_k\}$ and $\overline{B} = \{b_1, b_2, \ldots, b_{k-1}\}$. Note that $(B - \overline{B}) \cup \overline{A}$ is a stable set which contains one more element than does B. Thus B is not a maximum stable set.

2. Suppose that there exists no odd maximal alternating sequence, and suppose that B is not maximum (proof by contradiction).

Let T be a maximum stable set. Thus $|T - B| > |B - T|$. By the maximum stable set lemma, there is a matching of $B - T$ into T, and thus into $A = T - B$; let $a = \sigma(b)$ be the mapping that defines this matching. Let a_1 be a vertex of A that is not saturated by this matching. Since (a_1) is an odd alternating sequence, it is not maximal, and there exists some $b_1 \in B$ adjacent to a_1. Let $a_2 = \sigma(b_1)$; thus $a_2 \neq a_1$. Since (a_1, b_1, a_2) is an odd alternating sequence, which cannot be maximal, there exists some $b_2 \in B$, $b_2 \neq b_1$, adjacent to $\{a_1, a_2\}$. Let $a_3 = \sigma(b_2)$, etc. The sequence $(a_1, b_1, a_2, b_2, a_3, \ldots)$ can always be continued, which is a contradiction.

Q.E.D.

Call a vertex x whose removal from the graph G changes the stability number a *critical vertex*; it is defined by any of the following (equivalent) conditions:

   (i) $\alpha(G_{X-\{x\}}) \neq \alpha(G)$,

   (ii) $\alpha(G_{X-\{x\}}) = \alpha(G) - 1$,

   (iii) every maximum stable set of G contains x.

For example, isolated points of a graph are always critical vertices.

Call an edge $[a, b]$ whose removal changes the stability number a *critical edge*; it is defined by any of the following (equivalent) conditions:

   (i) $\alpha(G - [a, b]) \neq \alpha(G)$;

   (ii) $\alpha(G - [a, b]) = \alpha(G) + 1$;

   (iii) every maximum stable set of $G - [a, b]$ contains a and b.

Note that if $[a, b]$ is a critical edge, than a is not a critical vertex. In fact, there exists in G a set S of cardinality $\alpha(G) + 1$ which contains only one edge, namely $[a, b]$. Thus $S - \{a\}$ is a stable set which does not contain a, and a is not a critical vertex. Similarly, if a is a critical vertex, then it is not incident to a critical edge.

Theorem 3. *If G contains no critical vertex, then every stable set S can be matched into $X - S$.*

The proof is by induction on $|S|$. Suppose that $|S| = 1$. Let $S = \{x\}$. Since G has no isolated vertices, there must be some edge incident to x, yielding the desired matching.

Suppose that the Theorem holds for all stable sets with cardinality $< p$. Consider a stable set S with $|S| = p > 1$. Let $\underline{a} \in S$. Since $\underline{a}$ is not a critical vertex, there exists a stable set B which does not contain $\underline{a}$.

By the preceding Lemma, $S - B$ can be matched into B, and thus into $B - S$. By the induction hypothesis, the stable set $S \cap B$ can be matched into its complement, and thus into $X - (S \cup B)$. Combining these two matchings gives a matching of S into $X - S$.

$$Q.E.D.$$

Corollary 1. *If G has no critical vertices, then $\alpha(G) \leq \dfrac{n}{2}$. If equality holds, then G has a perfect matching.*

In fact, let S be any maximum stable set; it can be matched into $X - S$, and

$$\alpha(G) = |S| \leq |X - S| = n - \alpha(G).$$

Hence, $n \geq 2\alpha(G)$, with equality if and only if G has a perfect matching.

Corollary 2 (Folkman). *If for all $x \in X$, $\alpha(G_{X-x}) > \dfrac{n}{2}$, then $\alpha(G) > \dfrac{n}{2} + 1$.*

By Corollary 1, G has a critical vertex a. Thus $\alpha(G) = \alpha(G_{X-a}) + 1 > \dfrac{n}{2} + 1$.

$$Q.E.D.$$

With the following theorems, it is sometimes easy to recognize a maximum stable set; however, the problem of recognizing a maximum stable set of a graph is in general a difficult problem that is, in fact, 'N–P complete'. Often, a 'weight' $p(x) > 0$ is associated with each vertex, and one seeks a stable set of maximum total weight. By arranging the weights $p(x)$ to be integers, and by replacing each vertex x with a stable set of $p(x)$ points, a graph G' can be constructed in which a *maximum stable set* corresponds to a *maximum weight stable set* in G.

EXAMPLE: Find numbers $z_1, z_2, z_3, z_4 \in \{0, 1\}$ which maximize the 'pseudo-boolean' function

$$f(z_1, z_2, z_3, z_4) = 4z_1\bar{z}_2 + 3z_3z_4 + 6\,\bar{z}_1z_4 + \cdots, \qquad \text{where } \bar{z}_i = 1 - z_i.$$

Construct the graph G whose vertices $a_1, a_2, a_3, \ldots$ represent the monomials $z_1\bar{z}_2$, z_3z_4, $\bar{z}_1z_4$, $\ldots$ with respective weights 4, 3, 6, etc $\ldots$

Join points such as a_1 and a_3 whose corresponding monomials cannot be simultaneously equal to 1. Thus, a maximum weight stable set in G will yield immediately the values of the variables for which f is maximized.

2. Turan's theorem and related results

In this section we study bounds on the stability number as a function of the degrees or the number of edges.

Theorem 4 (Meyer [1972]). *Let G be a simple graph of order n with no isolated vertices, whose vertices $x_1, x_2, \ldots, x_n$ are indexed such that*

$$d_G(x_1) \leq d_G(x_2) \leq \cdots \leq d_G(x_n).$$

If for any integer k, $2 \leq k \leq n$,

(1) $$d_G(x_n) + \cdots + d_G(x_{n-k+2}) \leq n - k,$$

then every maximal (with respect to inclusion) stable set has at least k vertices.

If $k = 2$, (1) gives $\max_x d_G(x) = d_G(x_n) \leq n - 2$. Thus, each vertex x is non-adjacent to some vertex $a \neq x$, and the result is proved.

Suppose that the result is true for k, and consider the case for the value $k + 1$ (proof by induction). Let G be a graph that satisfies (1) for the value $k + 1$. Then

$$d_G(x_n) + \cdots + d_G(x_{n-k+2}) \leq d_G(x_n) + \cdots + d_{G(x_{n-k+1})} \leq n - (k+1) < n - k.$$

Hence, by the induction hypothesis, a stable set S with fewer than k vertices is contained in the stable set $S_0 = \{ y_1, y_2, \ldots, y_k \}$ with k vertices, and:

$$d_G(y_1) + \cdots + d_G(y_k) \leq d_G(x_n) + \cdots + d_G(x_{n-k+1}) \leq n - k - 1 < |X - S_0|.$$

It follows that there exists at least one vertex a in $X - S_0$ that is not adjacent to any vertex of S_0. Thus $S_0 \cup \{ a \}$ is a stable set with $k + 1$ vertices that contains S.

                                                                Q.E.D.

Corollary 1. *Let G be a simple graph of order n with maximum degree h. Let $\alpha'(G)$ denote the minimum cardinality of a maximal stable set; then $\alpha'(G) \geq \left[\dfrac{n}{h+1}\right]^*$, and this bound is the best possible.*

Let

$$(2) \qquad\qquad k = \left\lfloor \frac{n}{h+1} \right\rfloor^*.$$

Hence, $k - 1 < \dfrac{n}{h+1}$, from which

$$(3) \qquad\qquad h < \frac{n}{k-1} - 1 = \frac{n-k+1}{k-1}.$$

From (3), we have

$$d_G(x_n) + \ldots + d_G(x_{n-k+2}) \leqslant (k-1)h < n - k + 1.$$

Thus, condition (1) is satisfied, and consequently $\alpha'(G) \leqslant k$. To see that this bound is the best possible for all n, consider the following graph G: partition n vertices into the classes $C_1, C_2, \ldots, C_q$. Put $h + 1$ vertices into each of the first $q - 1$ classes, leaving at most $h + 1$ vertices in C_q, and thus $q = \left\lfloor \dfrac{n}{h+1} \right\rfloor^*$.

Join two vertices if and only if they belong to the same class. Thus G is a graph of order n, with maximum degree h, and $\alpha'(G) = \left\lfloor \dfrac{n}{h+1} \right\rfloor^*$.

Corollary 2. *Let G be a simple graph of order n with maximum degree h. Then $\alpha(G) \geqslant \left\lfloor \dfrac{n}{h+1} \right\rfloor^*$, and this bound is the best possible.*

In fact, $\alpha(G) \geqslant \alpha'(G) \geqslant \left\lfloor \dfrac{n}{h+1} \right\rfloor^*$.

As well, the example of the previous corollary proves the existence of a graph G of order n with maximum degree h, and $\alpha(G) = \left\lfloor \dfrac{n}{h+1} \right\rfloor^*$.

EXAMPLE. Consider 99 bridge players, and suppose that each of them will play only with the 3 players he or she already knows. If h is the maximum number of players unknown to any one player, for what value of h is one guaranteed to find at least one game of bridge?

Let G be the graph whose vertices represent the players, with two vertices adjacent if the players do not know each other. If $h = 32$, then

$$\left\lfloor \frac{n}{h+1} \right\rfloor^* = \left\lfloor \frac{99}{33} \right\rfloor^* = 3,$$

and it is possible that there will be no bridge four-some. On the other hand, if $h = 31$, a bridge game is guaranteed; what is more, any 3 players arbitrarily selected will be able to find a fourth.

Remark. Associated with these results are the following theorems, which we will not use.

The Hajnal–Szémérédi theorem [1970]: *If G is a graph of order n and of degree h, then its vertices can be partitioned into $n + 1$ stable sets, each with either $\left[\dfrac{n}{h+1}\right]^*$ or $\left[\dfrac{n}{h+1}\right]$ elements.*

For example, if $n = 100$ and $h = 24$ in the above bridge players' graph, then all people can play bridge simultaneously.

Erdös' theorem. *Let G be a graph of order n whose largest clique has k elements and whose vertices $x_1, x_2, \ldots, x_n$ are indexed so that*

$$d_G(x_1) \leq d_G(x_2) \leq \cdots \leq d_G(x_n).$$

Then there exists a complete k-partite graph H whose vertices $y_1, y_2, \ldots, y_n$ satisfy

$$d_G(x_i) \leq d_H(y_i) \qquad (i = 1, 2, \ldots, n).$$

Also, if G has the same degree sequences as H, then G and H are isomorphic.

(The proof is by induction on k).

We now prove a classic result due to Turán, which addresses the problem of finding the maximum number of edges in a graph of order n with no k-cliques. Although the result follows simply from Erdös' theorem, we present Turán's original proof, as it is quite interesting.

Many extensions of Turán's theorem have recently been announced, notably: Roman [1976]; Albertson, Bollobás, Tucker [1976]; Bollobás, Thomason [1981]; Bondy [1983].

Theorem 5 (Turán [1941]). *Let n and k be two integers, $n \geq k \geq 1$, and let $G_{n,k}$ be the graph consisting of k disjoint cliques of size $\left[\dfrac{n}{k}\right]$ or $\left[\dfrac{n}{k}\right]^*$ (an "equi-partition"). If G is a graph of order n with $\alpha(G) = k$, then $m(G) \geq m(G_{n,k})$, with equality if and only if G is isomorphic to $G_{n,k}$.*

Let $q = \left[\dfrac{n}{h}\right]^*$. Note that $\alpha(G_{n,k}) = k$, and that $n = k(q-1) + r, 0 < n \leqslant k$.

1. Consider the various values for n given below:

$$n = \begin{cases} k+1 & 2k+1 & 3k+1 & \ldots & qk+1 & \ldots \\ k+2 & 2k+2 & 3k+2 & \ldots & qk+2 & \ldots \\ \ldots & \ldots & \ldots & \ldots & \ldots & \ldots \\ 2k & 3k & 4k & \ldots & (q+1)k & \ldots \end{cases}$$

The proof is immediate for the values of n in the first column; we shall show by induction that if the theorem is true for the q-th column, then it is true for the $(q+1)$-st column, i.e. for graphs with order

$$n = k(q+1) + r, \qquad 0 < r \leqslant k.$$

2. Let $G = (X, E)$ be a graph of order n with $\alpha(G) \leqslant k$, and with a minimum number of edges. Hence $\alpha(G) = k$. Let $S = \{s_1, s_2, ..., s_k\}$ be a stable set with k vertices. Each vertex $x \in X - S$ is adjacent to S, since, otherwise, $\alpha(G) > k$.

Subgraph G_{X-S} has order $n - k$ and stability number $\leqslant k$; hence, by the induction hypothesis,

$$m(G_{X-S}) \geqslant m(G_{n-k,k}).$$

Since $G_{n,k}$ can be formed from $G_{n-k,k}$ by adding a vertex to each of the disjoint cliques in $G_{n-k,k}$,

$$m(G_{n,k}) - m(G_{n-k,k}) = n - k.$$

Furthermore, since $m(G) \leqslant m(G_{n,k})$ by the definition of G, it follows that

$$n - k = |X - S| \leqslant m_G(X - S, S) \leqslant$$
$$\leqslant m(G) - m(G_{X-S}) \leqslant m(G_{n,k}) - m(G_{n-k,k}) = n - k.$$

Hence,

$$\begin{cases} m(G) = m(G_{n,k}), \\ m(G_{X-S}) = m(G_{n-k,k}), \end{cases}$$

and, by the induction hypothesis,

$$G_{X-S} = G_{n-k,k}.$$

Thus, G is formed from k disjoint cliques $C_1, C_2, ..., C_k$ and one stable set S.

3. The above inequalities show also that

$$|X - S| = m_G(X - S, S).$$

Thus, each vertex $x \in X - S$ is joined to S by exactly one edge. Let $s(x)$ denote the neighbour of x in S.

If $x \in C_i$, $y \in C_j$, $i \neq j$, then $s(x) \neq s(y)$ since, otherwise, the set

$$\{x, y\} \cup (S - \{s(x)\})$$

would be a stable set with $k + 1$ vertices. If $x \in C_i$, $y \in C_i$, then $s(x) = s(y)$ since, otherwise, the number of cliques C_i would be $> |S| = k$. Thus, graph G is isomorphic to graph $G_{n,k}$.

Q.E.D.

Corollary 1. *If G is a simple graph with n vertices and m edges, and with $\alpha(G) = k$, then*

$$m \geqslant (q - 1)\left(n - \frac{kq}{2}\right)$$

where $q = \left[\dfrac{n}{k}\right]^*$. Equality holds if, and only if, G is isomorphic to graph $G_{n,k}$.

Graph $G_{n,k}$ consists of r cliques of q vertices and $(k - r)$ cliques of $q - 1$ vertices. Thus the number of edges equals

$$m(G_{n,k}) = r\,\frac{q(q - 1)}{2} + (k - r)\frac{(q - 1)(q - 2)}{2} =$$

$$= \frac{q - 1}{2}(rq + kq - rq - 2k + 2r) =$$

$$= \frac{(q - 1)(n - k + r)}{2} =$$

$$= \frac{1}{2}(q - 1)(n - k + n - k(q - 1)) =$$

$$= (q - 1)\left(n - \frac{k}{2}q\right).$$

Q.E.D.

Corollary 2. *If G is a simple graph with n vertices and m edges, then*

$$\alpha(G) \geqslant \frac{n^2}{2m + n}.$$

Equality holds if, and only if, the connected components of G are cliques with the same cardinality.

1. If $\alpha(G) = k$, $n = k(q - 1) + r$, then from Corollary 1,

$$m \geqslant \frac{1}{2}(q - 1)(n - k + r) = \frac{1}{2k}(n - r)(n - k + r).$$

It is easy to see that $(n - x)(n - k + x)$ has its minimum value in $[1, k]$ only for $x = k$. Hence

$$2km \geqslant (n - k)n.$$

or

$$k \geqslant \frac{n^2}{2m + n}.$$

2. If G consists of p cliques with cardinality n_0, then $\alpha(G) = p$, and

$$\frac{n^2}{2m + n} = \frac{p^2 n_0^2}{pn_0(n_0 - 1) + pn_0} = p = \alpha(G).$$

Therefore, equality holds in the inequality of the corollary.

Conversely, if equality holds, then $G = G_{n,k}$, and $r = k$. Thus, G has the required form.

Q.E.D.

Corollary 3. *If G is a simple graph with n vertices and m edges, then*

$$\alpha(G) \geqslant \frac{2n - m}{3}.$$

Equality holds if, and only if, each connected component of G is either a 2-clique or a 3-clique.

We may assume that G is connected (otherwise, the result could be demonstrated for each connected component).

1. If G is a 2-clique, then

$$1 = \alpha(G) \geqslant \frac{4 - 1}{3} = 1,$$

and equality holds throughout.

If G is a 3-clique, then

$$1 = \alpha(G) \geqslant \frac{6 - 3}{3} = 1,$$

and equality holds throughout.

2. Let G be a simple connected graph with n vertices and m edges. First suppose that $m \geqslant n$. From Corollary 2,

$$\alpha(G) \geqslant \frac{n^2}{2\,m + n} \;.$$

Note that the inequality

$$\frac{n^2}{2\,m + n} \geqslant \frac{2\,n - m}{3}$$

is equivalent to

$$n^2 - 3\,mn + 2\,m^2 \geqslant 0$$

or

$$(n - m)\,(n - 2\,m) \geqslant 0\;.$$

Thus, if $m \geqslant n$, we have

$$\alpha(G) \geqslant \frac{2n - m}{3}\;.$$

Equality can hold only if $m = n$ (because $m = \dfrac{n}{2}$ is not excluded since $m \geqslant n$) and if G is a clique. Thus, equality holds only if G is a 3-clique.

3. Now, suppose that $1 \leqslant m < n$; then $m = n - 1$ because G is connected. Thus, G is a tree. Since G is bipartite and $n \geqslant 2$,

$$\alpha(G) \geqslant \frac{n}{2} \geqslant \frac{n + 1}{3} = \frac{2\,n - (n - 1)}{3} = \frac{2\,n - m}{3}\;.$$

Equality holds only if

$$\alpha(G) = \frac{n}{2} = \frac{n + 1}{3},$$

i.e. if $n = 2$, i.e. if G is a 2-clique.

4. Finally, suppose that $m = 0$ and $n = 1$; then, clearly,

$$\alpha(G) > \frac{2\,n - m}{3}\;.$$

Q.E.D.

Corollary 4 (Zarankiewicz [1947]). *Let G be a simple graph with n vertices and with maximum degree h, and let $k = \left[\dfrac{n}{h + 1}\right]$. Then, $\alpha(G) \geqslant k$. Besides, if G does not consist of k disjoint cliques each with cardinality $\dfrac{n}{k}$, we have $\alpha(G) > k$.*

The number m of edges in G satisfies

$$2\,m = \sum_{x \in X} d_G(x) \leqslant hn \leqslant \left(\frac{n}{k} - 1\right) n\;.$$

Hence, from Corollary 2,

$$k \leqslant \frac{n^2}{2m + n} \leqslant \alpha(G) .$$

If G does not consist of k disjoint cliques with the same cardinality, then, from Corollary 2,

$$\alpha(G) > \frac{n^2}{2m + n} \geqslant k .$$

Q.E.D.

Remark. We shall show that Corollary 4 implies:

$$\alpha(G) \geqslant \left[\frac{n}{h + 1} \right]^* .$$

If n is a multiple of $h + 1$, let $k = \frac{n}{h + 1}$. Then,

$$\alpha(G) \geqslant \frac{n}{h + 1} .$$

If n is not a multiple of $h + 1$, let $k = \left[\frac{n}{h + 1} \right]$. Then,

$$\alpha(G) > \left[\frac{n}{h + 1} \right] .$$

In both cases,

$$\alpha(G) \geqslant \left[\frac{n}{h + 1} \right]^* .$$

Q.E.D.

3. α-Critical graphs

Call a graph G *α-critical* if every edge of G is critical. The odd cycle $C_{2p + 1}$, the complete graph K_n, and the graph in Figure 13.6 are all α-critical. Readers interested in the structure of α-critical graphs should consult Zykov [1949].

Property 1. *A graph G with $\alpha(G) = k$ has an α-critical partial graph H with $\alpha(H) = k$.*

If $\alpha(G) = k$, any graph H obtained from G by the removal of an edge satisfies $\alpha(H) = k$ or $k + 1$. Successively eliminate the edges of G whose removal does not change the stability number until no more such edges exist. The remaining graph is α-critical and is a partial graph of G.

Q.E.D.

Property 2. *In an α-critical graph G with $\alpha(G) = k$, for each vertex x, there* *is a stable set S_x with $| S_x | = k - 1$ whose union with x is a maximum stable set. Set S_x is called a "cell" of x.*

1. Let x be a vertex of G. If there exists a vertex a adjacent to x, then the removal of edge $[a, x]$ creates a stable set S with $k + 1$ vertices. Thus, a, $x \in S$, and the set $S - \{a, x\}$ is a cell of x.

2. If x has no neighbours, each maximum stable set S contains x, and therefore, $S - \{x\}$ is a cell of vertex x.

Q.E.D.

Property 3. *In an α-critical graph, two vertices a and b have a common cell if, and only if, they are adjacent.*

Let G be an α-critical graph with $\alpha(G) = k$.

1. If a and b are adjacent, the removal of edge $[a, b]$ creates a stable set S with $k + 1$ vertices. Thus, $S - \{a, b\}$ is a cell of both a and b.

2. If a and b are non-adjacent, they cannot have a common cell S because then $S \cup \{a, b\}$ would be a stable set with $k + 1$ vertices.

Q.E.D.

Property 4. *Let G be an α-critical graph with $\alpha(G) = k$. If S_0 is a stable set with $k - 1$ vertices, then the set C_0 of vertices that have S_0 as a cell is a clique; furthermore, C_0 is disjoint from S_0 and no edge goes from C_0 to S_0.*

The proof follows from Property 3.

Property 5. *In a connected, α-critical graph G of order $\geqslant 3$, each vertex has a degree $\geqslant 2$.*

Suppose that there exists a vertex a with $d_G(a) < 2$. Vertex a cannot be isolated since G is connected. Therefore, a is pendant and there is a vertex x with

$$[a, x] \in E, \qquad d_G(a) = 1 .$$

Since $n \geqslant 3$ and G is connected, edge $[a, x]$ is not an isolated edge, and therefore x has a neighbour $b \neq a$.

Let S_{bx} be a cell of both b and x. Then, $| S_{bx} | = k - 1$, and $a \notin S_{bx}$ (since a is a neighbour of x). Vertex a is not adjacent to S_{bx}, since $d_G(a) = 1$. But, then, $S_{bx} \cup \{a, b\}$ is a stable set with cardinality $k + 1$, which contradicts $\alpha(G) = k$.

Q.E.D.

Property 6. *Given two adjacent edges $[a, b]$ and $[b, x]$ in an α-critical graph, there is a maximum stable set S_0 such that*

(1) *$a, b \notin S_0$, $x \in S_0$,*

(2) *only edge $[b, x]$ joins b to S_0.*

Let S_{bx} be a cell of both b and x (which exists by Property 3), and let

$$S_0 = S_{bx} \cup \{ x \}.$$

Since $a \notin S_0$ (because a is a neighbour of b and is distinct of x), and since $b \notin S_0$, Property (1) follows. (2) is immediate.

Q.E.D.

The structure of α-critical graphs has been extensively studied. First, it has been proved by Beineke, Harary and Plummer [1967] that in an α-critical graph, any two adjacent edges lie on a common odd cycle; Andrásfai [1967] has proved that *in an α-critical graph each non-isolated edge lies on an odd cycle without chords.* The following theorem generalizes both of these results:

Theorem 6 (Berge [1970]). *In an α-critical graph G with $\alpha(G) = k$, any two adjacent edges $[a, b]$, $[b, x]$ lie on a common odd elementary cycle without chords.*

1. If edge $[b, x]$ is removed from G, a stable set S_{bx} with cardinality $k + 1$ is formed. Let $B = S_{bx} - \{ b \}$. Clearly, B is a maximum stable set in G, and $a, b \notin B$, $x \in B$. Only edge $[b, x]$ joins b to B.

2. Clearly, stable set B is not maximum in the partial graph $G - [a, b]$; therefore, by Theorem 2, there exists in $G - [a, b]$ a maximal alternating sequence

$$\sigma = (a_1, b_1, a_2, b_2, ..., a_q),$$

with $a_i \in X - B$ and $b_i \in B$ for all i. Since set $T = (B - \sigma) \cup (\sigma - B)$ is a maximum stable set in $G - [a, b]$, we have $a, b \in T$. Hence, $a, b \in \sigma - B$.

The subgraph of $G - [a, b]$ generated by σ is connected and has bicolouring $(\sigma \cap B, \sigma - B)$. Let μ be the shortest chain connecting a and b. Since $a, b \in \sigma - B$, μ together with edge $[a, b]$ form an odd elementary cycle without chords in G. This cycle contains $[a, b]$ and $[b, x]$ (since only $[b, x]$ joins b to $\sigma \cap B$).

Q.E.D.

Corollary 1. *A connected α-critical graph has no articulation vertices.*

Suppose that a is an articulation vertex. Let B and C denote the connected components obtained by removing a, and let $[a, b]$ and $[a, c]$ be two edges with $b \in B$ and $c \in C$.

No elementary cycle can pass through these two edges. This contradicts Theorem 6.

Q.E.D.

Corollary 2. *No clique of a connected α-critical graph is an articulation set.*

Suppose that A_0 is a clique that is an articulation set. We shall show that the graph is not α-critical. Let $A \subseteq A_0$ be a minimal articulation set. Then, $|A| > 1$ since, otherwise, the graph is not α-critical by Corollary 1. Let B and C be the two connected components resulting from the removal of A.

Each vertex $a \in A$ is adjacent to both B and C (otherwise, $A - \{a\}$ would be an articulation set, which contradicts the minimality of A).

Let $[a, b]$ and $[a, c]$ be two edges with $b \in B$ and $c \in C$. Each elementary cycle that contains these two edges contains a vertex $a' \in A$ distinct from a. Since A is a clique, this cycle has a chord $[a, a']$.

From Theorem 6, this contradicts the hypothesis that the graph is α-critical.

Q.E.D.

Corollary 3. *A connected α-critical graph G is either a clique, or contains an odd cycle of length $\geqslant 5$ without chords.*

If G is not a clique, then there exist two non-adjacent vertices a and b. Let $[a, x_1, x_2, \ldots, b]$ be a shortest chain connecting a and b. Clearly, $x_1 \neq a, b$. No triangle contains both $[a, x_1]$ and $[x_1, x_2]$. Thus, from Theorem 6, there is an odd cycle of length $\geqslant 5$ without chords.

Q.E.D.

The following theorem characterizes α-critical graphs G with $\alpha(G) = \theta(G)$.

Theorem 7 (Berge [1960]). *For an α-critical graph $G = (X, E)$, the following conditions are equivalent:*

(1) *the smallest number $\theta(G)$ of cliques that partition X satisfies*

$$\theta(G) = \alpha(G),$$

(2) *graph G consists of $\alpha(G)$ disjoint non-adjacent cliques;*

(3) *each cycle of length 5 in the complementary graph $\overline{G} = (X, \mathscr{P}_2(X) - E)$ has at least two chords.*

(1) $\Rightarrow$ (2) Let $\theta(G) = k$, and let $C_1, C_2, ..., C_k$ be k cliques that partition X. Suppose cliques C_1 and C_2 are adjacent and let $a_1 \in C_1$, $a_2 \in C_2$ with $[a_1, a_2] \in E$. The graph H obtained from G by removing edge $[a_1, a_2]$, satisfies

$$\alpha(H) = k + 1 .$$

Since $(C_1, C_2, ..., C_k)$ is also a partition into cliques, it follows that

$$\theta(H) \leqslant k .$$

Thus,

$$k + 1 = \alpha(H) \leqslant \theta(H) \leqslant k ,$$

which is a contradiction. Thus, condition (2) holds.

(2) $\Rightarrow$ (3) Let $\bar{\mu} = [x_1, x_2, x_3, x_4, x_5, x_1]$ be a cycle in $\bar{G}$. We have:

$$[x_1, x_2], [x_2, x_3], [x_3, x_4], [x_4, x_5], [x_5, x_1] \in \mathscr{P}_2(X) - E .$$

Let $C_1, C_2, ..., C_k$ be k disjoint non-adjacent cliques that form graph G. We shall show that $\bar{\mu}$ has at least two chords in $\bar{G}$.

If $k = 1$ or 2, cycle $\bar{\mu}$ cannot exist; therefore, we may assume that $k \geqslant 3$. Cycle $\bar{\mu}$ encounters at least three different C_i successively; for example, suppose

$$x_1 \in C_1, \qquad x_2 \in C_2, \qquad x_3 \in C_3 .$$

We may assume that either vertex x_4 or vertex x_5 belongs to C_2 Otherwise, $\bar{\mu}$ would have two chords in $\bar{G}$ and (3) would hold.

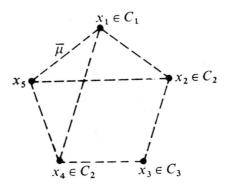

Fig. 13.3

For example, suppose $x_4 \in C_2$ (see Fig. 13.3). Then $[x_1, x_4]$ and $[x_2, x_5]$ are two chords of $\bar{\mu}$ in $\bar{G}$, and (3) holds.

(3) $\Rightarrow$ (1) First, we shall show that *if a, x, y are three distinct vertices of G with*

$$[a, x] \in E, \qquad [a, y] \in E,$$

then $[x, y] \in E$.

Suppose that $[x, y] \notin E$, and consider a common cell S_{ax} of a and x and a common cell S_{ay} of a and y. (From Property 3, these cells exist.)

Since $[a, y] \in E$, $y \notin S_{ax}$. Similarly, $x \notin S_{ay}$. Cell S_{ax} is not a cell of y, because $[x, y] \notin E$ and Property 3. Let b be a vertex of S_{ax} that is adjacent to y. Similarly, let c be a vertex of S_{ay} that is adjacent to x (see Fig. 13.4).

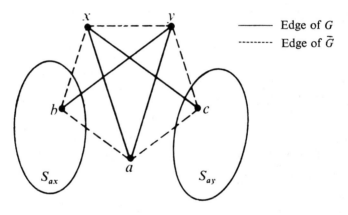

Fig. 13.4

We have $b \neq c$, because one of these two vertices is a neighbour of x and the other is not. Therefore, $[x, y, c, a, b, x]$ is an elementary cycle of $\bar{G}$ of length 5, and has at most one chord in $\bar{G}$, which contradicts (3).

Thus, the binary relation defined by "*x is adjacent to or identical to y*" is an equivalence relation. The classes of this equivalence relation are disjoint, non-adjacent cliques of G. Hence, $\alpha(G) = \theta(G)$.

Q.E.D.

Theorem 8. *If G is an α-critical graph with no isolated edge or vertex, then every stable set S satisfies $\mid \Gamma_G(S) \mid > \mid S \mid$.*

Suppose $\mid S \mid = 1$; then $S = \{ x \}$, and by Property 5, $d_G(x) \geq 2$. Thus $\mid \Gamma_G(S) \mid > \mid S \mid$.

Let $p > 1$. Suppose now that the theorem holds for all stable sets S with $\mid S \mid < p$.

Let B be a stable set with $\mid B \mid = p$ and let $b \in B$. By Property 5, there exist two distinct edges $[b, a]$ and $[b, x]$ incident to B; by Property 6 there is a maximum stable set S_0 such that $a, b \notin S_0$, and where b is joined to S_0 only by the edge $[b, x]$. Then

(1) $$\mid \Gamma_G(B) \cap S_0 \mid \geq \mid B - S_0 \mid,$$

for otherwise $[S_0 - \Gamma_G(B) \cap S_0] \cup (B - S_0)$ would be a stable set and with cardinality $> \mid S_0 \mid$, a contradiction.

CASE 1: $B \cap S_0 = \emptyset$.
This implies, by virtue of (1):

$$\mid \Gamma_G(B) \mid \geq \mid \Gamma_G(B) \cap S_0 \mid + \mid \{ a \} \mid > \mid \Gamma_G(B) \cap S_0 \mid \geq \mid B - S_0 \mid = \mid B \mid.$$

CASE 2: $B \cap S_0 \neq \emptyset$.
Here, $B \cap S_0$ is a stable set with cardinality $< p$, and by the induction hypothesis, $\mid \Gamma_G(B \cap S_0) \mid > \mid B \cap S_0 \mid$. This, together with (1), yields

$$\mid \Gamma_G(B) \mid \geq \mid \Gamma_G(B) \cap S_0 \mid + \mid \Gamma_G(B \cap S_0) \mid > \mid B - S_0 \mid + \mid B \cap S_0 \mid = \mid B \mid.$$

In both cases $\mid \Gamma_G(B) \mid > \mid B \mid$.

$$\text{Q.E.D.}$$

Corollary. *Every α-critical graph with no isolated vertex G is 'regularizable', that is, it is possible to obtain a regular multigraph by adding to G edges parallel to edges originally in G.*

Assume without loss of generality that G is connected and has more than one edge (otherwise the result if trivial). Let $A \subset X$, $S \neq \emptyset$, $A \neq X$, where S is the set of isolated vertices in the subgraph G_A. Thus

$$\Gamma_G(A) \supset (A - S) \cup \Gamma_G(S).$$

If $S = \emptyset$, then $A \neq X$ and the connectivity of G imply $\mid \Gamma_G(A) \mid > \mid A \mid$.

If $S \neq 0$, then $|\Gamma_G(S)| > |S|$ (by Theorem 8), and hence $|\Gamma_G(A)| > |A - S| + |S| = |A|$.

Consider the bipartite graph $H = (X, \bar{X}, E)$ formed by taking two copies X and $\bar{X}$ of the vertices of G, and joining $x \in X$ to $y \in \bar{X}$ if $[x, y]$ is an edge of G. Then

(1) $|\Gamma_H(A)| = |\Gamma_G(A)| \geqslant |A| + 1$ $(A \subset X, A \neq \varnothing, A \neq X)$.

We now show that H is regularizable. It suffices to show that for any edge $[a, \bar{b}]$ of H, the subgraph H' induced by $X \cup \bar{X} - \{a, \bar{b}\}$ has a perfect matching $F_{a,b}$ (otherwise $F_{a,b} \cap [a, \bar{b}]$ would form a regular multigraph). For $A \subset X - \{a\}$, $A \neq \varnothing$, (1) implies

$$|\Gamma_{H'}(A)| = |\Gamma_H(A) - \{\bar{b}\}| \geqslant |A| + 1 - 1 = |A|$$

Now, by Theorem 5 (Ch. 7) the matching $F_{a,b}$ exists and H is regularizable by multiplying each edge $[a, \bar{b}]$ by $k(a, \bar{b})$. Hence G is also regularizable, by multiplying each edge $[a, b]$ by $k(a, \bar{b}) + k(b, \bar{a})$.

Q.E.D.

(In fact, the condition of Theorem 8 is equivalent to: "G is regularizable and has no bipartite connected component.")

Lemma. *Consider a graph G with no stable set of cardinality $k + 1$. If $S_1, S_2, ..., S_p$ are stable sets with k elements, then*

$$\left| \bigcup_{i=1}^{p} S_i \right| + \left| \bigcap_{i=1}^{p} S_i \right| \geqslant 2k.$$

If $p = 1$, the result is evident. If the result is valid for $p - 1$ stable sets, we shall show that it is also valid for p stable sets $S_1, S_2, ..., S_p$.

For $k = 1, 2, ..., p$, let

$$A_k = \bigcup_{i=1}^{k} S_i$$

$$B_k = \bigcap_{i=1}^{k} S_i.$$

Clearly, set B_{p-1} is non-adjacent to set A_{p-1}; thus, no vertex of set $(A_{p-1} \cap S_p) \cup B_{p-1}$ is adjacent to another vertex of the set and, consequently, the set is stable. Hence

(1) $\left| (A_{p-1} \cap S_p) \cup B_{p-1} \right| \leqslant \alpha(G) = k.$

Furthermore,

(2) $| A_{p-1} \cap S_p | = | S_p | - | S_p - A_{p-1} | =$

$$= | S_p | - | A_p - A_{p-1} | = k - | A_p | + | A_{p-1} | .$$

By comparing equations (1) and (2), we have

$$\left| B_{p-1} - (A_{p-1} \cap S_p) \right| = \left| B_{p-1} \cup (A_{p-1} \cap S_p) \right| - | A_{p-1} \cap S_p | \leqslant$$

$$\leqslant k - (k - | A_p | + | A_{p-1} |) = | A_p | - | A_{p-1} | .$$

Since the induction hypothesis yields that $| A_{p-1} | + | B_{p-1} | \geqslant 2 k$,

$$| B_p | = | B_{p-1} \cap S_p | = | B_{p-1} | - | B_{p-1} - S_p | \geqslant$$

$$\geqslant | B_{p-1} | - \left| B_{p-1} - (A_{p-1} \cap S_p) \right| \geqslant$$

$$\geqslant (2 k - | A_{p-1} |) - (| A_p | - | A_{p-1} |) = 2 k - | A_p | .$$

$$\text{Q.E.D.}$$

Theorem 9 (Hajnal [1965]). *Let G be an α-critical graph without isolated vertices with $\alpha(G) = k$ and $| X | = n$. Then*

$$d_G(x) \leqslant n - 2 k + 1 \qquad (x \in X) .$$

Suppose there is a vertex a with $| \Gamma_G(a) | > n - 2 k + 1$. We shall show that there exists a stable set S_0 with $| \Gamma_G(S_0) | < | S_0 |$.

Let $S_1, S_2, ..., S_p$ be the stable sets with cardinality k that contain vertex a. If we let $A = \bigcup S_i$, $B = \bigcap S_j$, then, from the lemma,

$$| A | + | B | \geqslant 2 k .$$

Furthermore, each vertex $x \in A$ is non-adjacent to set $B - \{ a \}$. Besides, if $y \in \Gamma_G(a)$, there exists a common cell S_{ay} of a and y (from Property 3). From the definition of B,

$$S_{ay} \cup \{ a \} \supset B .$$

Thus, $S_{ay} \cup \{ y \}$ is a stable set that contains $B - \{ a \}$, and again, y is non-adjacent to set $B - \{ a \}$. Finally, by letting $S_0 = B - \{ a \}$,

$$x \in A \cup \Gamma_G(a) \qquad \Rightarrow \qquad x \notin \Gamma_G(S_0)$$

and so

$$\Gamma_G(S_0) \subset X - A - \Gamma_G(a) .$$

Hence,

$$| \Gamma_G(S_0) | \leqslant n - | A | - | \Gamma_G(a) | < n - (2k - | B |) - (n - 2k + 1) =$$
$$= | B | - 1 = | S_0 | .$$

Thus, set S_0 is stable and satisfies $| \Gamma_G(S_0) | < | S_0 |$, which contradicts Theorem 8.

Q.E.D.

Corollary 1. *Let G be an α-critical graph without isolated vertices. Then $\alpha(G) \leqslant \dfrac{n}{2}$. Equality holds if, and only if, each connected component of G is a 2-clique.*

Let $k = \alpha(G)$; then each vertex x satisfies

$$1 \leqslant d_G(x) \leqslant n - 2k + 1 .$$

Hence,

$$n - 2k \geqslant 0 .$$

Equality can hold only if each vertex has degree 1, i.e., if all edges of G are isolated edges.

Q.E.D.

Corollary 2 (Erdös, Gallai [1961]). *An α-critical graph G with $\alpha(G) = k$ and $| X | = n$ contains at least $2k - n$ isolated vertices.*

If G has n isolated vertices, then the result follows because $k = \alpha(G) = n$. If G has $p > n$ isolated vertices, the graph G' obtained from G by removing the isolated vertices satisfies

$$n' = n - p ,$$
$$\alpha(G') = k' = k - p .$$

From Corollary 1, graph G', which is α-critical and without isolated vertices, satisfies

$$0 \leqslant n' - 2k' = (n - p) - 2(k - p) .$$

Thus $p \geqslant 2k - n$.

Q.E.D.

Remark. The previous results can be generalized in the following manner. Let $\mathscr{A}(G)$ represent the set of graphs that can be obtained by replacing each edge of the simple graph G by an elementary chain of odd length. For G a simple graph of order n and stability number $\alpha(G)$, let $\delta(G) = n - 2\alpha(G)$.

Corollary 1 can be restated: *G is α-critical connected with $\delta(G) = 0$ if and only if $G = K_2$.*

This follows immediately from Theorem 9: *G is α-critical connected with*
$\delta(G) = 1$ *if and only if* $G \in \mathcal{A}(K_3)$, *i.e. G is an odd cycle.*

Andrasfai [1966] proved: *G is α-critical connected with* $\delta(G) = 2$ *if and only
if* $G \in \mathcal{A}(K_4)$.

Lovasz [1976] proved: *G is α-critical connected with* $\delta(G) = 3$ *if and only if*
$G \in \mathcal{A}(K_5)$, *or G is the graph with vertices a, b, c, d, e, f, g and edges ab, ac,
af, ag, bc, fg, bd, cd, fe, ge, de* (cf. Fig. 13.6). *More generally, α-critical
connected graphs with* $\delta(G) = \delta$ *are graphs of* $\mathcal{A}(K_{\delta + 2})$, *together with a finite
number of additional graphs.*

Theorem 10 (Erdös, Hajnal, Moon [1964]). *If G is an α-critical graph with
n vertices and m edges, and with* $\alpha(G) = k$, *then*

$$m \leqslant \binom{n - k + 1}{2}.$$

*Equality holds if, and only if, G consists of a stable set with $k - 1$ vertices
and a clique with $n - k + 1$ vertices.*

Let $f_k(n)$ denote the maximum possible number of edges in a graph G
with the properties:

(1) G has at least one stable set with cardinality k,
(2) the removal of any edge creates a new stable set with cardinality $k + 1$,
(3) G has order n.

1. First, we shall show that

$$f_k(n) \leqslant \binom{n - k + 1}{2}.$$

Note that

$$f_k(k + 1) = 1$$

because each graph satisfying (1), (2) and (3) with $k + 1$ vertices and with a
maximum number of edges consists of a 2-clique and a stable set of $k - 1$
vertices.

Consider a graph G that satisfies (1), (2) with order $n > k + 1$ and with
$f_k(n)$ edges. Let a and b be two adjacent vertices. From (2), the removal
of edge $[a, b]$ creates a stable set S with cardinality $k + 1$, and $a \in S$, $b \in S$.

Consider the graph G' obtained from G by removing vertex b and each
edge $[a, z]$ with $z \notin \Gamma_G(b)$. For G', set $S - \{b\}$ is stable with cardinality k.
The removal of an edge of G' creates a stable set with cardinality
$k + 1$, since this creates in G a stable set with cardinality $k + 1$ that either
contains a and not b or contains neither a nor b. Hence the number of edges

in G' is

$$m(G') \leqslant f_k(n - 1) .$$

Furthermore, the number of edges removed from G to form G' is

$$m(G) - m(G') \leqslant 1 + | X - S | = 1 + n - (k + 1) = n - k .$$

Hence

$$f_k(n) = m(G) \leqslant m(G') + n - k \leqslant f_k(n - 1) + (n - k) .$$

Thus,

$$\begin{cases} f_k(n) \leqslant f_k(n - 1) + (n - k) \\ f_k(n - 1) \leqslant f_k(n - 2) + (n - 1 - k) \\ \cdots \\ f_k(k + 1) = 1 . \end{cases}$$

Hence,

$$f_k(n) \leqslant 1 + 2 + \cdots + (n - k) = \frac{(n - k)(n - k + 1)}{2} = \binom{n - k + 1}{2} .$$

2. Denote by $G_k(n)$ a graph consisting of the union of a $(n - k + 1)$-clique and a stable set of $k - 1$ vertices. Clearly, graph $G_k(n)$ satisfies (1), (2) and (3), and has $\binom{n - k + 1}{2}$ edges.

Thus

$$f_k(n) = \binom{n - k + 1}{2} ,$$

and $G_k(n)$ is a graph with the maximum possible number of edges.

3. We shall show that each graph of order n with $f_k(n)$ edges that satisfies (1) and (2) is isomorphic to $G_k(n)$. Clearly, this is true for $n = k + 1$. By induction, we shall show that it is also true for a graph G of order $n > k + 1$.

Consider an edge $[a, b]$ of G, and construct a graph G' by removing vertex b and each edge $[a, z]$ with $z \notin \Gamma_G(b)$. As before, $m(G) - m(G') \leqslant n - k$. Hence

$$m(G') \geqslant f_k(n) - (n - k) = f_k(n - 1) .$$

From Part 1, G' satisfies conditions (1) and (2) since G has order $n - 1$ and has the maximum possible number of edges, it follows from the induction hypothesis that

$$G' \equiv G_k(n - 1) \equiv K_{n-k} \cup S_{k-1} ,$$

where K_{n-k} is a clique with $n - k$ vertices and S_{k-1} is a stable set with cardinality $k - 1$. We shall consider two cases.

CASE 1: $a \in S_{k-1}$ (see Fig. 13.5).
Since the removal of edge $[a, b]$ creates in G a stable set with cardinality $k + 1$, there exists a vertex y with

$$y \in K_{n-k}, \quad y \notin \Gamma_G(b), \quad \Gamma_G(y) \cap S_{k-1} = \varnothing.$$

Furthermore, if $c \in K_{n-k}$ and $c \neq y$, then c is not adjacent to a because, otherwise, a stable set with cardinality $k + 1$ could not be created in G by removing edge $[a, c]$. Furthermore, c is adjacent to b, since the number of edges removed from G to form G' is equal to

$$f_k(n) - f_k(n - 1) = n - k.$$

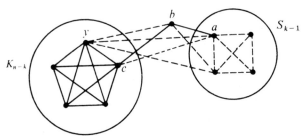

Fig. 13.5

Then, if edge $[b, c]$ is removed from G, a stable set with cardinality $k + 1$ cannot be created, which contradicts the hypothesis. Thus, Case 1 cannot occur.

CASE 2: $a \in K_{n-k}$.
Vertex b cannot be adjacent to set S_{k-1} in G (because the removal of $[a, b]$ would create a stable set with cardinality $k + 1$). On the other hand, b is adjacent to each vertex of K_{n-k} (because the number of edges removed from G to form G' equals $n - k$). Thus, G is the union of an $(n - k + 1)$-clique, $K_{n-k} \cup \{b\}$, and a stable set S_{k-1}.

$$\text{Q.E.D.}$$

4. The Gallai–Milgram theorem and related results

Let $G = (X, U)$ be a 1-graph. Gallai and Milgram showed that the least number of (elementary) paths that partition X is $\leqslant \alpha(G)$, and that this bound is the best possible, as equality occurs if all arcs incident with a stable set S_0 are

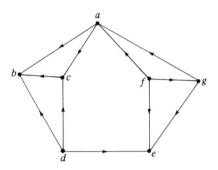

Fig. 13.6

directed out. For example, with the 1-graph in Figure 13.6, $\alpha(G) = 2$, and two paths are necessary to partition the vertices, namely (d, e) and (f, g, a, c, b).

In this section we generalize the preceding result. Recall that an *arborescence forest* H of G is a partial 1-graph of G each of whose components is an arborescence. Denote by $R(H)$ the set of roots of these arborescences and by $S(H)$ the set of terminal vertices. Thus, each isolated vertex of H belongs to $R(H) \cap S(H)$. Call a *terminal branch* of H a maximal path of H with terminal vertex $s \in S(H)$, all of whose vertices $z \neq s$ satisfy $d_H^+(z) = 1$.

Theorem 11. *Let H_0 be an arborescence forest, and let $R(H_0) = R_0$ and $S(H_0) = S_0$. Then for every arborescence forest H with $R(H) \subset R_0$, $S(H) \subset S_0$ and $|S(H)|$ minimum, there exists a stable set S which intersects every terminal branch of H.*

The proof is by induction. Let $G = (X, U)$ be a graph of order n, and assume that the result holds for all 1-graphs with order $< n$. Let H be an arborescence forest with $R(H) \subset R_0$, $S(H) \subset S_0$ and $|S(H)|$ minimum.

If $S(H)$ is a stable set, we are done. If not, there exist two adjacent vertices a, b in $S(H)$ such that $(b, a) \in U$. Then $a \notin R(H)$ (else $H' = H + (b, a)$ contradicts the minimality of H). Thus H contains an arc $(a_1, a) \in U$.

Also, $d_H^+(a_1) = 1$, for otherwise H would have at least two descendants of a_1 in $S(H)$, and $H' = H - (a_1, a) + (b, a)$ would contradict the minimality of H.

The subgraph $\bar{G}$ of G induced by $X - \{a\} = \bar{X}$ has an arborescent forest $\bar{H} = H_{\bar{X}}$ with $R(\bar{H}) \subset R_0$ and $S(\bar{H}) \subset (S_0 - \{a\}) \cup \{a_1\}$. We now show that $|S(\bar{H})|$ is minimum with these conditions.

Suppose (proof by contradiction) that there exists a forest $\bar{H}'$ with the same conditions and with $| S(\bar{H}') | < | S(\bar{H}) |$.

CASE 1: $a_1 \in S(\bar{H}')$. Then $\bar{H}'' = \bar{H}' + (a_1, a)$ is a forest of G with $R(\bar{H}'') \subset R_0$, $S(\bar{H}'') \subset S_0$ and $| S(\bar{H}'') | = | S(\bar{H}') | < | S(H) |$, which contradicts the minimality of H.

CASE 2: $a_1 \notin S(\bar{H}')$, $b \in S(\bar{H}')$. In this case, $\bar{H}'' = \bar{H}' + (b, a)$ is a forest that contradicts the minimality of H.

CASE 3: $a_1 \notin S(\bar{H}')$, $b \notin S(\bar{H}')$. In this case, $| S(\bar{H}') | \leq | S(\bar{H}) | - 2$; thus $\bar{H}'' = \bar{H}' + (a_1, a)$ is a forest of G with

$$| S(\bar{H}'') | = | S(\bar{H}') | + 1 \leq | S(\bar{H}) | - 1 = | S(H) | - 1,$$

contradicting the minimality of H.

Thus, $\bar{H}$ is a minimal forest of $\bar{G}$, and by the induction hypothesis there exists in $\bar{G}$ a stable set $\bar{S}$ meeting every terminal brench of $\bar{H}$. Thus, $\bar{S}$ meets every terminal branch of H.

Q.E.D.

Corollary 1. (Las Vergnas [1976]). *A quasi-strongly connected 1-graph G has a partial graph that is an arborescence with at most $\alpha(G)$ terminal vertices.*

Corollary 2 (Gallai–Milgram theorem [1960]). *If $M = \{ \mu_1, \mu_2, \ldots \}$ is a path partition of the vertices of G, then* min $| M | \leq \alpha(G)$.

In fact, a forest H satisfying the conditions of Theorem 11 can be decomposed into $| S(H) |$ disjoint paths successively obtained by removing from H the longest path with a terminal vertex in $S(H)$.

Corollary 3 (Linial [1978]). *There exists a path partition M and a stable set S such that S meets every path in M.*

(Same proof).

Corollary 4 (Rédei's theorem). *If G is a complete anti-symmetric 1-graph ('tournament'), then G has a path which meets every vertex exactly once ('hamiltonian path').*

In fact, $\alpha(G) = 1$.

Corollary 5 (Dilworth's theorem). *Let $G = (X, U)$ be a transitive 1-graph, i.e.*

$$(x, y) \in U, (y, z) \in U \quad \Rightarrow \quad (x, z) \in U.$$

If M is a partition of X into elementary paths, then $\min |M| = \alpha(G)$.

In fact, if G is transitive, then every path induces a clique; also, by Corollary 3, every clique induces a path. Thus

$$\alpha(G) \leq \theta(G) = \min |M| \leq \alpha(G)$$

(by the Gallai–Milgram theorem). Hence, $\min |M| = \alpha(G)$.

Corollary 6 (Camion's theorem [1959]). *Every strongly connected tournament has a hamiltonian circuit.*

Let μ be a longest circuit of G. If μ is not hamiltonian, there exists a path μ' with origin $z \in \mu$ and terminus $a \in \mu$ whose interior vertices are not in μ. The 1-graph obtained from $\mu + \mu'$ by removing arcs $(y, a) \in \mu$ and $(y', a) \in \mu'$ is an arborescence H with $R(H) = \{ a \}$, $S(H) = \{ y, y' \}$. Since the subgraph of G induced by $\mu + \mu'$ has stability number 1, it can be covered by an arborescence H' with $R(H') = \{ a \}$ and with the single terminal vertex $s \in \{ y, y' \}$.

A hamiltonian circuit is then created by adding the arc (s, a) to H'.

Q.E.D.

The following analogous statements are conjectured for a strongly connected 1-graph G with $\alpha(G) > 1$.

Conjecture 1 (Bermond). *The vertices of a strongly connected graph G can be covered with $\alpha(G)$ elementary circuits or singletons.*

Conjecture 2 (Las Vergnas). *A strongly connected graph G has a partial graph H that is an arborescence with $|S(H)| \leq \alpha(G) - 1$.*

For the case $\alpha(G) = 2$, these conjectures have been proved by Chen and Manalastas [1982]. It has also been proved by Berge [1983] for those 1-graphs that, like symmetric 1-graphs, have a circuit meeting all maximum stable sets.

Conjecture 3 (Berge). *In every 1-graph G there exists a maximum stable set S with $|S| = \alpha(G)$, and a path partition $\mu_1, \mu_2, \ldots, \mu_{\alpha(G)}$ with*

$$|\mu_i \cap S| = 1 \qquad (i = 1, 2, \ldots, \alpha(G)).$$

A graph G is called α-*diperfect* if every subgraph G_A has the following property (P): *for every maximum stable set there is a path partition of the vertices, each path meeting the stable set in one point.*

EXAMPLE 1. The 1-graph in Figure 13.6 does not have property (P); hence, it is not α-diperfect.

EXAMPLE 2. Every symmetric 1-graph is α-diperfect. Let S be a maximum stable set of G, and let G' be the partial graph obtained by removing arcs directed into S. Then G' can be covered with $\alpha(G') = \alpha(G)$ disjoint paths, which necessarily meet S in one point, satisfying (P).

EXAMPLE 3. Call a 1-graph G 'perfect' if $\alpha(G_A) = \theta(G_A)$ for every $A \subset X$. Then G is then also α-diperfect, for a maximum stable set of G meets all cliques of G_A forming an optimal partition, and each of these cliques is spanned by a path (by Corollary 3).

The characterization of α-diperfect graphs leads to the following conjecture (Berge [1981]):

Conjecture 4. *A 1-graph is α-diperfect if and only if it contains as induced subgraph no chordless odd cycle $\mu = (x_1, x_2, x_3, \ldots)$ of length $\geqslant 5$ with the following property: the longest path in μ has length 2, and for each vertex x in $\{ x_1, x_2, x_3, x_4, x_6, x_8, x_{10}, \ldots \}$, the two arcs of μ incident to x are directed either both towards x or both away from x.*

The reader can verify that for such a cycle, the maximum stable set $S = \{ x_1, x_4, x_6, x_8, \ldots, x_{2k} \}$ does not meet all the paths of an optimal partition M with $| M | = \alpha(G) = k$.

Theorem 11 shows that *the only critical vertices in an α-critical graph are the isolated vertices*. The following theorem applies this result.

Theorem 13. *If a graph G with order n has no critical vertices, then $\alpha(G) \leqslant \dfrac{n}{2}$. Furthermore, if equality holds, then*

$$\alpha(G) = \theta(G) = \frac{n}{2}.$$

Suppose $\alpha(G) \geqslant \dfrac{n}{2}$; we shall show that $\alpha(G) = \theta(G) = \dfrac{n}{2}$.

Since G has no critical vertices, each vertex of G belongs to the complement of a maximum stable set. After removing an edge without changing the stability number, this remains true a fortiori.

Suppose that enough edges are removed to form an α-critical graph H with $\alpha(H) = \alpha(G)$. Then, graph H has no critical vertices (and no isolated vertices) and satisfies $\alpha(H) \geqslant \dfrac{n}{2}$.

From Corollary 1 to Theorem 9, $\alpha(H) = \dfrac{n}{2}$, and H consists of $\dfrac{n}{2}$ pairwise disjoint edges. Hence,

$$\frac{n}{2} = \alpha(H) = \alpha(G) \leqslant \theta(G) \leqslant \theta(H) = \frac{n}{2} .$$

Therefore

$$\alpha(G) = \theta(G) = \frac{n}{2} .$$

Q.E.D.

Corollary. *If, for each vertex* x, *the subgraph* $G_{X - \{x\}}$ *contains a stable set* S *with* $|S| > \dfrac{n}{2}$, *then* G *contains a stable set* S_0 *with*

$$|S_0| > \frac{n}{2} + 1 .$$

From Theorem 13, G has at least one critical vertex a. Hence

$$\alpha(G) = \alpha(G_{X - \{a\}}) + 1 > \frac{n}{2} + 1 .$$

Q.E.D.

5. Stability number and vertex-coverings by paths

In this section, we shall assume that G is a 1-graph.

Theorem 14 (Gallai, Milgram [1960]). *In a 1-graph* $G = (X, U)$, *there exist* $\alpha(G)$ *elementary paths that partition* X.

Let $M = \{ \mu_1, \mu_2, ..., \mu_k \}$ be a family of elementary pairwise disjoint paths that cover X. Let $A(M) = \{ a_1, a_2, ..., a_k \}$ be the set of initial vertices of the paths in M. A family M always exists since all paths of length 0 can be chosen.

We shall show that *there is a family* M' *of elementary paths that partition* X *such that* $|M'| \leqslant \alpha(G)$ *and* $A(M') \subset A(M)$. (This proposition is in fact stronger than the theorem.)

1. Clearly, this proposition is true for a graph with one vertex. If it is valid for all graphs with less than n vertices, we shall show that it is also valid for a graph G with n vertices.

2. If $|M| > \alpha(G) + 1$, we shall show that M can be replaced by a family $\overline{M}$ with $|\overline{M}| \leqslant \alpha(G) + 1$ and $A(\overline{M}) \subset A(M)$. Consider the subgraph generated by $X_1 = X - \mu_1$. The stability number of this subgraph is

$$\alpha(G_{X_1}) \leqslant \alpha(G) ,$$

and $M_1 = \{ \mu_2, ..., \mu_k \}$ partitions the vertex set of G_{X_1}. Hence, from the induction hypothesis, there exists a family $\overline{M}_1$ in G_{X_1} such that

$$|\overline{M}_1| \leqslant \alpha(G_{X_1}) \leqslant \alpha(G), \qquad A(M_1) \subset A(M_1).$$

Thus, family $\overline{M} = \overline{M}_1 \cup \{\mu_1\}$ is the required family of G.

3. Clearly, if $|\overline{M}| \leqslant \alpha(G)$, the proposition holds.

If $|\overline{M}| = \alpha(G) + 1$, set $A(\overline{M}) = \{a_1, a_2, ...\}$ is not stable, because $|A(\overline{M})| > \alpha(G)$. Thus, there is an arc joining two of its vertices; let this arc be (a_1, a_2).

If the path $\bar{\mu}_1 \in \overline{M}$ which begins at a_1 has zero length, then the proposition holds. Otherwise, let

$$\bar{\mu}_1 = [a_1, b_1, c_1, ...].$$

The subgraph generated by $X_1 = X - \{a\}$ possesses a family M_1 with

$$|M_1| \leqslant \alpha(G), A(M_1) \subset \{b_1, a_2, a_3 ...\}.$$

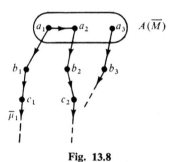

Fig. 13.8

APPLICATION ("Sperner's theorem"). *If X is a set of n elements, and if $\mathcal{F}$ is a family of distinct subsets of X such that no member of $\mathcal{F}$ is contained in another member of $\mathcal{F}$, then*

$$|\mathcal{F}| \leqslant \left(\begin{array}{c} n \\ \left[\dfrac{n}{2}\right] \end{array} \right).$$

This famous result has recently been extended in several directions (Erdös and Katona, Kleitman, Meshalkin, etc). We shall show that the Spencer theorem is an easy consequence of Corollary 2.

Consider the graph G whose vertices represent the different subsets of X. Place the vertices of G into $n + 1$ rows $0, 1, 2, ..., n$ where the h-th row contains the vertices corresponding to the subsets with $n - h$ elements. There are $\binom{n}{h}$ such subsets. Join vertex a to vertex b by an arc (a, b) if the corresponding sets A and B satisfy $A \supset B$.

The maximum cardinality of a family $\mathscr{F}$ with the required property equals the stability number $\alpha(G)$. Since the set of elements in a row is stable, there exists a stable set with $\left(\begin{array}{c} n \\ \left[\frac{n}{2} \right] \end{array} \right)$ members in row $h = \left[\frac{n}{2} \right]$.

From the Dilworth theorem, this stable set is maximum if and only if the vertex set of G can be partitioned into $\left(\begin{array}{c} n \\ \left[\frac{n}{2} \right] \end{array} \right)$ paths.

To show that this is possible, notice that, given two consecutive rows, the smaller can always be matched into the larger: Since two consecutive rows form a bipartite graph in which all vertices of the same class have the same degree, this follows from Corollary 4 to Theorem (5, Ch. 7). For all h, let E_h be a maximum matching between the rows h and $h + 1$. Then the union of the E_h define the edges of the required paths.

$$\text{Q.E.D.}$$

EXERCISES

1. Using the Turán theorem, show that the minimum number of edges in a graph G with n vertices and stability number k equals

$$\sum_{i=0}^{n-1} \left[\frac{i}{k} \right].$$

(Las Vergnas)

2. If $n \geqslant 4(p + 1)$, and if each subgraph G_A with $|A| = 2p + 1$ satisfies $|A| - \alpha(G_A) \leqslant p$, then show that $m \leqslant \left[\frac{n^2}{4} \right]$.   (Erdös, Gallai [1961])

3. Consider a simple graph $G = (X, E)$ and a partition $(X_1, X_2, ..., X_p)$ of X such that

$$\alpha(G) = \sum_{i=1}^{p} \alpha(G_{X_i}).$$

Show that $\alpha(G) = \theta(G)$ if

$$\alpha(G_{X_i}) = \theta(G_{X_i}) \qquad (i = 1, 2, ..., p).$$

4. If the partition $(X_1, X_2, ..., X_p)$ above has more than one class, then G is said to be *decomposable*. Show that a connected α-critical graph is not decomposable.

(Plummer)

5. For a non-decomposable graph G that is not a clique, show that $\alpha(G) < \theta(G)$.

6. Let $G = (X, E)$ be a connected graph, and let E_0 be a maximum matching. Suppose that E_0 is not a perfect matching. Let X_1 denote the set of vertices that can be reached from an unsaturated vertex by an even alternating chain (and by no odd alternating chain). Let X_2 denote the set of vertices that can be reached from an unsaturated vertex by an odd alternating chain (and by no even alternating chain). Show that if $X_1 \cup X_2 = X$, then X_1 is a maximum stable set of graph G.   (Berge [1957])

7. If $A \subset X$ is a set of vertices in G, let $S(A)$ denote a maximum stable set of the subgraph generated by A. Given a maximum matching E_0, let M_1, M_2, ..., denote the connected components of the subgraph generated by the "mixed" vertices. Let I_1, I_2, ..., denote the connected components of the subgraph generated by the "inaccessible" vertices. Let

$$S = X_1 \cup S(M_1) \cup S(M_2) \cup \cdots \cup S(I_1) \cup S(I_2) \cup \cdots$$

If there is at most one connected component of mixed vertices, show that S is a maximum stable set of G.   (Berge [1957])

8. Show that in a graph G without critical vertices, $|\Gamma_G(S)| \geq |S|$ for every stable set S.

9. Recall that if G is an α-critical graph without isolated vertices then G has no critical vertices. Show that the converse is not true by considering the following graphs:
   (1) $\overline{C}_7$, the complementary graph of a cycle of length 7 without chords,
   (2) C_5 augmented with a vertex that is joined to each vertex of C_5.

10. If G is a graph with n vertices and m edges with $m > \left[\dfrac{n^2}{4}\right]$, show that G contains a triangle.

Show that there exists a graph of order n with $\left[\dfrac{n^2}{4}\right]$ edges that contains no triangles.

11. Using Corollary 1 of Theorem 3, show that if G is a graph of order n odd with $\alpha(G) = \dfrac{n-1}{2}$ and $\alpha(G)_{X-\{x,y\}} = \alpha(G)$ for all $x, y \in X$, then G is a chordless odd cycle.

(Melknikov and Vizing [1971])

12. Show that through any pair of adjacent critical edges there passes a chordless odd cycle.

(George [1971])

13. Let G be a simple triangle-free graph of order n and maximum degree Δ. Show that $m(G) \leq \Delta(n - \Delta)$.

14. Let $T_{n,k}$ be the number of edges in the complement of the graph $G_{n,k}$ defined in Theorem 5 (Turán's graph). Show directly that for a graph of order n with $m(G) > T_{n,k}$, every vertex x_0 of maximum degree belongs to a $(k + 1)$-clique. (It suffices to count the number of edges in the subgraph induced by $\Gamma_G(x_0)$).

(Bondy [1983])

15. Using the Gallai–Milgram theorem, prove the following result, due to Erdös–Szekeres: let $\sigma = (a_1, a_2, ..., a_n)$ be a sequence of n distinct real numbers, and let p, q be integers with $pq < n$. Then either there exists an increasing subsequence σ' of σ with $|\sigma'| > p$, or else there exists a decreasing subsequence σ'' of σ with $|\sigma''| > q$. (Consider the graph G whose vertices are a_i, with the arc (a_i, a_j) in G if $a_i < a_j$ and $i < j$).

16. Show that if G does not consist of disjoint cliques then

$$\alpha(G) \geq \frac{n^3 + 2n + 1}{n(2m + n)}$$

(Koh Khee-Ming [1976])

CHAPTER 14

Kernels and Grundy Functions

1. Absorption number

For a 1-graph $G = (X, \Gamma)$, a set $A \subset X$ is defined to be *absorbant* if for each $x \notin A$,

$$\Gamma(x) \cap A \neq \emptyset .$$

Let $\mathscr{A}$ denote the family of all absorbant sets of graph G. Then

$$X \in \mathscr{A},$$
$$A \in \mathscr{A}, \quad A' \supset A \quad \Rightarrow \quad A' \in \mathscr{A} .$$

The *absorption number* of graph G is defined to be

$$\beta(G) = \min_{A \in \mathscr{A}} | A | .$$

This section studies the absorbant sets with minimal cardinality.

EXAMPLE 1. *Radar stations.* A number of strategic locations $x_1, x_2, \ldots$ (called cells) are kept under the surveillance of radar. Radar in cell x_4 can survey x_1 or x_2 or x_3, as shown in Fig. 14.1. Similarly, cell x_2 can be surveyed by radar located in x_3 or x_5, etc.

What is the minimum number of radar stations needed to survey all the cells? It is the absorption number of the graph in Fig. 14.1, which is equal to 2, since radar stations placed at x_2 and x_4 are sufficient.

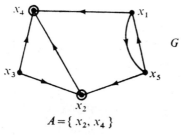

$$A = \{ x_2, x_4 \}$$

Fig. 14.1

303

EXAMPLE 2. *Chessboard controlled by five queens.* What is the minimum number of queens that can be placed on a standard chessboard so that each square is controlled by at least one queen? Five queens are sufficient because the absorption number $\beta(G)$ of the graph G, defined by the moves of a queen on a chessboard, is 5 (see Fig. 14.2).

Note that the same placement of the 5 queens also controls a 9×9 chessboard. The placement at the right in Fig. 14.2 shows how five queens can control an 11×11 chessboard. No general results are available for the maximum size $n(k)$ of a chessboard that can be controlled by k queens.

Fig. 14.2

Finding a minimum absorbant set is a special case of the minimum transversal problem, for which constructive algorithms are given in Chapter 18. Here, we shall give bounds on the absorption number.

Proposition 1. *If G is a 1-graph with n vertices and m arcs, then*

$$\beta(G) \geqslant n - m .$$

Let A be a minimum absorbant set. Each vertex of $X - A$ is the initial endpoint of an arc going into A. Thus,

$$n - |A| = |X - A| \leqslant m .$$

Hence,

$$\beta(G) = |A| \geqslant n - m .$$

Q.E.D.

Proposition 2. *If G is a 1-graph without loops with n vertices, then*

$$\beta(G) \leqslant n - \max_{x \in X} d_G^-(x) .$$

Let x_0 be a vertex with $d_G^-(x_0) = \max_{x \in X} d_G^-(x)$. Since the set

$$A = X - \Gamma_G^-(x_0)$$

is absorbant, we have

$$\beta(G) \leqslant |A| = n - \max_{x \in X} d_G^-(x) .$$

$$\text{Q.E.D.}$$

If G is a simple graph, its *dominance number* $\beta^*(G)$ is defined to be the absorption number $\beta(G^*)$ of the symmetric graph G^* obtained by replacing each edge in G by two oppositely directed arcs. As in the Turán theorem, we shall calculate the maximum number of edges possible in a simple graph with a given order and dominance number.

Theorem 1 (Vizing [1965]). *If G is a simple graph with n vertices, m edges and $\beta^*(G) = k \geqslant 2$, then*

$$m \leqslant \left[\frac{1}{2}(n - k)(n - k + 2) \right] .$$

For each n and each k, equality holds if and only if G is isomorphic to a graph $G_{n,k}$ obtained by removing from a $(n - k)$-clique the edges of a minimum covering and by joining its vertices to each vertex in a stable set of cardinality k.

1. First, we shall show that

$$m \leqslant \frac{1}{2}(n - k)(n - k + 2) .$$

Clearly, this inequality is satisfied for $n = 2$. We shall assume that it is satisfied for all graphs with order $< n$, and we shall show that it is satisfied for a graph G with order $n > 2$ and with $\beta^*(G) = k > 1$. Let G^* be the symmetric graph obtained by replacing each edge of G by two oppositely directed arcs. Let x_0 be a vertex with maximum degree. From Proposition 2, we have

$$\left| \Gamma_G(x_0) \right| = d_{G^*}^-(x_0) = \max_x d_{G^*}^-(x) \leqslant n - k .$$

Thus, we may write:

$$\left| \Gamma_G(x_0) \right| = n - k - r ; \qquad 0 \leqslant r \leqslant n - k .$$

Let $S = X - \{ x_0 \} - \Gamma_G(x_0)$ then

$$|S| = k + r - 1 .$$

If $y \in \Gamma_G(x_0)$, the set $(S - \Gamma_G(y)) \cup \{ x_0, y \}$ is absorbant, and, therefore,

$$\left| S - \Gamma_G(y) \right| + 2 \geqslant k .$$

Thus,

$$k + r - 1 - \left| \Gamma_G(y) \cap S \right| + 2 \geqslant k$$

or

$$\left| \Gamma_G(y) \cap S \right| \leqslant r + 1 .$$

Furthermore, if G_S is the subgraph generated by set S and if A is a minimum absorbant set of graph G_S, then

$$\left| A \cup \{ x_0 \} \right| \geqslant k$$

(because $A \cup \{ x_0 \}$ is absorbant in G). Hence,

$$\beta^*(G_S) \geqslant k - 1 .$$

By the induction hypothesis, the number of edges in G_S is

$$m(G_S) \leqslant \frac{1}{2} \left(k + r - 1 - (k - 1) \right) \left(k + r - 1 - (k - 1) + 2 \right) =$$

$$= \frac{1}{2} r(r + 2) .$$

Therefore, the number $m(G)$ of edges in G satisfies

$$2\, m(G) = 2\, m(G_S) + \left| \Gamma_G(x_0) \right| + \sum_{y \in \Gamma_G(x_0)} \left(|\, \Gamma_G(y) \cap S | + |\, \Gamma_G'(y) | \right) \leqslant$$

$$\leqslant r(r + 2) + (n - k - r) + (n - k - r)(r + 1) + (n - k - r)^2 =$$

$$= (n - k)(n - k + 2) - r(n - k - r) \leqslant$$

$$\leqslant (n - k)(n - k + 2) .$$

The required inequality follows.

2. We shall show that there exists a graph $G_{n,k}$ with order n and with $\beta^*(G_{n,k}) = k$, such that

$$m(G_{n,k}) = \left[\frac{1}{2} (n - k)(n - k + 2) \right] .$$

For $k = 2$, consider the graph $G_{n,2}$ defined by the union of a set $A_2 = \{ x_1, x_2 \}$ and a set $B_{n-2} = \{ x_3, ..., x_n \}$ where each vertex of A_2 is joined with each vertex of B_{n-2}, each vertex of B_{n-2} is joined with every other vertex of B_{n-2}, except for some pairs of vertices that constitute a minimum covering of B_{n-2} (see Ch. 7, § 2).

This graph satisfies $\beta^*(G_{n,2}) = 2$. Furthermore,

$$m(G_{n,2}) = m(B_{n-2}) + |A_2| \cdot |B_{n-2}| =$$

$$= \frac{1}{2}(n-2)(n-3) - \left[\frac{1}{2}(n-2)\right]^* + 2(n-2) =$$

$$= \left[\frac{1}{2}(n-2)n\right].$$

Thus, for this graph $G_{n,2}$ the required equality holds.

If $k > 2$, define graph $G_{n,k}$ by adding to graph $G_{n-k+2,2}$ a set of $k-2$ isolated vertices. Then,

$$m(G_{n,k}) = m(G_{n-k+2,2}) = \left[\frac{1}{2}(n-k+2-2)(n-k+2)\right]$$

$$= \left[\frac{1}{2}(n-k)(n-k+2)\right].$$

Again, equality holds. The inequalities in Part 1 can be used to show the uniqueness of graph $G_{n,k}$.

Q.E.D.

Corollary. *If G is a simple graph with n vertices and m edges, then*

$$\beta^*(G) \leqslant n + 1 - \sqrt{1 + 2m}.$$

From the inequality of Theorem 1, we have

$$(n-k)^2 + 2(n-k) - 2m \geqslant 0.$$

Since $n - k \geqslant 0$, then

$$n - k \geqslant -1 + \sqrt{1 + 2m}$$

or

$$k \leqslant n + 1 - \sqrt{1 + 2m}.$$

Q.E.D.

2. Kernels

For a 1-graph $G = (X, \Gamma)$, a set $S \subset X$ is defined to be a *kernel* if it is both stable and absorbant, i.e. if

(1)	$x \in S$	$\Rightarrow$	$\Gamma(x) \cap S = \varnothing$	(stable)
(2)	$x \notin S$	$\Rightarrow$	$\Gamma(x) \cap S \neq \varnothing$	(absorbant) .

Not every graph has a kernel (see graph in Fig. 14.3), and if a graph possesses a kernel, this kernel is not necessarily unique (see graph in Fig. 14.4). In this section we shall present existence and uniqueness theorems.

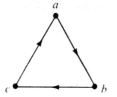

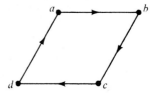

Fig. 14.3. Graph without kernels   **Fig. 14.4.** Graph with two kernels

EXAMPLE 1 (Von Neumann, Morgenstern [1944]). The concept of a kernel was first presented (under the name *solution*) in game theory.

Suppose that n players, denoted by $(1), (2), ..., (n)$, can discuss together to select a point x from a set X (the "situations"). If player (i) prefers situation a to situation b, we shall write $a \geqslant^i b$. The individual preferences might not be compatible, and, consequently, it is necessary to introduce the concept of *effective preference*. The situation a is said to be *effectively preferred* to b, or $a \succ b$, if there is a set of players who prefer a to b and who are all together capable of enforcing their preference for a. However, effective preference is not transitive, i.e., $a \succ b$ and $b \succ c$ do not necessarily imply that $a \succ c$.

Consider the 1-graph (X, Γ), where $\Gamma(x)$ denotes the set of situations effectively preferred to x. Let S be a kernel of the graph (if one exists). Von Neumann and Morgenstern suggested that the selection be confined to the elements of S. Since S is stable, no situation of S is effectively preferred to another situation of S. Since S is absorbant, for every situation $x \notin S$, there is a situation in S that is effectively preferred to x, so that x can be immediately discarded.

EXAMPLE 2. *Basis of axioms.* Consider a "theory", i.e., a set of propositions $a, b, c, ...$, that we shall represent by vertices; add arc (a, b) if proposition b implies proposition a. The resulting graph $G = (X, U)$ is transitive, i.e.:

$$(a, b) \in U, \ (b, c) \in U \ \Rightarrow \ (a, c) \in U.$$

A *basis of axioms* of the theory is a set B of propositions (called "axioms") such that:

(1) each proposition not in B follows from one of the axioms,
(2) no axiom follows from another axiom.

The problem of finding a basis of axioms reduces to finding a kernel of G. It will be shown later that each transitive graph has a kernel.

Proposition 1. *A necessary and sufficient condition for a set $S \subset X$ to be a kernel of a 1-graph $G = (X, \Gamma)$ is that its characteristic function $\varphi_s(x)$ satisfies*

$$\varphi_s(x) = 1 - \max_{y \in \Gamma(x)} \varphi_s(y) .$$

Recall that the characteristic function $\varphi_s(x)$ of set S is defined by

$$\varphi_s(x) \begin{cases} = 1 & \text{if} \quad x \in S , \\ = 0 & \text{if} \quad x \notin S . \end{cases}$$

If $\Gamma(x) = \varnothing$, we define

$$\max_{y \in \Gamma(x)} \varphi_s(y) = 0 .$$

1. Let S be a kernel. Since S is stable, we have

$$\varphi_s(x) = 1 \quad \Rightarrow \quad x \in S \quad \Rightarrow \quad \max_{y \in \Gamma(x)} \varphi_s(y) = 0 .$$

Since S is absorbant, we have

$$\varphi_s(x) = 0 \quad \Rightarrow \quad x \notin S \quad \Rightarrow \quad \max_{y \in \Gamma(x)} \varphi_s(y) = 1 .$$

Combining these, the required formula follows.

2. Let $\varphi_s(x)$ be the characteristic function of a set S which satisfies the formula; then we have

$$x \in S \quad \Rightarrow \quad \varphi_s(x) = 1 \quad \Rightarrow \quad \max_{y \in \Gamma(x)} \varphi_s(y) = 0 \quad \Rightarrow \quad \Gamma(x) \cap S = \varnothing$$

$$x \notin S \quad \Rightarrow \quad \varphi_s(x) = 0 \quad \Rightarrow \quad \max_{y \in \Gamma(x)} \varphi_s(y) = 1 \quad \Rightarrow \quad \Gamma(x) \cap S \neq \varnothing .$$

Thus, S is a kernel.

Q.E.D.

Proposition 2. *If S is a kernel, then S is a maximal stable set and a minimal absorbant set.*

Let S be the kernel of a 1-graph $G = (X, \Gamma)$. If $a \notin S$, the set $S \cup \{ a \}$ cannot be stable because $\Gamma(a) \cap S \neq \varnothing$. If $b \in S$, the set $T = S - \{ b \}$ cannot be absorbant because $b \notin T$ and $\Gamma(b) \cap T = \varnothing$.

Q.E.D.

Theorem 2. *If $G = (X, \Gamma)$ is a symmetric 1-graph, then G has a kernel. Furthermore, a set $S \subset X$ is a kernel if, and only if, S is a maximal stable set.*

1. Clearly, a maximal stable set S of G is absorbant (otherwise, there would exist a vertex $x \notin S$ non-adjacent to S, and S could not be a maximal stable set). Thus, S is a kernel.

2. Conversely, if S is a kernel of a symmetric graph G, then S is a maximal stable set (because, otherwise, S would not be absorbant).

Q.E.D.

Algorithms to construct all kernels of a graph have been presented by Roy [1970] and Rudeanu [1966].

Theorem 3. *If $G = (X, \Gamma)$ is a transitive 1-graph, each minimal absorbant set has cardinality $\beta(G)$. Furthermore, a set $S \subset X$ is a kernel if, and only if, S is a minimal absorbant set*

Let $G = (X, \Gamma)$ be a transitive 1-graph, i.e.:

$$y \in \Gamma(x), \quad z \in \Gamma(y) \quad \Rightarrow \quad z \in \Gamma(x).$$

Consider the strongly connected components of G. If C is a component with $m_G^+(C, X - C) = 0$ (i.e., no arcs leave C), C will be called a *terminal component*. Since the graph obtained from G by contracting each strongly connected component contains no circuits, G possesses terminal components. Let $C_1, C_2, ..., C_q$ be the terminal components of G. Each of these is a complete symmetric subgraph, because G is transitive.

1. If A is a minimal absorbant set, then A contains at least one vertex from each terminal component. Otherwise, $A \cap C_i = \emptyset$, and each $x \in C_i$ satisfies

$$x \notin A, \quad \Gamma(x) \cap A = \emptyset,$$

which contradicts that A is absorbant.

Let $a_i \in A \cap C_i$ for each i. Consider the set

$$A' = \{ a_1, a_2, ..., a_q \}.$$

A' is also an absorbant set because each $x \notin A'$ is the initial endpoint of an arc leading into A' (because of transitivity). Hence $A' = A$, and each minimal absorbant set is also a minimum absorbant set.

2. If S is a kernel, then S is a minimal absorbant set by Proposition 2. Conversely, if $A = \{ a_i, a_2, ..., a_q \}$ is a minimal absorbant set, then A is stable because no arc leaves a terminal component; therefore, A is a kernel.

Q.E.D.

Corollary 1. *A transitive 1-graph has a kernel, and all of its kernels have the same cardinality.*

Consequently, in Example 2, all the axiom bases have the same cardinality.

Corollary 2. *In a graph* $G = (X, U)$, *there is a set* $B \subset X$ *such that*

(1) *no paths join two distinct vertices of* B,
(2) *each vertex* $x \notin B$ *is the initial endpoint of a path leading into* B.

The corollary follows by applying Corollary 1 to the transitive 1-graph $(X, \hat{\Gamma})$ obtained by creating an arc (x, y) if there is a path from x to y in G.

Q.E.D.

Theorem 4. *A 1-graph without circuits possesses a kernel, and this kernel is unique.*

Given a 1-graph $G = (X, \Gamma)$ without circuits, consider the sets

$$X(0) = \{ x \mid x \in X, \ \Gamma(x) = \varnothing \}$$
$$X(1) = \{ x \mid x \notin X(0), \ \Gamma(x) \subset X(0) \}$$
$$X(2) = \{ x \mid x \notin X(0) \cup X(1), \ \Gamma(x) \subset X(0) \cup X(1) \}, \quad \text{etc...}$$

These sets are pairwise disjoint, and $x \in X(k)$ if, and only if, the longest path from x has length k. Since G contains no circuits, the sets $X(k)$ form a partition of X.

A characteristic function φ_S of a kernel S can be successively defined on the sets $X(0)$, $X(1)$, $X(2)$, etc. by the equality of Proposition 1. Furthermore, $\varphi_S(x)$ is uniquely defined for each x.

Q.E.D.

The following theorem due to Richardson shows that a graph without odd circuits possesses a kernel. The original proof of this result was long and involved. The less complicated proof presented here is due to Victor Neumann.

A *semi-kernel* of a 1-graph $G = (X, \Gamma)$ is defined to be a non-empty stable set $S \subset X$ such that each x adjacent to S has at least one successor in S.

Lemma. *If for each non-empty subset* $A \subset X$, *the subgraph* G_A *has a semi-kernel, then* G *has a kernel.*

Let S be a maximal semi-kernel of G, and let $A = X - S - \Gamma_G(S)$. If $A = \varnothing$, then S is a kernel. If $A \neq \varnothing$, then G_A has a semi-kernel T. Sets S and T are non-adjacent; thus, $S \cup T$ is stable. Each $x \notin S \cup T$ that is adjacent to $S \cup T$ has a successor in $S \cup T$. Thus, $S \cup T$ is a semi-kernel of G, which contradicts the maximality of S.

Q.E.D.

Theorem 5 (Richardson [1953]). *If* $G = (X, \Gamma)$ *is a 1-graph without odd circuits, then* G *has at least one kernel.*

From the lemma, it suffices to show that G possesses a semi-kernel. We may assume that G is connected.

Let X_1 be a strongly connected component of G with $\Gamma(X_1) \subset X_1$. If $|X_1| = 1$, then X_1 is a semi-kernel. If $|X_1| > 1$, and if $x_0 \in X_1$, let x be a vertex in X_1 distinct from x_0. Then, all paths $\mu[x_0, x]$ remain in the interior of X_1; and they have the same parity (since, otherwise, an odd circuit could be formed with a path $\mu[x, x_0]$).

Let S denote the set of all $x \in X_1$ such that all paths $\mu[x_0, x]$ are even. Set S is stable. If $z \in X$ is adjacent to S, then each successor of z belongs to S. Thus, S is a semi-kernel.

Q.E.D.

Note that the graph in Fig. 14.3 has no kernel; therefore it contains an odd circuit. Similarly, the graph in Fig. 14.4 does not have a unique kernel, therefore it contains a circuit.

3. Grundy functions

Consider a 1-graph $G = (X, \Gamma)$ without loops. A non-negative integer function $g(x)$ is called a *Grundy function* on G if for every vertex x, $g(x)$ is the smallest non-negative integer which does not belong to the set $\{ g(y) \mid y \in \Gamma(x) \}$. This concept originated with P. M. Grundy [1939] for graphs without circuits. It was extended to 1-graphs by Berge and Schützenberger [1956].

The Grundy function can also be defined as a function $g(x)$ such that

(1) $g(x) = k > 0$ *implies that for each $j < k$, there is a $y \in \Gamma(x)$ with $g(y) = j$,*

(2) $g(x) = k$ *implies that each $y \in \Gamma(x)$ satisfies $g(y) \neq k$.*

A *pseudo-Grundy function* is defined to be a function that satisfies property (1) above and

(2′) $g(x) = 0$ *implies that each $y \in \Gamma(x)$ satisfies $g(y) \neq 0$.*

We shall see in the following development that a Grundy function (or a pseudo-Grundy function) determines a kernel of a graph.

Remark 1. Some graphs have no Grundy function. Some graphs have more than one Grundy function. (See Fig. 14.5 where the value of a Grundy function is written next to each vertex.)

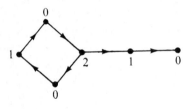

Fig. 14.5

Remark 2. If G has a Grundy function $g(x)$, then G has a kernel, since the set

$$S = \{ x \mid x \in X, \ g(x) = 0 \}$$

satisfies simultaneously:

(1) $\quad x \in S \quad \Rightarrow \quad g(x) = 0 \quad \Rightarrow \quad \min_{y \in \Gamma(x)} g(y) > 0 \quad \Rightarrow \quad \Gamma(x) \cap S = \varnothing$,

(2) $\quad x \notin S \quad \Rightarrow \quad g(x) > 0 \quad \Rightarrow \quad \min_{y \in \Gamma(x)} g(y) = 0 \quad \Rightarrow \quad \Gamma(x) \cap S \neq \varnothing$.

The converse is not true. It is left to the reader to verify that the graph G in Fig. 14.6 has a kernel $\{ d \}$ but has no Grundy function.

Theorem 6. *If G is a 1-graph such that each subgraph has a kernel, then G possesses a Grundy function.*

Let G be such a graph, and let S_0 be a kernel of G. Let S_1 be a kernel of $G_1 = G_{X-S_0}$; let S_2 be a kernel of $G_2 = G_{X-(S_0 \cup S_1)}$, etc.
The sets S_i form a partition of X. Let $g(x)$ be an integer defined by:

$$g(x) = k \quad \Leftrightarrow \quad x \in S_k .$$

We shall show that $g(x)$ is a Grundy function of G.

1. Let $g(x) = k$; we shall show that for each $j < k$, there exists a vertex $y \in \Gamma(x)$ such that $g(y) = j$.
Since $x \in S_k$, and $k > j$, vertex x is present in graph G_j. Since $x \notin S_j$, there is a vertex $y \in S_j$ such that $y \in \Gamma(x)$; thus, there is in $\Gamma(x)$ a vertex y with $g(y) = j$.

2. Let $g(x) = k$; there is no vertex $y \in \Gamma(x)$ such that $g(y) = k$, because then S_k would not be stable.

Thus $g(x)$ is a Grundy function.

$$\text{Q.E.D.}$$

Corollary 1. *A symmetric graph possesses a Grundy function.*

This follows from Theorem 2.

Corollary 2. *A transitive graph possesses a Grundy function.*

This follows from Theorem 3.

Corollary 3. *A graph without odd circuits possesses a Grundy function.*

This follows from Theorem 5.

Theorem 7 (Grundy [1939]). *A graph G without circuits possesses a unique Grundy function g(x). Moreover, for each vertex x, g(x) does not exceed the length of the longest path from x.*

As in Theorem 4, consider the sets:

$$X(0) = \{ x \mid x \in X, \; \Gamma(x) = \varnothing \}$$
$$X(1) = \{ x \mid x \notin X(0), \; \Gamma(x) \subset X(0) \}$$
$$X(2) = \{ x \mid x \notin X(0) \cup X(1), \; \Gamma(x) \subset X(0) \cup X(1) \}$$
etc.

Clearly, these sets partition X, and $x \in X(k)$ if, and only if, the longest path from x has length k. The values $g(x)$ can be successively defined on the sets $X(0)$, $X(1)$, etc.

If $x \in X(0)$, let $g(x) = 0$. If $x \in X(1)$, let $g(x) = 1$.

If for each $y \in X(k)$, the value $g(y)$ is uniquely defined and satisfies $g(y) \leqslant k$, then for $x \in X(k + 1)$, the value $g(x)$ is uniquely defined and $g(x) \leqslant k + 1$.

$$\text{Q.E.D.}$$

Before demonstrating the fundamental properties of Grundy functions, it is necessary to define the cartesian sum of 1-graphs:

Let $G_1 = (X_1, \Gamma_1)$, $G_2 = (X_2, \Gamma_2),..., G_p = (X_p, \Gamma_p)$ be 1-graphs, and let $P = \{ 1, 2, ..., p \}$. The *normal product* of these 1-graphs is defined to the 1-graph $G = G_1 . G_2 . .... . G_p$ with vertex set

$$X = X_1 \times X_2 \times \cdots \times X_p = \prod_{i \in P} X_i$$

and with correspondence

$$\Gamma(x_1, x_2, ..., x_p) = \bigcup_{\substack{I \subset P \\ |I| \neq 0}} \left(\prod_{i \in I} \Gamma_i(x_i) \times \prod_{j \in P-I} \{ x_j \} \right).$$

The *cartesian sum* of these graphs is defined to be the 1-graph $G = G_1 + G_2 + \cdots + G_p$ with vertex set $X = \prod_{i \in P} X_i$ and with correspondence

$$\Gamma(x_1, x_2, ..., x_p) = \bigcup_{i \in P} (\{ x_1 \} \times \cdots \times \{ x_{i-1} \} \times \Gamma_i(x_i) \times \{ x_{i+1} \} \times \cdots \times \{ x_p \}).$$

Finally, the *cartesian product* of these graphs is defined to be the 1-graph

$$G = G_1 \times G_2 \times \cdots \times G_p$$

with vertex set $X = \prod_{i \in P} X_i$ and with correspondence

$$\Gamma(x_1, x_2, ..., x_p) = \prod_{i \in P} \Gamma_i(x_i) .$$

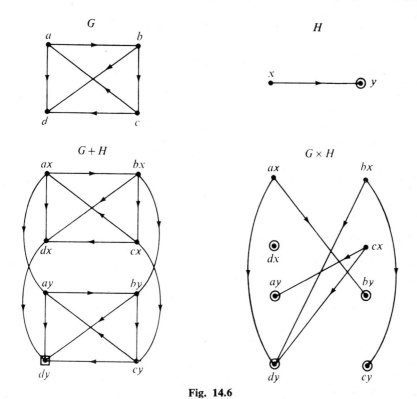

Fig. 14.6

EXAMPLE. Consider two machines, and let X_1 denote the set of possible states for the first machine. For $x_1, x_1' \in X_1$, let $x_1' \in \Gamma_1(x_1)$ if state x_1' can follow state x_1. Thus, the first machine defines a graph $G_1 = (X_1, \Gamma_1)$. Similarly, the second machine defines a graph $G_2 = (X_2, \Gamma_2)$.

Let (x_1, x_2) describe the state x_1 of the first machine and the state x_2 of the second machine. If the operator works both machines simultaneously, graph $G_1 \times G_2$ represents the possible changes of situations. If the operator works only one machine at a time, graph $G_1 + G_2$ represents the possible changes of situations. If the operator can work either one machine or two machines simultaneously, graph $G_1 \cdot G_2$ represents the possible changes of situations.

Remark. The connectivity of a cartesian product of 1-graphs has been studied by McAndrews [1963], who has shown the following result:

If G and H are two strongly connected graphs respectively with the sets $M(G)$ and $M(H)$ of circuits, then the cartesian product $G \times H$ is strongly *connected if, and only if, the lengths of the circuits in* $M(G) \cup M(H)$ *are relatively prime.*

Similar properties for the cartesian sum have been studied by Aberth [1964], who has shown that:

Graph G + H is strongly connected if, and only if, both G and H are strongly connected.

Other properties of $G + H$ and $G \times H$ have been studied by Picard [1970] and Pultr [1970].

Proposition 1. *If* 1-*graph* $G = (X, \Gamma)$ *possesses a kernel S, and if* 1-*graph* $H = (Y, \Delta)$ *possesses a kernel T, then the normal product G. H has the cartesian product* $S \times T$ *as a kernel.*

It is left to the reader to verify the stability and absorption of set $S \times T$.

Does the cartesian sum of two graphs, each with a kernel, also have a kernel? The example in Fig. 14.6, due to C.Y. Chao [1963], provides a counter example: graphs G and H respectively have kernels $\{ d \}$ and $\{ y \}$, but graph $G + H$ does not possess a kernel, because such a kernel would necessarily contain vertex dy, (since $\Gamma(d, y) = \varnothing$) and two of the three vertices ax, bx, or cx, which contradicts stability.

Later, it will be shown that if both graphs G and H possess a Grundy function, then graph $G + H$ has a kernel. Before demonstrating these results some preliminary developments are required.

The *binary expansion* of an integer p is a sequence $(p^1, p^2, ..., p^k)$ of digits such that

$$p = p^1 + 2 p^2 + 4 p^3 + \cdots + 2^{k-1} p^k .$$

Its *binary form* is

$$p = p^k p^{k-1} ... p^2 p^1 .$$

For example:

Decimal form		Binary form		Binary expansion	Decimal form		Binary form		Binary expansion
0	=	0	=	(0)	6	=	110	=	(0, 1, 1)
1	=	1	=	(1)	7	=	111	=	(1, 1, 1)
2	=	10	=	(0, 1)	8	=	1 000	=	(0, 0, 0, 1)
3	=	11	=	(1, 1)	9	=	1 001	=	(1, 0, 0, 1)
4	=	100	=	(0, 0, 1)	10	=	1 010	=	(0, 1, 0, 1)
5	=	101	=	(1, 0, 1)	11	=	1 011	=	(1, 1, 0, 1)

Let $[n]_{(2)}$ denote the integer n modulo 2 (which is the remainder of n divided by 2).

The *digital sum* of the integers $p_1, p_2, ..., p_n$ is defined to be an integer $p = p_1 \overset{\bullet}{+} p_2 \overset{\bullet}{+} \cdots \overset{\bullet}{+} p_n$ obtained from the binary expansions of the p_i by letting:

$$p = \left(\left[\sum_{i=1}^{n} p_i^1 \right]_{(2)}, \left[\sum_{i=1}^{n} p_i^2 \right]_{(2)}, \left[\sum_{i=1}^{n} p_i^3 \right]_{(2)}, ... \right).$$

For example, we obtain:

$$3 \overset{\bullet}{+} 7 = (1, 1) \overset{\bullet}{+} (1, 1, 1) = (0, 0, 1) = 4$$

$$1 \overset{\bullet}{+} 3 \overset{\bullet}{+} 11 = (1) \overset{\bullet}{+} (1, 1) \overset{\bullet}{+} (1, 1, 0, 1) = (1, 0, 0, 1) = 9, \quad \text{etc.}$$

Proposition 2. *The digital sum has the following properties*:

(1) *Associativity*: $(p \overset{\bullet}{+} q) \overset{\bullet}{+} r = p \overset{\bullet}{+} (q \overset{\bullet}{+} r)$,

(2) *Existence of an element* 0 *satisfying*: $p \overset{\bullet}{+} 0 = p$,

(3) *For each p, existence of an element* $(\overset{\bullet}{-} p)$ *satisfying*: $p \overset{\bullet}{+} (\overset{\bullet}{-} p) = 0$,

(4) *Commutativity*: $p \overset{\bullet}{+} q = q \overset{\bullet}{+} p$.

The proof follows immediately.

Thus, the first three properties imply that the integers under the binary sum operation form a *group*. The last property shows that this group is *abelian*.

Corollary. *Given two integers p and q, there exists a unique integer r such that* $p \overset{\bullet}{+} r = q$.

This is a well known property of groups.

Theorem 8. *If the 1-graphs $G_1 = (X_1, \Gamma)$, ..., $G_n = (X_n, \Gamma_n)$ respectively possess Grundy functions $g_1(x)$, ..., $g_n(x)$, then the function*

$$g(x_1, x_2, ..., x_n) = g_1(x_1) \overset{\bullet}{+} g_2(x_2) \overset{\bullet}{+} \cdots \overset{\bullet}{+} g_n(x_n)$$

is a Grundy function for the cartesian sum $G = G_1 + G_2 + \cdots + G_n$.

Consider a vertex $(x_1, x_2, ..., x_n)$ of G, and let

$$g_i(x_i) = p_i = (p_i^1, p_i^2, ...) \qquad (i = 1, 2, ..., n)$$

$$g(x_1, ..., x_n) = p_1 \overset{\bullet}{+} p_2 \overset{\bullet}{+} \cdots \overset{\bullet}{+} p_n = p = (p^1, p^2, ...)$$

where

$$p^k = \left[\sum_{i=1}^{n} p_i^k \right]_{(2)}.$$

1. We shall show that for each $q < p$, there is a vertex

$$(y_1, y_2, ..., y_n) \in \Gamma(x_1, x_2, ..., x_n)$$

such that $g(y_1, y_2, ..., y_n) = q$.

Let k_0 be the largest k for which $p^k \neq q^k$. Then, $p^{k_0} = 1, q^{k_0} = 0$, because, otherwise, we would have

$$0 + \sum_{\substack{k \geq 1 \\ k < k_0}} 2^{k-1} p^k > 2^{k_0-1} + \sum_{\substack{k \geq 1 \\ k < k_0}} 2^{k-1} q^k$$

and

$$2^{k_0-1} < \sum_{\substack{k \geq 1 \\ k < k_0}} 2^{k-1}(p^k - q^k),$$

which is impossible.

Thus,

$$\left[\sum_{i=1}^{n} p_i^{k_0} \right]_{(2)} = p^{k_0} = 1 .$$

Without loss of generality, suppose that $p_1^{k_0} = 1$. From the preceding corollary, there is an integer r such that $p \overset{\cdot}{+} r = q$, and, consequently,

$$p_1 \overset{\cdot}{+} r < p_1 .$$

Hence, there is a vertex $y_1 \in \Gamma_1(x_1)$ in graph G_1 with $g_1(y_1) = p_1 \overset{\cdot}{+} r$. Hence

$$g(y_1, x_2, x_3, ..., x_n) = \sum p_i \overset{\cdot}{+} r = q .$$

2. We shall show that if $(y_1, x_2, ..., x_n) \in \Gamma(x_1, x_2, ..., x_n)$, then

$$g(y_1, x_2, ..., x_n) \neq p .$$

If $y_1 \in \Gamma_1(x_1)$, then $g_1(y_1) \neq g_1(x_1)$. Therefore, if we had

$$g_1(y_1) \overset{\cdot}{+} g_2(x_2) \overset{\cdot}{+} \cdots \overset{\cdot}{+} g_n(x_n) = g_1(x_1) \overset{\cdot}{+} g_2(x_2) \overset{\cdot}{+} \cdots \overset{\cdot}{+} g_n(x_n)$$

then, from the preceding corollary, $g_1(y_1) = g_1(x_1)$, which is a contradiction. Therefore, g is a Grundy function for G.

Q.E.D.

4. Nim games

Given two players A and B and a 1-graph (X, Γ), the following game can be defined: Starting from some initial vertex x_0, player A selects a vertex x_1 from $\Gamma(x_0)$. Player B selects any vertex x_2 from $\Gamma(x_1)$. Next, player A selects any vertex $x_3 \in \Gamma(x_2)$, etc. If a player selects a vertex x_i with $\Gamma(x_i) = \emptyset$, then that player wins, and his opponent loses. Clearly, if there are circuits in the graph, the game need not terminate.

This game is called a *Nim-type* game. We shall study characterizations of its winning positions, i.e., those vertices which must be chosen by a player in order to win no matter how his opponent plays.

EXAMPLE 1. *Fan Tan*. There are p piles of matches. Each of two players alternately select a pile and remove one or more matches from it. The player who removes the last match wins.

EXAMPLE 2. Two players play with the same rules, but the last player to remove matches is the loser. This game was popularized by the film "Last Year at Marienbad" by Alain Resnais; it reduces to a Nim game by adding one vertex to the graph.

EXAMPLE 3. Each of two players alternately place a tile that covers three squares in a straight line on an $n \times n$ chessboard. No square can be covered twice. The last player to place a tile wins.

If n is odd, the first player can win by placing his first tile in the centre of the chessboard. Then, he should place his second tile in a position that is symmetric with respect to the centre of the chessboard to the first tile placed by his opponent, etc. (This is always possible since the centre of the chessboard has been covered.) Thus, his opponent must lose.

EXAMPLE 4 (Withoff [1907]). Let X be the set of non-negative integral points in the plane. From an integral point (p, q) in X, a player can select any

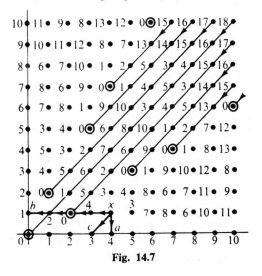

Fig. 14.7

point such that the value of one of the co-ordinates decreases and the value of the other co-ordinate remains unchanged, or such that the value of both co-ordinates decrease by an equal amount. The players take alternate turns. The first player to select the origin (0, 0) wins.

For example, from point (4, 1), the following choices are available: (4, 0), (3, 1), (2, 1), (1, 1), (0, 1), (3, 0).

The graph (X, Γ) defined by this game contains no circuits, and, consequently, it possesses a Grundy function, indicated in Fig. 14.7. The winning positions are circled. Note that these positions are distributed symmetrically around the main diagonal.

The above game is due to R. Isaacs [1958], but it was noticed by J. Kenyon [1967] that it is equivalent to a game invented by Withoff in 1907.

Theorem 9. *If the graph (X, Γ) possesses a kernel S, and if a player chooses a vertex in S, this choice assures him of a win or a draw.*

If player A chooses a vertex x_1 in S, either $\Gamma(x_1) = \emptyset$ (victory), or his opponent B will be forced to choose a vertex x_2 in $X - S$. Then, player A can choose again a vertex x_3 in S, etc. The game ends when one of the players chooses a vertex x_k with $\Gamma(x_k) = \emptyset$. Clearly, $x_k \in S$, and the winning player cannot be B.

$$\text{Q.E.D.}$$

To win games of this form, a player could calculate a Grundy function $g(x)$ (if there exists one) and then play on the kernel

$$S = \{ x \mid g(x) = 0 \}.$$

If the initial vertex x_0 satisfies $g(x_0) = 0$, then player A is in a dangerous position because his opponent can get a win or a draw. If $g(x_0) \neq 0$, then player A can assure himself of a win or a draw by choosing any successor x_1 of x_0 with $g(x_1) = 0$.

Consider n Nim games $(X_1, \Gamma_1), (X_2, \Gamma_2), ..., (X_n, \Gamma_n)$. Suppose now that each player can play only one of the Nim games during his turn; the first player who cannot play at all is the loser.

This situation is in fact a Nim game on the cartesian sum of the graphs (X_i, Γ_i). The following theorem gives a winning strategy.

Theorem 10. *Consider n Nim-games $G_1 = (X_1, \Gamma_1), G_2 = (X_2, \Gamma_2), ..., G_n = (X_n, \Gamma_n)$ with Grundy functions $g_1, g_2, ..., g_n$. A player will not lose the game on the cartesian sum $G = \sum G_i$ if he chooses a position*

$$x = (x_1, x_2, ..., x_n)$$

such that

$$g_1(x_1) \overset{\bullet}{+} g_2(x_2) \overset{\bullet}{+} \cdots \overset{\bullet}{+} g_n(x_n) = 0 \,.$$

The proof follows immediately from Theorem 8 and Theorem 9.

EXAMPLE. *Fan Tan.* Each of two players alternately selects one of p piles of matches and removes at least one match from it. The player who removes the last match wins.

This game is the cartesian sum of the games $(X_1, \Gamma_1), (X_2, \Gamma_2), \ldots, (X_n, \Gamma_n)$, where $x_k \in X_k$ represents the state of the k-th pile. Clearly $g_k(x_k)$ equals the number of matches in the k-th pile.

From Theorem 10, a position is winning if, and only if, the digital sum of the number of matches in the piles is zero.

Consider a Nim game $G = (X, \Gamma)$ without circuits and with a Grundy function $g(x)$. Let $\overset{\bullet}{+}$ be any operation that associates a vertex z to every ordered pair of vertices (x, y). This is written as $z = x \overset{\bullet}{+} y$. For $S \subset X$ and $T \subset X$, let

$$S \overset{\bullet}{+} T = \{ s \overset{\bullet}{+} t \,/\, s \in S, \quad t \in T \} \,.$$

The following result produces a winning strategy:

Theorem 11 (Grundy [1939]). *Given a 1-graph $G = (X, \Gamma)$ without circuits and an operation $\overset{\bullet}{+}$ such that*

$$\Gamma(x \overset{\bullet}{+} y) = (\Gamma(x) \overset{\bullet}{+} y) \cup (x \overset{\bullet}{+} \Gamma(y)) \,,$$

then the Grundy function g of G satisfies for all $x, y \in X$,

(1) $$g(x \overset{\bullet}{+} y) = g(x) \overset{\bullet}{+} g(y) \,.$$

We shall prove the theorem by induction. Consider the sets

$$\begin{aligned} X(0) &= \{ x \,/\, \Gamma(x) = \varnothing \} \\ X(1) &= \{ x \,/\, \Gamma(x) \subset X(0) \} \\ X(2) &= \{ x \,/\, \Gamma(x) \subset X(1) \}, \text{ etc...} \end{aligned}$$

Since the graph contains no circuits, $X = X(p)$ for some value of p. Furthermore, we have:

$$x \overset{\bullet}{+} y \in X(0) \quad \Rightarrow \quad (\Gamma(x) \overset{\bullet}{+} y) \cup (x \overset{\bullet}{+} \Gamma(y)) = \varnothing \quad \Rightarrow \quad x, y \in X(0) \,.$$

Thus, $x \overset{\bullet}{+} y \in X(0)$, implies that:

$$g(x \overset{\bullet}{+} y) = 0 = 0 \overset{\bullet}{+} 0 = g(x) \overset{\bullet}{+} g(y) \,.$$

Therefore, equation (1) is satisfied on $X(0)$.

Suppose that equation (1) is satisfied on $X(k-1)$; we shall show that it is satisfied for a vertex $z = x \overset{\cdot}{+} y$ in $X(k)$.

If this were not true, then either

$$g(z) > g(x) \overset{\cdot}{+} g(y)$$

or

$$g(z) < g(x) \overset{\cdot}{+} g(y) .$$

CASE 1. Suppose $g(z) > g(x) \overset{\cdot}{+} g(y)$. There is a vertex $z_1 \in \Gamma(z)$ such that

$$g(z_1) = g(x) \overset{\cdot}{+} g(y) .$$

Without loss of generality, we may write $z_1 = x_1 \overset{\cdot}{+} y$, $x_1 \in \Gamma(x)$. Since, $z_1 \in \Gamma(z) \subset X(k-1)$, we have

$$g(z_1) = g(x_1 \overset{\cdot}{+} y) = g(x_1) \overset{\cdot}{+} g(y) .$$

Hence,

$$g(x) \overset{\cdot}{+} g(y) = g(x_1) \overset{\cdot}{+} g(y) .$$

Therefore, $g(x) = g(x_1)$, which contradicts that $x_1 \in \Gamma(x)$.

CASE 2. Suppose $g(z) < g(x) \overset{\cdot}{+} g(y)$. Then, as in Theorem 8, there is an integer r such that

$$g(z) = r \overset{\cdot}{+} g(y) , \qquad r < g(x) .$$

Since there is a vertex $x_1 \in \Gamma(x)$ with $g(x_1) = r$, then

$$g(z) = g(x_1) \overset{\cdot}{+} g(y) .$$

Since $x_1 \overset{\cdot}{+} y \in \Gamma(z) \subset X(k-1)$, we have

$$g(x_1 \overset{\cdot}{+} y) = g(x_1) \overset{\cdot}{+} g(y) = g(z) ,$$

which contradicts that $x_1 \overset{\cdot}{+} y \in \Gamma(z)$.

Thus, in both cases, we have a contradiction.

$$\text{Q.E.D.}$$

Corollary. *Let $G = (X, \Gamma)$ be a 1-graph without circuits, and let $\overset{\cdot}{+}$ be an operation on X such that*

(1) $x \overset{\cdot}{+} y = y \overset{\cdot}{+} x ,$

(2) $x \overset{\cdot}{+} (y \overset{\cdot}{+} z) = (x \overset{\cdot}{+} y) \overset{\cdot}{+} z ,$

(3) $x \overset{\bullet}{+} z = y \overset{\bullet}{+} z \quad \Rightarrow \quad x = y$,

(4) *there exists a vertex $e \in X$ such that $x \overset{\bullet}{+} e = x$ for all $x \in X$ and such that*
$\Gamma(e) = \varnothing$,

(5) $\Gamma(x \overset{\bullet}{+} y) = \left(\Gamma(x) \overset{\bullet}{+} y\right) \cup \left(x \overset{\bullet}{+} \Gamma(y)\right)$.

A vertex z is defined to be "irreducible" if there exist no vertices x and y distinct from e such that

$$z = x \overset{\bullet}{+} y.$$

Each vertex $x \in X$ can be uniquely written as a digital sum of irreducible vertices. If the value of the Grundy function $g(z)$ is known for all the irreducible vertices, then $g(x)$ is defined for x by

$$g(x) = g(z_1 \overset{\bullet}{+} z_2 \overset{\bullet}{+} \cdots) = g(z_1) \overset{\bullet}{+} g(z_2) \overset{\bullet}{+} \cdots.$$

The uniqueness of this decomposition is a well known result from the theory of semi-groups.

EXAMPLE. Two players alternately choose one of n piles of matches and divide the chosen pile into two unequal piles. The last player who can make such a division is the winner.

A vertex x is represented by the numbers $\bar{x}_1, \bar{x}_2, \ldots, \bar{x}_k$ that denote the numbers of matches in each pile. Let

$$x \overset{\bullet}{+} y = (\bar{x}_1, \bar{x}_2, \ldots, \bar{x}_k) \overset{\bullet}{+} (\bar{y}_1, \bar{y}_2, \ldots, \bar{y}_l) = (\bar{x}_1, \bar{x}_2, \ldots, \bar{x}_k, \bar{y}_1, \ldots, \bar{y}_l).$$

The operation $\overset{\bullet}{+}$ satisfies:

(1) $x \overset{\bullet}{+} y = y \overset{\bullet}{+} x$,

(2) $x \overset{\bullet}{+} (y \overset{\bullet}{+} z) = (x \overset{\bullet}{+} y) \overset{\bullet}{+} z$,

(3) $x \overset{\bullet}{+} z = y \overset{\bullet}{+} z \quad \Rightarrow \quad x = y$,

(4) There exists a position e (no matches) such that $x \overset{\bullet}{+} e = x$ for all x and such that $\Gamma(e) = \varnothing$,

(5) $\Gamma(x \overset{\bullet}{+} y) = \left(\Gamma(x) \overset{\bullet}{+} y\right) \cup \left(x \overset{\bullet}{+} \Gamma(y)\right)$.

Thus, the value of the Grundy function g for the irreducible vertices permits the calculation of the other values by means of a simple digital sum. An irreducible vertex corresponds to a single pile with $\bar{x}_1$ matches, and some values of $g(\bar{x}_1)$ are given below

$\bar{x}_1 =$	1	2	3	4	5	6	7	8	9	10	11	12	13	14	15	16	17	18	19	20
$g(\bar{x}_1) =$	0	0	1	0	2	1	0	2	1	0	2	1	3	2	1	3	2	4	3	0

EXERCISES

1. Let λ be an integer such that

$$\max \{ n - m, 1 \} \leqslant \lambda \leqslant [n + 1 - \sqrt{1 + 2m}].$$

Show that there exists a simple graph G with n vertices and m edges such that the corresponding directed symmetric graph G^* satisfies $\beta(G^*) = \lambda$.

2. Show that in the Nim game in Example 4, the abscissa of the n-th winning position below the diagonal is

$$\left[\frac{3 + \sqrt{5}}{2} n \right].$$

3. Show that for the Marienbad game (Example 2) there is a Grundy function $g'(x)$ that can be constructed from the Grundy function for Fan Tan (Example 1) as follows:

$g'(x) = g(x)$ if in position x, there is at least one pile with more than one match
$g'(x) = 1 - g(x)$ if no pile has more than one match.

4. The graph in Fig. 14.5, known as $\vec{C}_7(1, 2)$, is obtained from a circuit of length 7 by joining all arcs of the form $(i, 1 + 1)$ and $(i, i + 2)$. Show that the graphs $\vec{C}_7(1, 2)$ and $\vec{C}_{11}(1, 2, 4)$ are minimal kernel-free. Do there exist other kernel-free graphs of the form $\vec{C}_n(p, q, r)$?

5. Show that the graph with vertices 1, 2, ..., 9 and arcs (2, 3), (3, 4), (4, 5), (5, 6), (6, 7), (7, 8), (1, 8), (8, 9), (9, 2), (8, 3), (2, 4), (3, 5), (4, 6), (5, 7), (6, 1), (7, 2), (1, 3) is the smallest minimal kernel-free graph G such that G^{-1}, obtained by reversing all arcs in G, is not minimal kernel-free.

(Duchet and Meyniel [1981])

6. Using Richardson's theorem, show that if every arc in an odd circuit of the graph G belongs to a double-edge, then G has a kernel.

7. More generally, show that if every odd circuit has at least 2 arcs belonging to double edges, then G has a kernel.

(Duchet [1980])

For this, show that all components of the graph G' obtained by removing all double edges from G are bipartite and strongly connected, and that it is possible to colour these graphs such that no double edge of G has two endpoints of the same colour.

8. If every odd circuit $[x_1, x_2, ..., x_{2k+1}, x_1]$ has two crossing chords of the form (x_i, x_{i+2}) and (x_{i+1}, x_{i+3}), then the graph has a kernel (Duchet and Meyniel [1983]). However, it is not true that if every odd circuit has two chords, then the graph has a kernel. A counter-example was found by Galéana-Sanchez [1982]. Neumann and Galéana-Sanchez [1984] showed that if every odd circuit has two chords whose endpoints occur consecutively on the circuit, then the graph has a kernel.

9. Let the vertices of a graph G correspond to a set of mathematical properties P, with (p, p') an arc of G if p implies p'. To show that there are no new 'theorems' to be discovered, i.e. that all arcs not in G are 'bad', it is necessary to construct a set of counter-examples. Show that every minimum set of counter-examples is given by the kernel of a certain graph H easily obtained from G.

(Berge and Ramachandra Rao [1977]).

10. Use the graph with vertices 1, 2, ..., 9 and arcs (1, 8), (2, 1), (2, 5), (2, 9), (3, 8), (4, 2), (4, 3), (5, 4), (5, 7), (6, 1), (6, 3), (7, 6), (7, 9), (8, 1), (8, 3), (9, 2), (9, 5), (9, 8) to show that a minimal kernel-free graph is not necessarily strongly connected.

(Duchet and Meyniel [1981])

11. Berge and Duchet conjectured that if G is a perfect graph (cf. Ch. 16) and if every maximal clique has a kernel (thus a singleton), then G has a kernel. Show that the conjecture holds for triangulated graphs.

(Maffray [1984])

CHAPTER 15

Chromatic Number

1. Vertex colourings

The chromatic number of a graph G is defined to be the smallest number of colours needed to colour the vertices of the graph so that no two adjacent vertices have the same colour. The chromatic number of graph G is denoted by $\gamma(G)$.

A graph G with $\gamma(G) \leqslant k$ is called k-colourable. A k-colouring of the vertices is a partition $(S_1, S_2, ..., S_k)$ of the vertex set into k stable sets, each representing the vertices of one of the k different colours.

In this chapter, we shall assume that graph G is simple.

EXAMPLE 1. The chromatic index (see Chapter 12) of G is equal to the chromatic number of a graph $L(G)$ formed in the following way: The vertices of $L(G)$ represent the edges of G, and two vertices are adjacent in $L(G)$ if, and only if, the corresponding edges in G are adjacent. Thus, results for the chromatic number yield results for the chromatic index.

EXAMPLE 2. Is it possible to colour all the countries on a geographic map with four colours so that no two countries with a common border have the same colour? This famous problem, unsolved since 1890, is equivalent to asking whether every planar graph is 4-chromatic. This has recently been proven to be true, by using more than 1000 hours of computer time.

(Appel and Haken [1979])

EXAMPLE 3. Let X represent the inhabitants of an aquarium. Consider the 1-graph $G = (X, \Gamma)$ with $d_G^-(x) \leqslant 2$ for all $x \in X$. Is it possible that this graph represents a family tree, i.e., $y \in \Gamma(x)$ implies that y is the child of x? Clearly, a necessary condition is that the graph contains no circuits and its vertices can be coloured with two colours α_1 (male) and α_2 (female) such that $\Gamma_G^- x$ contains at most one element of each colour, for each x.

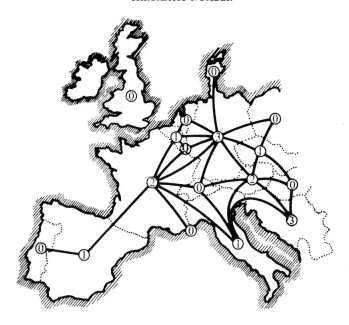

Fig. 15.1. Geographic map corresponding to a 4-colourable graph G. The colours 0, 1, 2, 3 correspond to the values of a Grundy function for the symmetric (directed) graph G^*

It is also necessary that the simple graph $H = (X, E)$, where $[x, y] \in E$ if, and only if,

$$\Gamma x \cap \Gamma y \neq \varnothing ,$$

is bicolourable.

EXAMPLE 4. *Examination schedule.* At the end of the term, each student must take an examination in each of his courses. Let X be the set of different courses and let Y be the set of students. Since the examination is written, it is convenient that all students in a course take the examination at the same time. What is the minimum number of examination periods needed?

For each examination x, let $S(x) \subset Y$ denote the set of students who must write examination x. Form graph $G = (X, E)$ where $[x, x'] \in E$ if $S(x) \cap S(x') \neq \varnothing$, i.e., if examinations x and x' cannot be given simultaneously. Each vertex colouring of this graph corresponds to a possible examination schedule, and vice versa. Thus, the examination scheduling problem is solved by finding $\gamma(G)$.

EXAMPLE 5. The *schoolgirl problem* (Kirkman). This famous problem can be stated as follows: Each day the fifteen schoolgirls $a, b, c, \ldots, m, n, o$ from a

boarding school take a walk in 5 rows of 3. Is it possible for them to take 7
walks such that no two girls walk together more than once?

By considering symmetry, Cayley found the following solution:

Sunday	Monday	Tuesday	Wednesday	Thursday	Friday	Saturday
afk	abe	alm	ado	agn	ahj	aci
bgl	cno	bcf	bik	bdj	bmn	bho
chm	dfl	deh	cjl	cek	cdg	dkm
din	ghk	gio	egm	fmo	efi	eln
ejo	ijm	jkn	fhn	hil	klo	fgj

The schoolgirl problem is related to the well known *Steiner problem*: Is it
possible for the 15 schoolgirls to form successively 35 distinct triples such
that no two girls appear together twice in a triple? To solve this problem,
form a graph G whose vertices are the $\binom{15}{3} = 455$ possible triangles. Two
vertices are joined together if the corresponding triangles have two girls in
common. Thus, the problem reduces to finding a maximum stable set S.
Clearly, $|S| \leqslant 35$, because no girl can appear in more than 7 distinct
triples of S, and therefore there are at most $15 \times 7 \times \frac{1}{3} = 35$ triangles).
If a stable set S has 35 vertices, then S is a maximum stable set.

To check if a solution S of the Steiner problem yields a solution of the
schoolgirl problem, construct a graph G' whose vertices are the 35 selected
triples, and with two vertices of G' triples joined by an edge if the corresponding triples have one girl in common. If the chromatic number $\gamma(G') = 7$,
the schoolgirl problem is solved. If $\gamma(G') > 7$, it is necessary to select another
set S.

A set of triples which meets the conditions of the Steiner problem is also
called a *Steiner triple system of order 15*, or a (15, 3, 1)-*block design*. It was
proved by the Reverend T. J. Kirkman, in 1847, that a Steiner triple system of
order n exists if and only if $n \equiv 1$ or $3 \pmod 6$.

Now, if $n = 6k + 3$ girls take a walk each day in $2k + 1$ rows of three,
can an arrangement be made for $3k + 1$ different days such that any pair of
girls belong to the same row on exactly one day of the $3k + 1$ days? The
existence of such an arrangement for all $k \geqslant 0$ was proved by Ray-Chaudhuri
and Wilson, [1971].

Algorithms to determine chromatic number. Several algorithms exist for the determination of a minimum vertex colouring of a graph. The most efficient of these algorithms use the following two principles:

I. *Principle of contraction-connection.*

Consider a simple graph G with two non-adjacent vertices a and b. The *connection $G \mid ab$* of graph G is obtained by joining two non-adjacent vertices a and b with an edge. The *contraction $G : ab$* of graph G is obtained by shrinking set $\{a, b\}$ into a single vertex $c(a, b)$ and by joining it to each neighbour in G of vertex a and of vertex b.

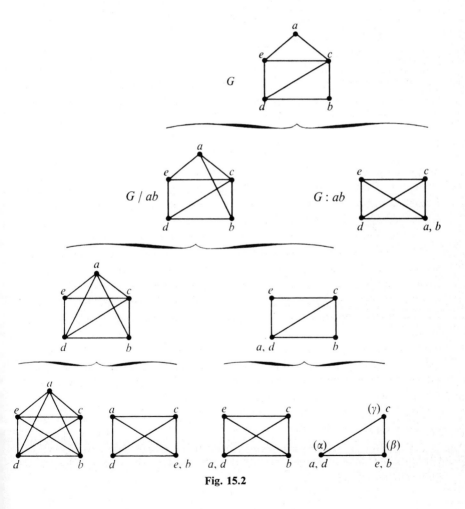

Fig. 15.2

A colouring of G in which a and b have the same colour yields a colouring of $G : ab$. A colouring of G in which a and b have different colours yields a colouring of $G \mid ab$.

Repeat the operations of connection and contraction, until the resulting graphs are all cliques. If the smallest resulting clique is a k-clique, then $\gamma(G) = k$.

For example, the graph G in Fig. 15.2 yields via the connection-contraction process a 3-clique. A 3-colouring of the original graph is obtained by recovering the original from the 3-clique. Vertices a and d will be coloured α; vertices e and b will be coloured β; vertex c will be coloured γ.

II. *Principle of separation into pieces*

Suppose that during the above contraction–connection process, we encounter a graph with an articulation set A that is a clique; the removal of A creates connected components $C_1, C_2, C_3, \ldots$, and, clearly, the subgraphs $G_{A \cup C_1} = H_{1_2}, G_{A \cup C_2} = H_2, G_{A \cup C_3} = H_3, \ldots$ can be coloured separately so that the colours of the vertices of A do not conflict.

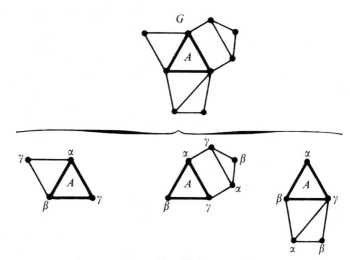

Fig. 15.3

For example, graph G in Fig. 15.3 is 3-colourable because each subgraph H_i is 3-colourable.

A *piece of graph G relative to the articulation set A* is defined to be a subgraph of G of the form $H_i = G_{A \cup C_i}$. Clearly, the pieces of a graph can be coloured separately.

Theorem 1. *If G is a graph of order n, then*

$$\begin{cases} \gamma(G)\alpha(G) \geqslant n \,, \\ \gamma(G) + \alpha(G) \leqslant n + 1 \,. \end{cases}$$

Let us colour the vertices of G with $q = \gamma(G)$ colours $\alpha_1, \alpha_2, ..., \alpha_q$. Let S_i denote the set of vertices with colour α_i. Since S_i is a stable set, we have

$$n = |\, S_1 \cup S_2 \cup \cdots \cup S_q \,| = \sum_{i=1}^{q} |\, S_i \,| \leqslant q\alpha(G) = \gamma(G)\,\alpha(G) \,.$$

Furthermore, if a maximum stable set S is coloured with the first colour, then $n - \alpha(G)$ colours can be used to colour $X - S$. Hence,

$$\gamma(G) \leqslant \big(n - \alpha(G)\big) + 1 \,.$$

$$\text{Q.E.D.}$$

Theorem 2 (Gaddum, Nordhaus [1960]). *Let $\overline{G}$ be the complementary graph of a simple graph G. Then*

$$\gamma(G) + \gamma(\overline{G}) \leqslant n + 1 \,.$$

Recall that the complementary graph of $G = (X, E)$ is defined to be $\overline{G} = (X, \mathscr{P}_2(X) - E)$. Clearly, the theorem is true for $n = 1$ and $n = 2$.

We shall suppose that the theorem is valid for every graph with $n - 1$ vertices, and we shall show that the theorem holds for a graph G with n vertices, $n > 2$. Let G_0 be the subgraph obtained by the removal of vertex x_0. Clearly, we have

(1) $$\gamma(G) \leqslant \gamma(G_0) + 1 \,.$$

(2) $$\gamma(\overline{G}) \leqslant \gamma(\overline{G_0}) + 1 \,.$$

1. If in both (1) and (2) equality holds, then

$$d_G(x_0) \geqslant \gamma(G_0)$$

$$d_{\overline{G}}(x_0) \geqslant \gamma(\overline{G_0}) \,.$$

Hence,

$$\gamma(G_0) + \gamma(\overline{G_0}) \leqslant d_G(x_0) + d_{\overline{G}}(x_0) = n - 1 \,.$$

Hence

$$\gamma(G) + \gamma(\overline{G}) \leqslant (n - 1) + 2 = n + 1 \,.$$

2. If either (1) or (2) is not satisfied with equality, then, by the induction hypothesis,

$$\gamma(G) + \gamma(\overline{G}) \leqslant \gamma(G_0) + \gamma(\overline{G_0}) + 1 \leqslant n + 1 \,.$$

Hence, the theorem is valid in both cases.[1]

Q.E.D.

Remark. This bound is the best possible: A *graph of type* $F_1(n, p)$ is defined to be a graph with n vertices formed from a stable set S_p with p vertices and a clique K_{n-p+1} with $n - p + 1$ vertices such that

$$| K_{n-p+1} \cap S_p | = 1 .$$

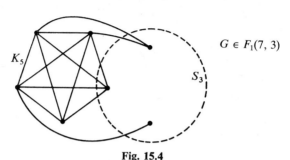

$G \in F_1(7, 3)$

Fig. 15.4

If graph G is of type F_1, then

$$\gamma(G) = n - p + 1 ,$$
$$\gamma(\overline{G}) = p .$$

Hence
$$\gamma(G) + \gamma(\overline{G}) = n + 1 .$$

Similarly, a *graph of type* $F_2(n, p)$ is defined to be a graph with n vertices formed from a cycle C_5 of length 5 without chords, a stable set S_p with $p \leqslant n - 5$ vertices, and a $(n - p - 5)$-clique K_{n-p-5}, such that these three sets are disjoint and

(1) each vertex C_5 is adjacent to each vertex of K_{n-p-5},
(2) no vertex of C_5 is adjacent to a vertex of S_p.

If G is a graph of type F_2, then

---

[1] A stronger inequality has been obtained by R. P. Gupta [1968].

A *quasi-colouring with k colours* is defined to be a mapping $f : X \rightarrow (\alpha_1, \alpha_2, \cdots, \alpha_k)$ such that $i \neq j$ implies the existence of two adjacent vertices x and y with $f(x) = \alpha_i$ and $f(y) = \alpha_j$. A colouring with a minimum number of colours is a quasi-colouring because, otherwise, two colours α_i and α_j are not adjacent and the vertices with any one of these two colours could be coloured by using only one colour.

Let $\psi(G)$ denote the maximum number of colours needed for a quasi-colouring of G. Thus, $\gamma(G) \leqslant \psi(G)$. Gupta has shown that

$$\gamma(G) + \psi(\overline{G}) \leqslant n + 1 .$$

Clearly, this result is stronger than Theorem 2.

$$\gamma(G) = (n - p - 5) + 3 = n - p - 2,$$

$$\gamma(\overline{G}) = p + 3.$$

Hence $\gamma(G) + \gamma(\overline{G}) = n + 1.$

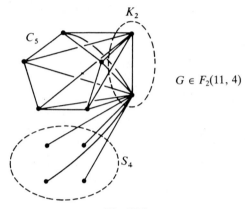

$G \in F_2(11, 4)$

Fig. 15.5

The bound given in Theorem 2 is attained by graphs of type F_1 and F_2. Furthermore, H.J. Finck [1966] has shown that only the graphs of types F_1 and F_2 attain this bound.

Corollary. *If $\overline{G}$ is the complementary graph of a simple graph G of order n, then*

$$\gamma(G)\,\gamma(\overline{G}) \leqslant \left[\left(\frac{n+1}{2} \right)^2 \right].$$

By Theorem 2, we have

$$4\,\gamma(G)\,\gamma(\overline{G}) \leqslant (\gamma(G) - \gamma(\overline{G}))^2 + 4\,\gamma(G)\,\gamma(\overline{G}) = (\gamma(G) + \gamma(\overline{G}))^2 \leqslant (n+1)^2.$$

Q.E.D.

Remark. This bound is the best possible. If n is odd, then each graph G of type

$$F_1\left(n, \frac{n+1}{2}\right) \qquad \text{or} \qquad F_2\left(n, \frac{n-5}{2}\right)$$

satisfies

$$\gamma(G)\,\gamma(\overline{G}) = \frac{n+1}{2} \cdot \frac{n+1}{2} = \left[\left(\frac{n+1}{2} \right) \right]^2.$$

If n is even, then each G of type

$$F_1\left(n, \frac{n}{2}\right), \quad \text{or} \quad F_1\left(n, \frac{n+2}{2}\right), \quad \text{or} \quad F_2\left(n, \frac{n-4}{2}\right), \quad \text{or} \quad F_2\left(n, \frac{n-6}{2}\right),$$

satisfies

$$\gamma(G)\,\gamma(\overline{G}) = \frac{n}{2} \cdot \frac{n+2}{2} = \frac{(n+1)^2}{4} - \frac{1}{4} = \left[\left(\frac{n+1}{2}\right)^2\right].$$

Theorem 3. *If G is a simple graph with n vertices and m edges, then*

$$\gamma(G) \geqslant \frac{n^2}{n^2 - 2m}.$$

Let $\gamma(G) = q$. Let $S_1, S_2, \ldots, S_q$ denote the sets of vertices of colour α_1, $\alpha_2, \ldots, \alpha_q$, respectively. The adjacency matrix of graph G has the form:

Let $n_i = |S_i|$. Let N_0 denote the number of 0 entries in this matrix, and let N_1 denote the number of 1 entries in this matrix. Then,

$$N_1 = 2m,$$

$$N_0 \geqslant n_1^2 + n_2^2 + \cdots + n_q^2 \geqslant \frac{(n_1 + n_2 + \cdots + n_q)^2}{q} = \frac{n^2}{q},$$

This follows from the Cauchy-Schwartz inequality $|\mathbf{x}|\,|\mathbf{y}| \geqslant |\langle \mathbf{x}, \mathbf{y}\rangle|$, with $\mathbf{x} = (n_1, n_2, \ldots, n_q)$ and $\mathbf{y} = (1, 1, \ldots, 1)$.

Thus, the total number of entries in the matrix satisfies:

$$n^2 = N_0 + N_1 \geqslant 2m + \frac{n^2}{q}.$$

Hence,

$$(n^2 - 2m)\, q \geqslant n^2.$$

Hence the required formula.

Q.E.D.

Remark. In order to have equality, it is necessary that $N_0 = \dfrac{n^2}{4}$, and that

$$|\mathbf{x}|.|\mathbf{y}| = |\langle \mathbf{x}, \mathbf{y} \rangle|,$$

which implies that $(n_1, n_2, ..., n_q)$ is proportional to $(1, 1, ..., 1)$. Thus, equality holds only if

$$n_1 = n_2 = \cdots = n_q,$$

and if the matrix has only 1 entries outside of the blocks $S_i \times S_i$, i.e., the graph is formed from q stable sets $S_1, S_2, ..., S_q$ of the same cardinality with two vertices joined together if, and only if, they belong to distinct S_i.

We now direct our interests to the study of the chromatic number of a graph G of order n as a function of the number of edges $m(G)$. If $m(G) = 0$, then G is the graph S_n consisting of n isolated vertices, and $\gamma(G) = 1$. If $m(G) = \binom{n}{2}$, G is the complete graph K_n, and $\gamma(G) = n$. As the number of edges increases from 0 to $\binom{n}{2}$, the chromatic number also is likely to increase from 1 to n.

Theorem 4. *Let k be an integer $1 < k \leqslant n$, and let G be any graph of order n other than the disjoint union of K_k and S_{n-k}. Then $\gamma(G) < k$ if $m(G) \leqslant \binom{k}{2}$.*

The result holds for $k = n$, for if $G \neq K_n$, then obviously $\gamma(G) < n = k$. Consider an integer $n \geqslant k + 1$, and suppose that the result holds for all graphs with order $< n$ (proof by induction).

Assume then that G is a graph of order n such that

(1) $$n(G) = n \geqslant k + 1,$$

(2) $$m(G) \leqslant \binom{k}{2},$$

(3) $$G \neq K_k + S_{n-k}.$$

G has a vertex x_0 with $d_G(x_0) < k - 1$, for otherwise

$$m(G) = \frac{1}{2}\sum_x d_G(\dot{x}) \geqslant \frac{n}{2}(k-1) > \binom{k}{2}$$

which contradicts (2). Let G_0 be the subgraph of G obtained by removing x_0.

Then $G_0 \neq K_k + S_{n-k-1}$ for otherwise $m(G_0) = \binom{k}{2} \geq m(G) \geq m(G_0)$, hence $d_G(x_0) = 0$ and $G = K_k + S_{n-k}$, contradicting (3). Thus:

(1') $n(G_0) = n - 1 \geq k,$

(2') $m(G_0) \leq \binom{k}{2},$

(3') $G_0 \neq K_k + S_{n-k-1}.$

Hence, by the induction hypothesis, $\gamma(G_0) < k$, Also, any $(k-1)$-colouring of G_0 extends to a $(k-1)$-colouring of G, since $d_G(x_0) < k$. Thus $\gamma(G) < k$.

Theorem 4'. *Let k be an integer, $1 < k \leq n$. If G is a graph of order n other than the complement of $G_{n,k}$ (as described in Theorem 5, section 2, Ch. 13), and if $m(G) \geq m(\bar{G}_{n,k})$, then $\gamma(G) > k$.*

(Recall that $\bar{G}_{n,k}$ is formed by joining k stable sets of cardinality $\left[\dfrac{n}{k}\right]$ or $\left[\dfrac{n}{k}\right]^*$ in all possible ways. This graph is sometimes called 'Turán's graph'.)

We have:

$$n(G) = n, \qquad G \neq \bar{G}_{n,k}, \qquad m(G) \geq m(\bar{G}_{n,k})$$

which implies:

$$n(\bar{G}) = n, \qquad \bar{G} \neq G_{n,k}, \qquad m(\bar{G}) \leq m(G_{n,k})$$

which, by Turán's theorem, implies $\alpha(\bar{G}) > k$, thus $\omega(G) > k$, and hence $\gamma(G) > k$.   Q.E.D.

Theorems 4 and 4' are summarized in the following table:

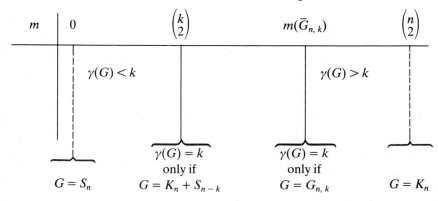

m	0	$\binom{k}{2}$	$m(\bar{G}_{n,k})$	$\binom{n}{2}$
		$\gamma(G) < k$	$\gamma(G) > k$	
	$G = S_n$	$\gamma(G) = k$ only if $G = K_n + S_{n-k}$	$\gamma(G) = k$ only if $G = G_{n,k}$	$G = K_n$

The unique graph with chromatic number $\leqslant k$ and maximum number of edges is $G_{n,\,k}$; the unique graph with chromatic number $\geqslant k$ and minimum number of edges is $K_k + S_{n-k}$.

The following theorems give bounds on the chromatic number as a function of the degree sequence. This next theorem is due to Welsh [1968]:

Theorem 5. *Let $(S_1, S_2, .. , S_q)$ be a q-colouring (not necessarily minimum) of a simple graph G, and let*

$$d_k = \max_{x \in S_k} d_G(x) .$$

Then,

$$\gamma(G) \leqslant \max_{k \leqslant q} \min \{ k , d_k + 1 \} .$$

1. If S_1 is not a maximal stable set, add vertices to S to obtain a maximal stable set S_1'. If $S_2 - S_1'$ is not a maximal stable set in $X - S_1'$, then add vertices to it to form a maximal stable set S_2' in $X - S_1'$, etc. This process defines a new colouring

$$(S_1', S_2', ..., S_r') ,$$

with

$$\bigcup_{i=1}^{\min \{ k,r \}} S_i' \supset S_k \qquad (k = 1, 2, ..., q).$$

2. Let $i(x)$ denote the index i such that $x \in S_i'$. Let x_0 be a vertex with $i(x_0) = k$; since x_0 is adjacent to each S_j' with $j \leqslant k - 1$ (by the maximality of S_j'), then

$$d_G(x_0) \geqslant k - 1 .$$

Thus, for all vertices x,

$$i(x) \leqslant d_G(x) + 1 .$$

3. Let $x_0 \in S_k$. From Part 1, it follows that $i(x_0) \leqslant k$ and, consequently,

$$i(x_0) \leqslant \max_{x \in S_k} i(x) \leqslant \max_{x \in S_k} (d_G(x) + 1) = d_k + 1 .$$

Hence,

$$\gamma(G) \leqslant \max_{x_0} i(x_0) \leqslant \max_{k \leqslant q} \min \{ k, d_k + 1 \} .$$

Q.E.D.

Corollary 1. *Let G be a simple graph. If for some integer q, the number of vertices with degree $\geqslant q$ is $\leqslant q$, then G is q-colourable.*

Assume that the vertices x_i are indexed with decreasing degrees.

Consider the n-colouring $(\{x_1\}, \{x_2\}, ..., \{x_k\}, ..., \{x_n\})$. Thus, $d_k = d_G(x_k)$. If $k \leqslant q$, we have

$$\min\{d_k + 1, k\} \leqslant q.$$

If $k \geqslant q + 1$, we have

$$\min\{d_k + 1, k\} \leqslant \min\{d_{q+1} + 1, k\} \leqslant \min\{q, k\} \leqslant q.$$

Thus, from Theorem 5, $\gamma(G) \leqslant q$.

Q.E.D.

Corollary 2. *A simple graph with maximum degree h, is $(h + 1)$-colourable.* The proof follows by letting $q = h + 1$ in Corollary 1.

Corollary 2 can be improved by the following result, known as the " Brooks Theorem"; the proof presented here is due to Melnikov and Vizing [1969].

Theorem 6 (Brooks [1941]). *Let G be a connected simple graph with maximum degree h. Then G is h-colourable, unless*
(1) *$h \neq 2$, and G is a $(h + 1)$-clique, or*
(2) *$h = 2$, and G is an odd cycle.*

Clearly, the theorem is true for $h = 0, 1, 2$. Let $G = (X, E)$ be a simple graph with maximum degree $h \geqslant 3$, with $\gamma(G) = h + 1$, and that contains no $(h + 1)$-clique. We shall show that this leads to a contradiction.

1. Since vertices that are not essential to the above properties can be removed, we may assume that G is minimal with respect to these properties.

Let $x_0 \in X$. The subgraph G_0 generated by $X - \{x_0\}$ contains no $(h + 1)$-clique, and therefore $\gamma(G_0) < h + 1$. Hence, $\gamma(G_0) = h$. This implies that $d_G(x_0) \geqslant h$ because, otherwise, one of the h colours used to colour G_0 could be used to colour x_0, which contradicts $\gamma(G) = h + 1$.

Thus, $d_G(x_0) = h$. We may assume that, *if $y_1, y_2, ..., y_h$ denote the h vertices adjacent to x_0, they are coloured with the colours $\alpha_1, \alpha_2, ..., \alpha_h$ respectively in a given h-colouring of G_0.*

2. Let $G(\alpha_i, \alpha_j)$ be the subgraph of G_0 generated by the vertices with colours α_i or α_j in the h-colouring of G_0. *Vertices y_i and y_j belong to the same connected component of $G(\alpha_i, \alpha_j)$* because, otherwise, after interchanging colours α_i and α_j in the connected component containing y_i, x_0 could be coloured α_i, which contradicts $\gamma(G) = h + 1$.

3. We shall show that *the connected component of $G(\alpha_i, \alpha_j)$ that contains y_i and y_j is an elementary chain $\mu[y_i, y_j]$ going from y_i to y_j.*

Vertex y_i is adjacent to only one vertex with colour α_j (otherwise, since $d_G(y_i) \leqslant h$, vertex y_i could be recoloured with a colour α_k, $k \neq i, j$, and vertex x_0 could be coloured α_i).

Consider a chain in $G(\alpha_i, \alpha_j)$ from y_i to y_j. We shall show that this chain is unique. Let x be the first vertex of this chain with $d_{G(\alpha_i, \alpha_j)}(x) > 2$. If x has colour α_i, for example, then there are three vertices with colour α_j adjacent to x; since $d_G(x) \leqslant h$, vertex x can be recoloured α_k, $k \neq i, j$. This would disconnect y_i and y_j and contradict Part 2.

4. We shall show that *the two chains* $\mu[y_i, y_j]$ *and* $\mu[y_i, y_k]$ *that constitute components* $G(\alpha_i, \alpha_j)$ *and* $G(\alpha_i, \alpha_k)$ *cannot contain a common vertex* $z \neq y_i$.

If such a vertex z existed, then it would have colour α_i and would be adjacent to two vertices with colour α_j and to two vertices with colour α_k. Thus, $h \geqslant d_G(z) \geqslant 4$, and there is a fourth colour $\alpha_l \neq \alpha_i, \alpha_j, \alpha_k$ to recolour z. This would disconnect y_i and y_j and contradict Part 2.

5. Since G does not contain any $(h + 1)$-cliques, there exist in $\Gamma_G(x_0)$ two non-adjacent vertices, say y_1 and y_2. Consider the connected component $\mu[y_1, y_2]$ of $G(\alpha_1, \alpha_2)$ and the connected component $\mu'[y_1, y_3]$ of $G(\alpha_1, \alpha_3)$. Let x be the first vertex after y_1 in chain $\mu[y_1, y_2]$. Since y_1 and y_2 are non-adjacent, $x \neq y_2$. If colours α_1 and α_3 are interchanged in the component of $G(\alpha_1, \alpha_3)$ that contains y_1 and y_3, then vertex y_1 is recoloured α_3 and vertex y_3 is recoloured α_1. Furthermore, the new component H_{12} with colours α_1 and α_2 that contains y_2 satisfies

$$H_{12} \supset \mu[x, y_2],$$

since from Part 4, chains $\mu[y_1, y_2]$ and $\mu'[y_1, y_3]$ have no common vertex (except y_1).

On the other hand, the component H_{23} with colours α_2 and α_3 that contains y_1 satisfies

$$H_{23} \supset \mu[y_1, x],$$

since x has colour α_2 and is adjacent to y_1. Hence, x is a vertex common to the connected components H_{12} and H_{23}. This contradicts Part 4 and completes the proof.

Q.E.D.

The next section will present extensions of the Brooks theorem.

2. γ-Critical graphs

A simple graph G is defined to be *γ-critical* if for each vertex x_0 the subgraph G_0 generated by $X - \{x_0\}$ has chromatic number $\gamma(G_0) < \gamma(G)$.

Note that in this case, $\gamma(G_0) = \gamma(G) - 1$ because if $\gamma(G) = q + 1$ and $\gamma(G_0) < q$, then G_0 can be coloured with $q - 1$ colours and vertex x_0 can be coloured with a q-th colour, which contradicts that $\gamma(G) > q$.

We shall now study the properties of γ-critical graphs.

Property 1. *A graph G with $\gamma(G) = q + 1$ has a γ-critical subgraph with* $\gamma(G) = q + 1$.

If G is not γ-critical, there is a vertex x_0 whose removal does not decrease the chromatic number. If the subgraph G_0 generated by $X - \{x_0\}$ is not γ-critical, then again there is a vertex x_1 whose removal from G_0 does not decrease the chromatic number, etc. Eventually, this process locates a γ-critical subgraph.
                                                                    Q.E.D.

Property 2. *If G is a simple γ-critical graph with $\gamma(G) = q + 1$, then* $d_G(x) \geqslant q$ *for all x.*

If $d_G(x_0) < q$, the subgraph G_0 generated by $X - \{x_0\}$ can be coloured with q colours. At least one of these q colours is not adjacent to vertex x; thus, x can be coloured with this colour, which contradicts $\gamma(G) = q + 1$.
                                                                    Q.E.D.

Property 3. *A γ-critical graph is connected.*

The proof is obvious.

Property 4. *A γ-critical graph contains no articulation set that is a clique.*

Let G be a γ-critical graph with $\gamma(G) = q + 1$, and let A be a clique that is an articulation set of G. Since G is γ-critical,

$$| A | = \gamma(G_A) \leqslant q \ .$$

Let $B_1, B_2, B_3, \ldots$ be the connected components of the subgraph generated by $X - A$, and let $B_1' = B_1 \cup A$, $B_2 = B_2' \cup A$, ... be the corresponding pieces. Since G is γ-critical, each piece can be coloured with q colours. Thus, G can be coloured with q colours, which contradicts $\gamma(G) = q + 1$.
                                                                    Q.E.D.

Property 5. *A γ-critical graph has no articulation points.*

This follows from property 4 since an articulation point is a 1-clique.

Property 6. *If G is a γ-critical graph with $\gamma(G) = q + 1$, and if $A = \{a, b\}$ is an articulation set of G, then there are exactly two pieces B_1' and B_2' relative to this articulation set and*

$$\gamma(B_1') = \gamma(B_2') = q \ .$$

Clearly, a piece B' relative to A can be coloured with q colours. Three cases must be considered:

(1) B' cannot be coloured with q colours so that a and b have the same colour,

(2) B' cannot be coloured with q colours so that a and b have different colours,

(3) There is a colouring of B' in which a and b have the same colour, and there is also a colouring of B' in which a and b have different colours.

Case (3) cannot apply to a piece B' because the corresponding component B can be removed without changing the chromatic number of G. If Case (1) applies to B', then $\gamma(B') = q$ because if $\gamma(B') = q - 1$, vertices a and b can be coloured with a q-th colour, which contradicts that Case (1) applies to B'. Similarly, if Case (2) applies to B', then $\gamma(B') = q$.

Since $\gamma(G) = q + 1$, there is a piece satisfying Case (1) and a piece satisfying Case (2). Since G is γ-critical, there is only one piece of each type.

Q.E.D.

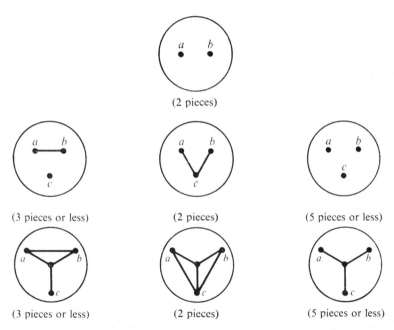

Fig. 15.6. Various types of articulation sets in a γ-critical graph

Property 7. *If G is γ-critical with $\gamma(G) = q + 1 \geqslant 4$, and if $A = \{a, b, c\}$ is an articulation set, then one of the following cases occurs:*

I. G_A *contains only one edge and A has at most three pieces, each with chromatic number q.*

II. G_A *contains exactly two edges, and A has at most two pieces, each with chromatic number q.*

III. G_A *contains no edges, and A has at most five pieces, each with chromatic number q or $q - 1$.*

We shall show only Case I. For example, suppose that a and b are the adjacent vertices in A. If piece B' is coloured with colours 1, 2, ..., q, then vertices a, b, c may be coloured 121, or 122, or 123. However, not all of these three colourings are possible for B' because G is γ-critical; it follows that $\gamma(B') = q$ (because if $\gamma(B') \leqslant q - 1$, then using a q-th colour, all three of these colourings would be possible).

If two of these colourings are possible for piece B'_1, then there is a piece B'_2 for which only the third colouring is possible (and there are only two pieces since G is γ-critical).

If no piece admits two of the three colourings, then there exist three pieces B'_1, B'_2, B'_3 that respectively admit one of each of the three colourings.

Q.E.D.

Property 8. *A γ-critical graph $G = (\dot{X}, E)$ with $\gamma(G) = q + 1$ cannot be disconnected by the removal of $q - 1$ edges, i.e.,*

$$m_G(A, X - A) \geqslant q \qquad (A \subset X, A \neq \varnothing, X).$$

We shall show that a contradiction results if there are two nonempty sets A and B of vertices G with

$$A \cup B = X, \quad A \cap B = \varnothing, \quad m_G(A, B) \leqslant q - 1.$$

Since G is γ-critical, the subgraph G_A is q-colourable; consider a *q-colouring function* for G_A, i.e., a mapping $f(a)$ from A into $\{1, 2, ..., q\}$ such that

$$x, y \in A, \ [x, y] \in E \quad \Rightarrow \quad f(x) \neq f(y).$$

Let $a_1, a_2, ..., a_r$ be the vertices of A adjacent to B; choose function f so that

$$f(a_i) \leqslant i \qquad (i = 1, 2, ..., r).$$

Let $e_1, e_2, ..., e_p$ denote the edges that join A and B, with indices chosen so that

$$e_s = [a_i, x], \quad e_t = [a_j, y], \quad i < j \quad \Rightarrow \quad s < t.$$

Let g^1 be a q-colouring function for the subgraph G_B. We shall show that by a sequence of transformations, this q-colouring of G_B can become compatible with the q-colouring of G_A, which contradicts $\gamma(G) = q + 1$. If edge e_1 is of the form $[a_1, b_i]$, define a q-colouring $g^2(y)$ of G_B by

$$g^2(y) \begin{cases} = g^1(y) & \text{if} \quad g^1(y) \neq 2, \, g^1(b_i) \\ = 2 & \text{if} \quad g^1(y) = g^1(b_i) \\ = g^1(b_i) & \text{if} \quad g^1(y) = 2 \, . \end{cases}$$

Thus

$$g^2(b_i) = 2 > f(a_1) \, .$$

Now consider $e_2 = [a_k, b_j]$, and similarly, define a q-colouring $g^3(y)$ by

$$g^3(y) \begin{cases} = g^2(y) & \text{if} \quad g^2(y) \neq 3, \, g^2(b_j) \\ = 3 & \text{if} \quad g^2(y) = g^2(b_j) \\ = g^2(b_j) & \text{if} \quad g^2(y) = 3 \, . \end{cases}$$

Thus,

$$g^3(b_i) \geqslant g^2(b_i) > f(a_1) \, ,$$
$$g^3(b_j) = 3 > f(a_k) \, .$$

This process can be continued, and since $p = m_G(A, B) \leqslant q - 1$, the last edge e_p gives a q-colouring g^{p+1} with

$$[a_s, b_t] \in E \quad \Rightarrow \quad g^{p+1}(b_t) > f(a_s) \, .$$

Thus G can be coloured with q colours, which contradicts $\gamma(G) = q + 1$.
$$\text{Q.E.D.}$$

Consider a γ-critical graph G with $\gamma(G) = q + 1$; from Property 2, graph G satisfies

$$\min_{x \in X} d_G(x) \geqslant q \, .$$

To study the structure of G, we shall consider the set

$$M = \{ x \mid x \in X, \, d_G(x) = q \} \, ,$$

and study the structure of the subgraph G_M generated by M.

Lemma 1. *Let* $\mu = [a_0, a_1, a_2, ..., a_{k-1}, a_0]$ *be a cycle of* G_M *and let* f *be a* $(q + 1)$-*colouring of* G *with only vertex* a_0 *having the* $(q + 1)$-*st colour; then there exists a* $(q + 1)$-*colouring* g *of* G *with only vertex* a_0 *having the* $(q + 1)$-*st colour such that*:

$$\begin{cases} g(x) &= f(x) \qquad (x \notin \mu) \\ g(a_0) &= f(a_0) \\ g(a_1) &= f(a_2) \\ g(a_2) &= f(a_3) \\ \cdots\cdots\cdots\cdots \\ g(a_{k-1}) &= f(a_1) \,. \end{cases}$$

Let $\alpha_1, \alpha_2, ..., \alpha_{q+1}$ be the colours in colouring f, where $f(a_0) = \alpha_{q+1}$.

Since $d_G(a_i) = q$, each colour $\alpha_1, \alpha_2, ..., \alpha_q$ appears exactly once in $\Gamma_G(a_i)$. Hence, interchanging the colours of a_0 and of a_1 gives a $(q + 1)$-colouring with only vertex a_1 having the $(q + 1)$-st colour. This procedure can be repeated around cycle μ until vertex a_0 has colour α_{q+1}. This produces the required colouring g.

Q.E.D.

Lemma 2. *Let* $\mu = [a_0, a_1, ..., a_{k-1}, a_0]$ *be a cycle of* G_M *with no chord incident to vertex* a_0; *then* μ *is an odd cycle.*

1. Since graph G is $(q + 1)$-critical, there exists a $(q + 1)$-colouring f of G with only vertex a_0 having the $(q + 1)$-st colour α_{q+1}.

Since $d_G(a_0) = q$, the set $\Gamma_G(a_0) - \{ a_1, a_{k-1} \}$ contains exactly $q - 2$ vertices, and none of these vertices lies on μ. Each of these vertices take a different colour in colouring f; let $\alpha_1, \alpha_2, ..., \alpha_{q-2}$ be these colours.

2. We shall show that only the colours α_{q-1} and α_q are assigned to vertices $a_1, a_2, ..., a_{k-1}$ in colouring f.

If this is not true, then there exists in cycle μ a vertex a_i, $i > 0$, with colour $\alpha_j, j \neq q - 1, q$. After repeating $i - 1$ times the operation of Lemma 1, a colouring g is obtained with only vertex a_0 having colour α_{q+1} and with

$$\begin{aligned} g(x) &= f(x) \qquad (x \notin \mu) \\ g(a_0) &= f(a_0) \\ g(a_1) &= f(a_i) = \alpha_j \\ g(a_2) &= f(a_{i+1}) \\ \cdots\cdots\cdots\cdots \\ g(a_{k-1}) &= f(a_{i-1}) \,. \end{aligned}$$

But, then, the set $\Gamma_G(a_0)$ has two vertices with colour α_j in colouring g. Thus, a_0 can be recoloured with a colour other than α_{q+1}, which contradicts $\gamma(G) = q + 1$.

3. In colouring f, all vertices of cycle μ, except a_0, have either colour α_{q-1} or α_q. Since vertices a_1 and a_{k-1} belong to $\Gamma_G(a_0)$ and cannot have the same colour, cycle μ is necessarily odd.

Q.E.D.

Theorem 7. *Let G be a γ-critical graph with $\gamma(G) = q + 1$; then $d_G(x) \geqslant q$ for all x, and each block of the subgraph G_M generated by set $M = \{ x \mid x \in X, d_G(x) = q \}$ is either a clique or an odd cycle without chords.*

From Lemma 2, each even cycle of graph G_M has at least two chords. Thus, from Theorem (7, Ch. 9), G_M has the required form.

Q.E.D.

Remark. If q equals

$$h = \max_{x \in X} d_G(x) \, ,$$

the Brooks theorem follows from Theorem 7: Let G be a graph with

$$\max d_G(x) = h \quad \text{and} \quad \gamma(G) = h + 1 \, .$$

Consider the γ-critical subgraph G' of G with $\gamma(G') = h + 1$ (which exists by Property 1). Since $d_{G'}(x) \geqslant h$ (Property 2), and since $d_{G'}(x) \leqslant d_G(x) \leqslant h$, graph G' is regular of degree h. In other words, the vertex set X' of graph G' is equal to the set

$$M = \{ x/x \in X' \, ; \, d_{G'}(x) = h \} \, .$$

Furthermore, G' is connected and has no articulation points (Properties 3 and 5). Since $G'_M = G'$ is a block, Theorem 7 shows that G' is either a $(h + 1)$-clique or an odd cycle without chords.

Hence graph G has either a connected component that is a $(h + 1)$-clique, or, if $h = 2$, an odd cycle without chords.

The Brooks theorem follows.

The next theorem is also an extension of the Brooks theorem.

Lemma. *Let G be a γ-critical graph with $\gamma(G) = q + 1 \geqslant 4$. If G contains a clique $C = \{ c_1, c_2, ..., c_q \}$ with q vertices such that*

$$\sum_{i=1}^{q} (d_G(c_i) - q) \leqslant q - 3 \, ,$$

then in a q-colouring of $X - C$, the sets

$$A_i = \Gamma_G(c_i) - C$$

contain a common colour.

Suppose that

$$d_G(c_1) \leqslant d_G(c_2) \leqslant \cdots \leqslant d_G(c_q) \, .$$

Since G is γ-critical, then by Property 2, $d_G(c_1) \geqslant q$. Since $d_G(c_i) - q \geqslant 1$

for at most $q - 3$ of the c_i, then

$$d_G(c_1) = d_G(c_2) = d_G(c_3) = q \, .$$

Thus $| A_1 | = 1$. Let α_1 be the colour of the unique vertex of A_1 in the q-colouring g of $X - C$. Suppose that there is a set A_{i_0} that does not contain colour α_1. (If several such sets exist, take index i_0 to be as small as possible.)

We shall now show that this is impossible, i.e., clique C can be coloured with the q colours used for $X - C$, which will contradict $\gamma(G) = q + 1$.

Let $\alpha_1, \alpha_2, ..., \alpha_q$ be the q colours of $g(X - C)$. Define the q-colouring $g(x)$ of C successively in the following way:

— let $g(c_{i_0})$ be colour α_1;
— if $q \neq i_0$, let $g(c_q)$ be any colour different from the colours of $A_q \cup \{ c_{i_0} \}$;
— if $q - 1 \neq i_0$, let $g(c_{q-1})$ be any colour different from the colours of $A_{q-1} \cup \{ c_{i_0}, c_q \}$;
— etc....

Note that, for $j \leqslant q$,

$$\sum_{i=j}^{q} \left(d_G(c_i) - q \right) = \sum_{i=j}^{q} \left(| A_i | - 1 \right) \leqslant q - 3 \, .$$

Thus,

$$(q - j + 1) \left(| A_j | - 1 \right) \leqslant q - 3 \, ,$$

or

$$| A_j | \leqslant \frac{q - 3}{q - j + 1} + 1 \, .$$

Let $g(c_j)$ take a value in the set $\{ \alpha_1, \alpha_2, ..., \alpha_q \}$ if the set

$$A_j \cup \{ c_{i_0}, c_q, c_{q-1}, ..., c_{j+1} \}$$

has less than q vertices. This is possible if

$$\frac{q - 3}{q - j + 1} + 1 + (q - j + 1) \leqslant q - 1 \, ,$$

i.e., if

$$- j + 3 + \frac{q - 3}{q - j + 1} \leqslant 0 \, ,$$

i.e.,

$$j^2 - (q + 4) j + 4 q \leqslant 0 \, .$$

Since the roots of this quadratic equation are $j' = 4$ and $j'' = q$, vertex c_j can be coloured for:

$$4 \leqslant j \leqslant q.$$

It remains to colour c_j for $j = 1, 2, 3$. Since

$$d_G(c_1) = d_G(c_2) = d_G(c_3) = q,$$

we may take: $A_1 = \{\, a_1 \,\}, \; A_2 = \{\, a_2 \,\}, \; A_3 = \{\, a_3 \,\}.$

If $i_0 \neq 2, 3$, then $g(a_1) = g(a_2) = g(a_3) = \alpha_1$. Moreover, $C - \{\, c_1, c_2, c_3 \,\}$ has been coloured with $q - 3$ colours, including α_1. Thus, c_1, c_2, c_3 can be coloured with 3 colours different from α_1.

If $i_0 = 3$, then $g(a_1) = g(a_2) = \alpha_1$. Moreover, $q - 3$ colours have been used (including colour α_1 for c_3). Thus, c_1 and c_2 can be coloured with 2 colours different from α_1.

If $i_0 = 2$, a similar result follows.

In all cases, C can be coloured with the q colours that have already been used for $X - C$.

Q.E.D.

Theorem 8 (Dirac [1952]). *Let G be a graph with chromatic number $\gamma(G) = q + 1$ and without any $(q + 1)$-cliques. Let $S = \{\, x \mid d_G(x) > q \,\}$. Then,*

$$\sum_{x \in S} (d_G(x) - q) \geqslant q - 2.$$

If $q = 1$, the theorem is trivial because G cannot have both chromatic number 2 and no 2-cliques.

If $q = 2$, then

$$\sum_{x \in S} (d_G(x) - q) \geqslant 0 = q - 2.$$

Suppose that the theorem is true for all graphs with chromatic number $\leqslant q$; we shall show that it is also true for a graph G with chromatic number $\gamma(G) = q + 1$, where $q \geqslant 3$. (From the Brooks Theorem, we already know that $S \neq \varnothing$.) Suppose that G contains no $(q + 1)$-cliques and that

$$\sum_{x \in S} (d_G(x) - q) < q - 2.$$

We shall show that this leads to a contradiction.

We may assume that G is γ-critical. (Otherwise, replace G by a γ-critical subgraph.) Since $d_G(x) \geqslant q$ for all $x \in X$, then

(1) $$\sum_{x \in X} (d_G(x) - q) < q - 2.$$

We shall now produce a contradiction for each of the following two possible cases:

CASE 1. *G contains no q-cliques.*

Let x_0 be a vertex of S. (By the Brooks theorem, S contains at least one vertex.) Let $X - T$ be a maximal stable set that contains x_0, and let G_T be the subgraph generated by T. Clearly, $\gamma(G_T) < q + 1$ (because $T \neq X$ and G is γ-critical).

If $\gamma(G_T) < q$, then G could be coloured with only q colours. Thus

$$\gamma(G_T) = q .$$

Let

$$S' = \{ x / x \in T, \; d_{G_T}(x) > q - 1 \} .$$

We have

$$d_{G_T}(x) + 1 \leqslant d_G(x) \qquad (x \in T) ,$$

(because each vertex $x \in T$ is adjacent to $X - T$, since $X - T$ is a maximal stable set.)

Hence $S' \subset S$.

In fact, S' is a proper subset of S since $x_0 \in S$ and $x_0 \notin S'$. Hence,

$$\sum_{x \in S'} [d_{G_T}(x) - (q - 1)] \leqslant \sum_{x \in S'} (d_G(x) - q) \leqslant$$

$$\leqslant \sum_{x \in S} (d_G(x) - q) - 1 < q - 3 .$$

But, from the induction hypothesis, graph G_T satisfies

$$\sum_{x \in S'} [d_{G_T}(x) - (q - 1)] \geqslant (q - 1) - 2 ,$$

which is the required contradiction.

CASE 2. *G contains a q-clique.*

Let $C = \{ c_1, c_2, ..., c_q \}$ be this clique. Set $A_i = \Gamma_G(c_i) - C$ is not empty (because $d_G(c_1) \geqslant q$ since G is γ-critical). Set $A = \Gamma_G(C) - C$ contains at least two vertices (because G contains no $(q + 1)$-cliques).

Renumbering the vertices if necessary, we may assume that

$$d_G(c_1) \leqslant d_G(c_2) \leqslant \cdots \leqslant d_G(c_q) .$$

Since $d_G(c_i) - q \geqslant 1$ for at most $q - 3$ of the c_i, from inequality (1), it follows that

$$d_G(c_1) = q .$$

Thus, let a_1 be the unique vertex of $A_1 = \Gamma_G(c_1) - C$. Since G contains no

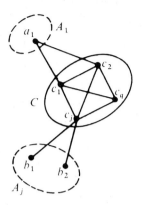

Fig. 15.7.

$(q + 1)$-cliques, there is a vertex c_j that is non-adjacent to a_1. (If several such vertices exist, let j be the smallest possible index.) Let

$$A_j = \{ b_1, b_2, ... \} .$$

Let G' be the graph obtained from G_{X-C} by joining vertex a_1 to vertices $b_1, b_2, \, ...$ Since G_{X-C} is q-colourable the graph $G' - \{ a_1 \}$ is q-colourable; hence

$$\gamma(G') \leqslant q + 1.$$

Furthermore, from the lemma, graph G' cannot be coloured with q colours. Hence,

$$\gamma(G') = q + 1 .$$

Let G'' be the γ-critical subgraph of G' with $\gamma(G'') = q + 1$. Let graph H be obtained from G'' by removing the edges of the form $[a_1, b_i]$.

Graph H contains vertex a_1 and one of the vertices of A_j, because if G'' did not contain an edge of the form $[a_1, b_i]$, then $\gamma(G'') \leqslant \gamma(G_{X-c}) \leqslant q$.

Note that

$$d_G(a_1) \geqslant d_{G''}(a_1) - | A_j | + (j - 1) \geqslant q - | A_j | - 1 + j .$$

If $| A_j | \geqslant 2$, we would have

$$\sum_{x \in S} [d_G(x) - q] \geqslant \sum_{i=j}^{q} (d_G(c_i) - q) + (d_G(a_1) - q) \geqslant$$

$$\geqslant (q - j + 1)(| A_j | - 1) + (- | A_j | - 1 + j) =$$

$$= (q - j + 1)(| A_j | - 1) - (| A_j | - 1) + j - 2 =$$

$$= (q - j)(|A_j| - 1) + j - 2 \geqslant$$
$$\geqslant (q - j) + j - 2 = q - 2.$$

This contradicts inequality (1).

Hence, $|A_j| = 1$.

Let b be the unique vertex of A_j. Graph $H = (Y, F)$ can be formed from the $(q + 1)$-critical graph G'' by the removal of the only edge $[a_1, b]$. Thus,

$$d_H(a_1) = d_{G''}(a_1) - 1 \geqslant q - 1$$
$$d_H(b) = d_{G''}(b) - 1 \geqslant q - 1$$
$$d_H(x) = q \quad (x \in Y, x \neq a_1, b).$$

Combining this with Property 8 gives

$$\sum_{x \in S} (d_G(x) - q) \geqslant \sum_{x \in Y} (d_G(x) - q) =$$
$$= m_G(Y, X - Y) + \sum_{x \in Y} (d_H(x) - q) \geqslant q - 2.$$

This contradicts inequality (1).

Q.E.D.

Corollary. *Let G be a γ-critical graph with $\gamma(G) = q + 1$, without $(q + 1)$-cliques, and with n vertices and m edges; then*

$$2m \geqslant (n + 1)q - 2.$$

Since $d_G(x) \geqslant q$ for each vertex x, we have

$$2m = \sum_{x \in S} d_G(x) + (n - |S|)q \geqslant q - 2 + |S|q + nq - |S|q =$$
$$= (n + 1)q - 2.$$

Q.E.D.

Remark. The inequalities of Theorem 8 and its corollary are the best possible for $q \geqslant 3$. To construct a graph G with

$$\gamma(G) = q + 1,$$

such that the inequality of Theorem 8 holds with equality,

$$C = \{c_1, c_2, ..., c_{q+1}\}$$

and

$$D = \{c_1, d_2, ..., d_{q+1}\},$$

and remove edges

$$[c_1, c_2], [c_1, d_{q+1}],$$

and add edge $[c_2, d_{q+1}]$. See Fig. 15.8.

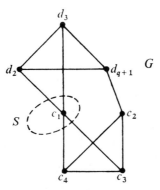

Fig. 15.8.

$$m = 11, \quad n = 7, \quad q = 3$$

Clearly, $\gamma(G) = q + 1$, and G contains no $(q + 1)$-clique. Furthermore, c_1 is the only vertex of S, and

$$d_G(c_1) = 2q - 2 .$$

Thus,

$$\sum_{x \in S} (d_G(x) - q) = d_G(c_1) - q = q - 2 .$$

Q.E.D.

3. The Hajós theorem

Let G be a simple graph. An *elementary contraction* on G is defined to be any operation that replaces two adjacent vertices a and b of G by a single vertex c and joins c to each vertex of

$$\Gamma_G(a) \cup \Gamma_G(b) .$$

If the contraction operation is repeated enough times, a clique will be obtained. See Fig. 15.9.

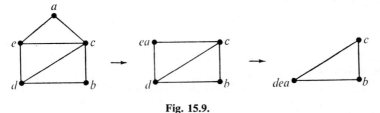

Fig. 15.9.

One of the most interesting conjectures concerning the chromatic number of a connected graph is due to Hadwiger:

Conjecture (Hadwiger [1943]). *Every connected graph G with $\gamma(G) = q$ can be transformed into a q-clique by a sequence of elementary contractions.*

For $q \leqslant 4$, Dirac [1952] and Halin have validated the conjecture. For $q = 5$, Wagner [1964] has shown that Hadwiger's conjecture is equivalent to the four colour theorem.

Let $\mathscr{G}_q$ be the class of all graphs with $\gamma(G) > q$. Consider the following three operations in $\mathscr{G}_q$:

(I) Add vertices and edges.

(II) Let G_1 and G_2 be two disjoint graphs. Let a_1 and b_1 be two adjacent vertices in G_1, and let a_2 and b_2 be two adjacent vertices in G_2. Remove edges $[a_1, b_1]$ and $[a_2, b_2]$. Add an edge $[b_1, b_2]$, and contract set $\{a_1, a_2\}$ into a single vertex. See Fig. 15.10.

(III) Contract a set of two non-adjacent vertices into a single vertex.

Clearly, if we perform these operations on graphs in $\mathscr{G}_q$, the resulting graph belongs also to $\mathscr{G}_q$.

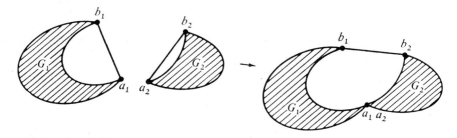

Fig. 15.10

Theorem 9 (Hajós [1961]). *Every graph G with $\gamma(G) > q$ can be obtained from $(q + 1)$-cliques by means of operations* (I), (II) *and* (III).

Suppose that $G = (X, E)$ is a graph of order n which contradicts the theorem, and assume that G has a maximal number of edges with respect to this property.

1. We shall show that the relation "vertices x and y are identical or non-adjacent" is transitive. Suppose that this is not true: there exist three vertices x, y, z such that $[x, y] \notin E$, $[y, z] \notin E$ and $[x, z] \in E$. From the maximality of G, the graphs $G + [x, y]$ and $G + [y, z]$ respectively contain graphs G_1 and G_2 that can be constructed from K_{q+1} by the operations. We may assume that

G_1 and G_2 are disjoint (by duplicating). Let $[x_1, y_1]$ be the edge in G_1 that corresponds to $[x, y]$, and let $[y_2, z_2]$ be the edge in G_2 that corresponds to $[y, z]$ (since $[x, y]$ appears in G_1 and $[y, z]$ appears in G_2). Remove edge $[x_1, y_1]$ from G_1 and edge $[y_2, z_2]$ from G_2, contract $\{y_1, y_2\}$ into a single vertex and join vertices x_1 and z_2 (i.e., perform operation (II) on graphs G_1 and G_2). In the resulting graph G', identify all pairs of vertices that correspond to the same vertex in G, which are necessarily non-adjacent in G' (operation (III)). The resulting graph G'' is a partial subgraph of G. Hence graph G can be constructed from graph G'' by operation (I). This contradicts the definition of G.

2. The above equivalence relation divides X into classes $S_1, S_2, ..., S_k$ such that any two vertices belonging to distinct classes are adjacent and such that any two vertices belonging to the same class are non-adjacent. Since G cannot be constructed from K_{q+1} by operation (I), we have $G \not\supset K_{q+1}$. Thus $k \leqslant q$ and, consequently,

$$\gamma(G) \leqslant q .$$

which contradicts $\gamma(G) > q$.

Q.E.D.

A variation of Hadwiger's conjecture was proposed by Hajós, who suggested that every q-chromatic graph has as a partial subgraph the graph obtained by replacing each edge of K_q by an elementary chain. For $q = 1, 2, 3$ this is trivial. For $q = 4$, this follows from a theorem due to Dirac.

However, this conjecture was disproved by Catlin [1979], who found counter-examples for $\gamma(G) \geqslant 7$.

Analogous conjectures have been proposed by Bollobas, Catlin and Erdös [1980].

4. Chromatic polynomials

Let $G = (X, E)$ be a graph with vertices $x_1, x_2, ..., x_n$. In this section we shall enumerate the distinct λ-colourings possible for graph G, i.e. the mappings $f(x)$ from X into $\{1, 2, ..., \lambda\}$ such that

$$[x, y] \in E \quad \Rightarrow \quad f(x) \neq f(y) .$$

The chromatic polynomial of G is defined to be a function $P(G; \lambda)$ that expresses for each integer λ the number of distinct λ-colouring possible for graph G. This number was originally expressed by G. Birkhoff [1912] with determinants. An excellent treatment of chromatic polynomials, due to

Read [1968], serves as the basis for this presentation.

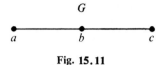

Fig. 15.11

EXAMPLE 1. Colour the graph G in Fig. 15.11 with λ colours; vertex b can be coloured first with any of the λ colours, then vertex a or c can be coloured with any of the remaining $(\lambda - 1)$ colours. Hence, $P(G; \lambda) = \lambda(\lambda - 1)^2$.

EXAMPLE 2. Let G be the n-clique K_n. Vertex x_1 can be coloured first with any one of the λ colours. Then vertex x_2 can be coloured with any one of the remaining $\lambda - 1$ colours. Then vertex x_3 can be coloured with any one of the remaining $\lambda - 2$ colours, etc. Hence,

$$P(G ; \lambda) = \lambda(\lambda - 1)(\lambda - 2)\dots(\lambda - n + 1).$$

This polynomial is often denoted by $[\lambda]_n$, where $[\lambda]_n = 0$ for $\lambda < n$.

Property 1. *Let a and b be two non-adjacent vertices in graph G. Let $\tilde{G}$ be the graph obtained from G by joining a and b by an edge, and let $\overline{G}$ be the graph obtained from G by contracting a and b into a single vertex. Then,*

$$P(G ; \lambda) = P(\tilde{G} ; \lambda) + P(\overline{G} ; \lambda).$$

Recall the principle of contraction–connection mentioned in Section 1: Clearly, $P(G ; \lambda)$ equals the number of λ-colourings in G for which a and b have different colours plus the number of λ-colourings of G for which a and b have the same colour. The formula follows.

Q.E.D.

If the connection and contraction operations performed on graph G terminate with complete graphs $H_1, H_2, H_3, \dots$, then

$$P(G ; \lambda) = \sum_i P(H_i ; \lambda).$$

For example, the contractions and the connections on the graph G in Fig. 15.2 terminate with five distinct cliques: one 5-clique, three 4-cliques and one 3-clique. Hence

$$P(G ; \lambda) = [\lambda]_5 + 3[\lambda]_4 + [\lambda]_3.$$

Corollary. *If G is a graph of order n, the chromatic polynomial $P(G ; \lambda)$*

is a polynomial of degree n in λ*. The coefficient of* λ^n *is* $+$ 1, *and the coefficient of* λ^0 *is* 0.

Clearly, $P(G \,;\, \lambda)$ is the sum of terms of the form $[\lambda]_p$, where p assumes the value n exactly once. Q.E.D.

Property 2. *If graph G has p connected components* $H_1, H_2, ..., H_p$, *then*

$$P(G \,;\, \lambda) = P(H_1 \,;\, \lambda)\, P(H_2 \,;\, \lambda) ... P(H_p \,;\, \lambda) \,.$$

Clearly, each connected component can be coloured separately, and the formula follows. Q.E.D.

Property 3. *If graph G contains an articulation set A that is a k-clique with q pieces* $H_1, H_2, ..., H_q$ *relative to A, then*

$$P(G \,;\, \lambda) = \left([\lambda]_k\right)^{1-q} P(H_1 \,;\, \lambda) ... P(H_q \,;\, \lambda) \,.$$

Colour A in one of the $[\lambda]_k$ possible ways. Then there are $\dfrac{P(H_i \,;\, \lambda)}{[\lambda]_k}$ different ways to colour piece H_i, and the formula follows.

Q.E.D.

Theorem 10. *The coefficients of* $P(G \,;\, \lambda)$ *are alternately non-negative and non-positive integers.*

This result is a direct consequence of the Möbius function theory developed by G. C. Rota [1964]. We shall give here an inductive proof on the order n of G.

If $n = 1, 2$, the theorem is obvious. If the theorem is valid for all graphs of order $< n$, we shall show that it is also valid for a graph G of order n.

Clearly, this is trivial if G has only one edge. If the theorem is true for all graphs of order n with less than m edges, we shall show that it is also true for a graph $\tilde{G}$ of order n with m edges.

If a and b are two adjacent vertices in $\tilde{G}$, let $G = \tilde{G} - [a, b]$, and denote by $\overline{G}$ the graph obtained from G by contracting $\{\, a, b \,\}$ into a single vertex.

From Property 1,

$$P(\tilde{G} \,;\, \lambda) = P(G \,;\, \lambda) - P(\overline{G} \,;\, \lambda) \,.$$

Since the theorem is assumed to be valid for all graphs of order n with less than m edges, there exist integers $a_i \geqslant 0$ such that

$$P(G \,;\, \lambda) = \lambda^n - a_1 \lambda^{n-1} + a_2 \lambda^{n-2} - \cdots$$

Furthermore, since the theorem is assumed to be valid for all graphs of order $< n$, there exist integers $b_i \geqslant 0$ such that

$$P(\overline{G}\,;\lambda) = \lambda^{n-1} - b_1\,\lambda^{n-2} + b_2\,\lambda^{n-3} - \cdots$$

Hence,

$$P(\widetilde{G}\,;\lambda) = \lambda^{n} - (a_1 + 1)\,\lambda^{n-1} + (a_2 + b_1)\,\lambda^{n-2} - \cdots$$

Thus, the coefficients are alternately non-negative and non-positive integers.

Q.E.D.

Corollary. *If G is a graph of order n with m edges, then the coefficient of* λ^{n-1} *in the polynomial* $P(G;\lambda)$ *equals* $- m$.

If $m = 1$, then $P(G\,;\lambda) = \lambda^{n-1}(\lambda - 1)$, and the result follows. If $m > 1$, it suffices to observe in the proof of theorem 10 that the addition of an edge increases the absolute value of the coefficient of λ^{n-1} by one unit.

Q.E.D.

Theorem 11. *A graph G of order n is a tree if, and only if,*

$$P(G\,;\lambda) = \lambda(\lambda - 1)^{n-1}\,.$$

1. First we shall show that if G is a tree of order n, then the formula is valid. Clearly, this is true for $n = 1$ and $n = 2$. If the formula is valid for all trees of order $< n$, we shall show that it is also valid for a tree G of order n.

Since G has a pendant edge (Theorem 2, Ch. 3), Property 3 can be invoked with the pendant edge as H_1 and with the remaining tree of order $n - 1$ as H_2. Thus,

$$P(G\,;\lambda) = (\lambda)^{1-2}\,\lambda(\lambda - 1)\,\lambda(\lambda - 1)^{n-2} = \lambda(\lambda - 1)^{n-1}\,.$$

2. Let G be a graph of order n with $P(G\,;\lambda) = \lambda(\lambda - 1)^{n-1}$. Graph G is connected: Otherwise, it has $p \geqslant 2$ components $H_1, H_2, ..., H_p$ and no constant terms appear in the polynomials $P(H_i\,;\lambda)$ by the corollary to Property 1; by Property 2, $P(G\,;\lambda)$ equals their product and, therefore, could not have a non-zero coefficient for λ.

Since the coefficient of λ^{n-1} is $- (n - 1)$, graph G has $n - 1$ edges (by the corollary to Theorem 10). Hence, G is a tree.

Q.E.D.

The roots of chromatic polynomials have been notably studied by Berman and Tutte [1969] and by Tutte [1970].

5. The Gallai–Roy theorem and related results

Gallai and Roy showed (independently) that the longest (elementary) path in a graph $G = (X, U)$ has at least $\gamma(G)$ vertices. This bound is the best possible, for equality occurs whenever every edge is directed towards the colour with the largest index. For example, the graph in Fig. 13.6 has $\gamma(G) = 4$ and also has a path with 4 vertices. In the following section we extend this result in several different directions.

Let $(S_1, S_2, ..., S_k)$ be a k-colouring of the vertices of G; the number of colours encountered by a path μ is both $\leq k$ and $\leq |\mu|$. We shall say that the k-colouring *strongly colours μ* if this path meets exactly min $\{ k, |\mu| \}$ colours.

Theorem 12. *Let $k = \max |\mu|$, i.e. the maximum cardinality of a path in G. Then for every path partition $M = \{ \mu_1, \mu_2, ... \}$ of the vertices of G, there exists a k-colouring $(S_1, S_2, ..., S_k)$ of G which strongly colours all paths in M.*

Construct a partial graph H of G by first taking all arcs from all paths in M and then adding as many arcs from G as possible without creating any circuits. For each vertex x, denote by $t(x)$ the maximum cardinality of a path in H with origin x. Note that H has no circuits, and that $t(x) \leq \max |\mu| = k$. Thus, if (x, y) is an arc of H then $t(x) > t(y)$. If (x, y) is an arc of $G - H$, then $H + (x, y)$ contains a circuit, thus H has a path from y to x, and thus $t(y) > t(x)$. Hence, every pair of adjacent vertices x and y satisfy $t(x) \neq t(y)$, and t defines a k-colouring of G. As well, every path in M is strongly coloured.

Q.E.D.

Corollary 1 (The Gallai–Roy theorem). *For every graph G, $\max |\mu| \geq \gamma(G)$.*

Take any partition $(\mu_1, \mu_2, ...)$ where μ_1 is a longest path, and all other μ_i are singletons.

Corollary 2 (Rédei's theorem). *Every complete anti-symmetric 1-graph G ('tournament') has a hamiltonian path.*

Since $\gamma(G) = n$, there is a path with n vertices.

Corollary 3. *For every transitive 1-graph G, $\max |\mu| = \gamma(G)$.*

From Corollary 1, $\gamma(G) \leq \max |\mu| = \omega(G) \leq \gamma(G)$.

Corollary 4 (The Chvátal–Komlös theorem). *Let G be a 1-graph whose arcs are partitioned in partial 1-graphs $\hat{G}_1, G_2, ..., G_q$, and let $p_1, p_2, ..., p_q$ be integers with $p_1 p_2 ... p_q < \gamma(G)$. Then there is some graph G_i which contains a path with cardinality $> p_i$.*

If not, then max $\{ \, | \, \mu \, | \, / \, \mu \text{ is a path in} \leqslant G_i \, \} \leqslant p_i$ for each i. Thus, by Corollary 1, $\gamma(G_i) \leqslant p_i$. Let $t_i(x)$ be a colouring of G_i with colours $1, 2, \ldots, p_i$. Then the q-tuple $t(x) = (t_1(x), t_2(x), \ldots, t_q(x))$ gives a colouring of G in $p_1 p_2 \ldots p_q$ colours. Hence, $\gamma(G) \leqslant p_1 p_2 \ldots p_q$, a contradiction.

Remark 1. If the 1-graph G is strongly connected, the Gallai–Roy theorem can be strengthened. Bondy [1976] proved that there exists a circuit μ with $| \, \mu \, | \geqslant \gamma(G)$.

Remark 2. An open question is whether or not every 1-graph G has an optimal colouring and a path μ such that μ meets each colour exactly once. This is obviously the case for $\gamma(G) = 2$, and can be easily proved if $\gamma(G) = 3$ (Berge [1983]).

In an attempt to unify the Gallai–Roy and Gallai–Milgram theorems, Berge [1982] proposed the following conjecture: call a family of k disjoint stable sets a 'partial k-colouring' of G. A path partition $M = (\mu_1, \mu_2, \ldots, \mu_p)$ of the vertices of G will be called k-optimal if it minimizes

$$B_k(M) = \sum_{i=1}^{p} \min \, \{k, | \, \mu_i \, |\}$$

For example, if there exists a hamiltonian path μ_0, then $M = \{ \, \mu_0 \, \}$ is a k-optimal partition for all k.

Conjecture. *Let k be an integer, $1 \leqslant k \leqslant \max | \, \mu \, |$. Then for every k-optimal partition M, there is a partial k-colouring of G which strongly colours each path in M.*

For $k = 1$, the conjecture yields the Gallai–Milgram theorem. For $k = \max | \, \mu \, |$, the conjecture gives the Gallai–Roy theorem (Theorem 12).

The conjecture has been proved for bipartite 1-graphs, transitive 1-graphs, circuit-free 1-graphs (Cameron), and symmetric 1-graphs (Payan). For a transitive 1-graph, it is an extension of a famous theorem due to Greene and Kleitman.

A weaker conjecture had been proposed by Linial [1981]. It would even be interesting to show the existence of any path partition M that could be strongly coloured by a k-colouring.

EXERCISES

1. Use the Brooks theorem to show that a regular graph G with

$$\gamma(G) + \gamma(\overline{G}) = n + 1 ,$$

is either
  (1) n isolated vertices, or
  (2) an n-clique, or
  (3) an elementary cycle of length 5.

2. A graph G with n vertices is said to be of type $T_3(n, p, q)$, where $n \geqslant pq$, if there exist two partitions $(C_1, C_2, ..., C_q)$ and $(D_1, D_2, ..., D_p)$ of the vertices such that
  (1) $\max | D_i | = q$,
  (2) $\max | C_i | = p$,
  (3) $| C_i \cap D_j | \leqslant 1$ for all i, j,
  (4) two vertices belonging to the same C_i are adjacent,
  (5) two vertices belonging to the same D_j are non-adjacent.
Show that if G is of type $T_3(n, p, q)$, then $\overline{G}$ is of type $T_3(n, p, q)$ and that

$$\gamma(G) = p \quad \Rightarrow \quad \gamma(\overline{G}) = q .$$

Show that for any two integers p and q such that

$$pq \geqslant n ,$$
$$p + q \leqslant n + 1 ,$$

there exists a graph G of type $T_3(n, p, q)$ with $\gamma(G) = p$ and $\gamma(\overline{G}) = q$.

(Finck [1966])

3. Show that in a 4-colourable simple graph G, the edges can be coloured red and blue such that each triangle contains two blue edges and one red edge.

4. Show that in a simple planar graph the edges can be coloured red and blue such that each triangle contains two blue edges and one red edge. (Schäuble [1968])

5. Consider an infinite graph whose vertices are the pairs (p, q) of integers with $p < q$, and with an edge linking (p, q) and (r, s) if $q = r$ or if $p = s$. Show that this graph contains no triangles and that its chromatic number is infinite. (Erdös, Hajnal [1960])

6. Consider an infinite family of sets $(A_i / i \in I)$ where $A_i \subset \{1, 2, ...\}$, $| A_i | < \infty$. Form a graph G whose vertices are the sets A_i and with an edge joining every pair of vertices whose corresponding sets meet.
  Show that for each i, the complementary graph of the subgraph of G generated by $\Gamma_G(i)$ has chromatic number $\leqslant | A_i |$.
  Show that an infinite graph $G = (I, \Gamma)$ represents a family of finite subsets of $\{ 1, 2, 3, ... \}$ if, and only if, for each $i \in I$,

$$\gamma(\overline{G}_{\Gamma_G(i)}) < \infty \qquad \text{(Kreweras [1946])}$$

7. Consider the graph of the queen's moves in a chess game: the vertices correspond to the squares of an $n \times n$ chessboard, and two vertices are joined by an edge if a queen placed on the square corresponding to the first vertex controls the square corresponding to the second vertex. Call this graph G_n^Q. Show that

$$\gamma(G_2^Q) = 4 \,,$$
$$\gamma(G_3^Q) = 5 \,,$$
$$\gamma(G_4^Q) = 5 \,,$$

$\gamma(G_n^Q) = n$ if n is not divisible by 2 or by 3,

$\gamma(G_n^Q) = n$ or $n + 1$ if $n + 1$ is not divisible by 2 or by 3,

$n \leqslant \gamma(G_n^Q) \leqslant n + 3$ in all other cases.

It has been conjectured that $\gamma(G_n^Q) = n$ or $n + 1$ for $n > 3$.

(M. R. Iyer, V. V. Menon [1966])

8. If G_n^K is the graph of the king's moves on an $n \times n$ chessboard where $n \geqslant 2$, show that $\gamma(G_n^K) = 4$.

9. If G_n^R is the graph of the rook's moves on an $n \times n$ chessboard, show that

$$\gamma(G_n^R) = n \,,$$

10. If G_n^N is the graph of the knight's moves on an $n \times n$ chessboard where $n > 2$, show that $\gamma(G_n^N) = 2$.

11. If G_n^B is the graph of the bishop's moves on an $n \times n$ chessboard, show that

$$\gamma(G_n^B) = n \quad \text{if } n \text{ is odd.}$$

12. Show that if G is an elementary cycle with n vertices, then

$$P(G \,;\, \lambda) = (\lambda - 1)^n + (-1)^n (\lambda - 1) \,.$$

13. Show that if G is connected, then the absolute value of the coefficient of λ^r in $P(G \,;\, \lambda)$ is $\geqslant \binom{n-1}{r-1}$.

(Read [1968])

14. Show that the smallest integer r such that λ^r has a non-zero coefficient in $P(G \,;\, \lambda)$ equals the number of connected components in G.

15. Denote by G_k the simple graph formed by joining the stable sets $X_i = \{x_i^1, x_i^2, ..., x_i^k\}$ for $i = 1, 2, ..., k + 1$ so that (for all i and j) the vertex x_i^j is adjacent to all vertices in X_{i+j}. Show that G_k has no triangles. Show also that there exists a sufficiently large integer p such that $k \geqslant 2^p$ implies $\gamma(G_k) > p$.

(Gyarfas [1980])

16. Let G be a simple graph with maximum degree h. The following are various generalizations of Brooks' theorem.
– There exists a k-colouring such that the vertices with colour 1 form a maximum stable set.

(Catlin [1976]; simple proof by Mitchem [1978])

– If h_i are integers such that $\displaystyle\sum_{i=1}^{p} (h_i + 1) = h + 1$, then there is a vertex partition $(S_1, S_2, ..., S_p)$ with $\Delta(G_{S_i}) = h_i$.

(Lovasz [1977])

– If G contains as induced subgraph neither $K_{1,3}$ nor K_5 minus an edge nor a certain unique graph H, then $\gamma(G) \leqslant h - 1$ if $h \geqslant 6$.

(Dhurandhar [1982])

17. Let G be a simple graph and let $p_1, p_2, ..., p_k$ be positive integers. Show that G is the union $\displaystyle\sum_{i=1}^{k} H_i$ of edge-disjoint partial graphs with $\gamma(H_i) \leqslant p_i$ if and only if $\gamma(G) \leqslant \pi p_i$.

CHAPTER 16

Perfect Graphs

1. Perfect graphs

For a simple graph $G = (X, E)$, let

$\alpha(G)$ denote the stability number,

$\theta(G)$ denote the minimum number of cliques that partition X,

$\gamma(G)$ denote the chromatic number,

$\omega(G)$ denote the maximum cardinality of a clique.

Clearly, $\alpha(G) \leqslant \theta(G)$ since a stable set S can have at most one vertex in each clique of the partition. Similarly, $\omega(G) \leqslant \gamma(G)$.

Graph G is defined to be α-*perfect* if

$$\alpha(G_A) = \theta(G_A) \qquad (A \subset X).$$

Graph G is defined to be γ-*perfect* if

$$\gamma(G_A) = \omega(G_A) \qquad (A \subset X).$$

EXAMPLE 1. If G is a bipartite graph, then we know that

$$\alpha(G) = \theta(G).$$

from Corollary 2 to the König theorem, Chapter 7. Thus, a bipartite graph is α-perfect. A bipartite graph G is also γ-perfect because if it has an edge, then

$$\gamma(G) = 2 = \omega(G)$$

(since from Theorem 4, Chapter 7, G contains no triangles), and if G has no edges, then $\gamma(G) = 1 = \omega(G)$.

EXAMPLE 2. If G consists of an odd cycle of length $2k + 1 > 3$ without chords, then G is not α-perfect because $\alpha(G) = k$ and $\theta(G) = k + 1$. (A minimum partition of G consists of k 2-cliques and one 1-clique.)

Moreover, G is not γ-perfect because $\gamma(G) = 3$ and $\omega(G) = 2$.

In section 7 we shall show that the definitions 'α-perfect' and 'γ-perfect' are equivalent. This is in fact the 'perfect graph theorem', a conjecture we made in 1961 and which was proved by Lovász in 1971. Thus, from this point on we only need to speak of 'perfect' graphs. However, we shall temporarily preserve the distinction of our former definitions in order to make the following results more precise.

Hence:

Theorem 1. *A simple graph G is α-perfect if, and only if, its complementary graph $\overline{G}$ is γ-perfect.*

Clearly,

$$\alpha(G_A) = \omega(\overline{G}_A) ,$$

$$\theta(G_A) = \gamma(\overline{G}_A) .$$

Thus, $\alpha(G_A) = \theta(G_A)$ is equivalent to $\omega(\overline{G}_A) = \gamma(\overline{G}_A)$.

Q.E.D.

Corollary. *If either graph G or its complementary graph $\overline{G}$ contain an odd elementary cycle of length > 3 without chords, then G is neither α-perfect nor γ-perfect.*

Let A be the vertex set of such a cycle of G. Then, from Example 2, $\alpha(G_A) \neq \theta(G_A), \omega(G_A) \neq \gamma(G_A)$. Thus, G is neither α-perfect nor γ-perfect.

If the complementary graph $\overline{G}$ contains such a cycle, then it is neither α-perfect nor γ-perfect, and, from Theorem 1, G is neither α-perfect nor γ-perfect.

Q.E.D.

This result, and the study of various classes of perfect graphs (Berge [1963], [1967], [1969]) suggest the following conjecture:

The strong perfect graph conjecture. *For a graph G, the following conditions are equivalent:*

(1) *G is α-perfect,*
(2) *G is γ-perfect,*
(3) *G contains no set A such that G_A or $\overline{G}_A$ is an odd elementary cycle of length > 3 without chords.*

We have shown above that (1) $\Rightarrow$ (3), (2) $\Rightarrow$ (3). If (3) $\Rightarrow$ (1), then (3) $\Rightarrow$ (2). If (3) $\Rightarrow$ (2), then (3) $\Rightarrow$ (1). This conjecture is still unproved.

Theorem 2. *If G is a connected graph with an articulation set A that is a clique, and if each piece relative to A is a γ-perfect graph, then G is γ-perfect.*

Let G be a graph that satisfies the hypothesis of the theorem. It suffices to show that

$$\omega(G) = \gamma(G) .$$

If $\omega(G) = k$, then there exists a k-clique in at least one piece G' relative to A, and $\gamma(G') = \omega(G') = k$. Each other piece G'' relative to A satisfies

$$\gamma(G'') = \omega(G'') \leqslant k .$$

Thus, G is k-colourable, and

$$k = \omega(G) \leqslant \gamma(G) \leqslant k .$$

Hence, $\omega(G) = \gamma(G)$.

Q.E.D.

Theorem 3. *If G is a connected graph with an articulation set A that is a clique, and if each piece relative to A is an α-perfect graph, then G is α-perfect.*

Let G be a graph that satisfies the hypothesis of the theorem. It suffices to show that

$$\alpha(G) = \theta(G) .$$

Let $C_1, C_2, ..., C_p$ denote the connected components of subgraph G_{X-A}, and let

$$A_i = \left\{ a \mid a \in A , \alpha(G_{C_i \cup \{a\}}) = \alpha(G_{C_i}) \right\} .$$

Two cases must be considered:

CASE 1. $\bigcup_{i=1}^{p} A_i \neq A$. Then, there is a vertex $a \in A$ with

$$\alpha(G_{C_i \cup \{a\}}) = \alpha(G_{C_i}) + 1 \qquad (i = 1, 2, ..., p) .$$

Therefore, a maximum stable set S_i in $G_{C_i \cup \{a\}}$ satisfies

$$\begin{cases} |S_i| = \alpha(G_{C_i}) + 1 , \\ \{a\} \subset S_i \subset C_i \cup \{a\} . \end{cases}$$

Set $S_0 = \bigcup_{i=1}^{p} S_i$ is stable in G, and

$$|S_0| = \sum_{i=1}^{p} \alpha(G_{C_i}) + 1 .$$

Moreover, since a partition $\mathscr{C}$ of G into cliques can be formed with A and with the $\theta(G_{C_i})$ cliques of a minimum partition of C_i, for $i = 1, 2, ..., p$, we have

$$\theta(G) \leqslant |\mathscr{C}| = \sum_{i=1}^{p} \theta(G_{C_i}) + 1 = \sum_{i=1}^{p} \alpha(G_{C_i}) + 1 =$$

$$= |S_0| \leqslant \alpha(G) \leqslant \theta(G) .$$

Hence, $\alpha(G) = \theta(G)$.

CASE 2. $\bigcup_{i=1}^{p} A_i = A$. Then, for all i,

$$\alpha(G_{C_i \cup A_i}) = \alpha(G_{C_i}).$$

Otherwise, there would exist a stable set $S_i \subset A_i \cup C_i$ with

$$|S_i| = \alpha(G_{C_i}) + 1,$$

and there would exist a vertex $a \in A_i$ that is non-adjacent to some maximum stable set of C_i, which contradicts the definition of A_i.

Hence,

$$\theta(G) \leqslant \sum_i \theta(G_{C_i \cup A_i}) = \sum_i \alpha(G_{C_i \cup A_i}) = \sum_i \alpha(G_{C_i}) \leqslant \alpha(G) \leqslant \theta(G).$$

Hence, $\alpha(G) = \theta(G)$.

Q.E.D.

The next three sections treat particular classes of α-perfect and γ-perfect graphs.

2. Comparability graphs

A simple graph $G = (X, E)$ is called a *comparability graph* if it is possible to direct its edges so that the resulting 1-graph (X, U) satisfies:

$$(x, y) \in U, \; (y, z) \in U \quad \Rightarrow \quad (x, z) \in U \qquad \text{(transitivity)}$$
$$(x, y) \in U \quad \Rightarrow \quad (y, x) \notin U \qquad \text{(anti-symmetry)}.$$

Clearly, a bipartite graph is a comparability graph. Furthermore, each subgraph of a comparability graph is a comparability graph.

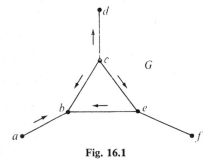

The graph in Fig. 16.1 is not a comparability graph because edges ab, bc, cd, be, ce can be directed appropriately, but then it is impossible to direct edge ef correctly.

Fig. 16.1

Theorem 4. *Every comparability graph is α-perfect.*

Consider a 1-graph $G = (X, U)$ whose arcs represent an order relation. From Corollary 2, to Theorem (14, Ch. 13), $\alpha(G)$ equals the smallest number of elementary paths that partition the vertex set. Because of transitivity, each path generates a clique of G, and, by the corollary to Theorem (4, Ch. 10) each clique is the vertex set of some elementary path of G. Hence, $\alpha(G) = \theta(G)$.

$$Q.E.D.$$

Theorem 5. *Each comparability graph is γ-perfect.*

It suffices to show that if $G = (X, U)$ is the 1-graph of an order relation, then $\gamma(G) = \omega(G)$.

Let $t(x)$ denote the length of the longest path from x plus one. Since G has no circuits, $t(x) < \infty$. If max $t(x) = k$, then there exists a k-clique. There exist no $(k + 1)$-cliques (because this clique would contain a path passing through all its vertices, and the longest path contains only k vertices). Thus

$$\omega(G) = k \, .$$

Consider k colours denoted by $1, 2, ..., k$, and colour each vertex x with colour $t(x)$. Two adjacent vertices cannot have the same colour, because if there is an arc directed from x to y, then $t(x) > t(y)$. Thus

$$\gamma(G) \leqslant k \, .$$

Since $\gamma(G) \geqslant \omega(G) = k$, we have

$$\gamma(G) = k = \omega(G) \, .$$

$$Q.E.D.$$

Theorem 6. *Let $G = (X, U)$ be a transitive 1-graph, and let E denote the set of all pairs of adjacent vertices. Then, the simple graph (X, E) is a comparability graph.*

We shall remove from G all the loops and one arc in each multiple edge, so that the resulting 1-graph (X, V) satisfies the transitivity property.

Let $x_1, x_2, ..., x_n$ be the vertices of G, and let $x_i \equiv x_j$ if either $i = j$ or

$$i \neq j \, , \qquad (x_i, x_j) \in U \, , \qquad (x_j, x_i) \in U \, .$$

From the transitivity of G, the relation $\equiv$ is an equivalence. Define set V as follows:

(1) If x_i and x_j are in the same equivalence class, and if $i > j$, let $(x_i, x_j) \in V$,

(2) If x_i and x_j are in different equivalence classes, and if $(x_i, x_j) \in U$, let $(x_i, x_j) \in V$.

Hence, we have:

$$(x, y) \in V, (y, z) \in V \quad \Rightarrow \quad (x, z) \in V \quad \text{(transitivity)}$$
$$(x, y) \in V \quad \Rightarrow \quad (y, x) \notin V \quad \text{(anti-symmetry)}.$$

Q.E.D.

Theorem 7 (Ghouila-Houri [1962]). *A relation $\succ$ is said to be a semi-order relation if we have*

$$a \succ b, \quad b \succ c \quad \Rightarrow \quad a \succ c \quad \text{or} \quad c \succ a \quad \text{(semi-transitivity)}$$
$$a \succ b \quad \Rightarrow \quad \text{not } b \succ a \quad \text{(anti-symmetry)}$$

A simple graph can be directed so that its arcs represent a semi-order relation if, and only if, it is a comparability graph.

It suffices to show that *if $G = (X, U)$ is a 1-graph whose arcs represent a semi-order relation, then there exists an orientation of the edges of G that represent an order relation.*

Assume that the theorem is true for all graphs of order $< n$; we shall show that it is valid for a 1-graph $G = (X, U)$ of order n whose arcs represent a semi-order.

First, note that if three distinct vertices a, b, and c satisfy

$$(a, b) \in U, \quad (b, c) \in U, \quad (c, a) \in U,$$

then each other vertex is adjacent with either 0 or 2, or 3 of these vertices.

Suppose that G is not transitive; then there exist three vertices x_1, x_2, x_3 satisfying

$$(x_1, x_2) \in U, \quad (x_2, x_3) \in U, \quad (x_3, x_1) \in U.$$

Two cases must be considered:

CASE 1. *Each vertex $x \neq x_1, x_2, x_3$ adjacent with one of these three vertices is adjacent with all three.*

Remove vertices x_2 and x_3 and direct the edges of the resulting subgraph so that it is transitive. (This is possible from the induction hypothesis.)

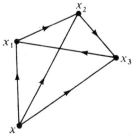

Fig. 16.2

Then, direct each edge of the form $[x, x_2]$ or $[x, x_3]$, where $x \neq x_1, x_2, x_3$, in the same direction as $[x, x_1]$, and direct transitively the triangle formed by x_1, x_2, x_3. Thus, the obtained graph is transitive.

CASE 2. *There exists at least one vertex $a \neq x_1, x_2, x_3$ that is adjacent with exactly two of these vertices, say, x_2 and x_3.*

Let A be the set of vertices y that satisfy

$$(y, x_2) \in U \quad \text{and} \quad (x_3, y) \in U \,.$$

Thus, $x_1 \in A$. Furthermore, since a is not joined to x_1, then $a \in A$. We shall show that *if $x \notin A$, then either x is adjacent with all the vertices of A, or x is not adjacent with set A.*

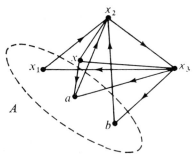

Fig. 16.3

Let $b \in A$, and suppose that x is joined to a but not to b. By considering the triangle formed by a, x_2, x_3, we see that x is joined to at least two of these three vertices and thus to x_2 or x_3. By considering the triangle formed by b, x_2, x_3, we see that x is necessarily joined to x_2 and x_3.

Since x is not adjacent with b, it follows that

$$(x, x_2) \in U \quad \text{and} \quad (x_3, x) \in U \,.$$

This contradicts $x \notin A$.

Now, direct transitively the subgraph generated by A and the subgraph obtained by removing from G the vertices of A other than x_1 (this is possible from the induction hypothesis), then direct each edge of the form $[x, a]$, with $x \notin A$ and $a \in A$, in the same direction as $[x, x_1]$. The resulting graph is transitive.

Q.E.D.

A conjecture due to A. Hoffman can now be proved:

Theorem 8 (Gilmore, Hoffman [1964]; Ghouila-Houri, [1962]). *A necessary and sufficient condition for a simple graph $G = (X, E)$ to be a comparability* *graph, is that for each pseudo-cycle* $[a_1, a_2, \ldots, a_{2q}, a_{2q+1}, a_1]$ *of odd length, there exists an edge of the form* $[a_i, a_{i+2}]$ *(where the addition is modulo* $2q + 1$*)*.

A *pseudo-cycle* is a sequence of vertices starting and ending with the same vertex such that any two consecutive vertices are adjacent. For example, graph G in Fig. 16.1 contains a pseudo-cycle

$$[a, b, c, d, c, e, f, e, b, a]$$

of length 9, but no edges of the form indicated. Thus, this graph is not a comparability graph.

1. *Necessity.* If the graph (X, U) of an order relation contains an odd pseudo-cycle $[a_1, a_2, \ldots, a_{2q+1}, a_1]$ without any edge of the form $[a_i, a_{i+2}]$, then when this pseudo-cycle is traversed, arcs are alternately encountered along and against their direction. But, this is impossible if the pseudo-cycle is odd.

2. *Sufficiency.* Consider a simple graph $G = (X, E)$, and associate with G the simple graph $H = (Y, F)$ defined by

$$(1) \quad \begin{cases} y \in Y & \Leftrightarrow \quad y = (a, b), \quad a, b \in X, \quad [a, b] \in E \\ [y, y'] \in F & \Leftrightarrow \quad \{y, y'\} = \{(a, b), (b, c)\}, \quad [a, c] \notin E. \end{cases}$$

It is easy to show that the hypothesis implies that H contains no odd cycles, and therefore, by Theorem (4, Ch. 7), H is bipartite. Thus, the vertices of H can be partitioned into two classes Y_0 and $Y - Y_0$, with

$$\begin{cases} y, y' \in Y_0 & \Rightarrow \quad [y, y'] \notin F, \\ y, y' \in Y - Y_0 & \Rightarrow \quad [y, y'] \notin F. \end{cases}$$

Direct each edge $[a, b]$ of G from a to b (and write $a \succ b$) if $(a, b) \in Y_0$:

From (1), note that in graph H the vertices $y = (a, b)$ and $y' = (b, a)$ are adjacent (because $[a, a] \notin E$, i.e. G has no loops); thus, these vertices belong to two distinct classes, and each edge of G has received exactly one direction.

Furthermore, we have

$$\begin{rcases} a \succ b \\ \text{or} \\ b \succ c \end{rcases} \Rightarrow \begin{cases} (a, b) \in Y_0 \\ \text{or} \\ (b, c) \in Y_0 \end{cases} \Rightarrow [(a, b), (b, c)] \notin F \Rightarrow$$

$$\Rightarrow [a, c] \in E \Rightarrow \begin{cases} a \succ c \\ \text{or} \\ c \succ a. \end{cases}$$

Therefore, the relation $\succ$ is a semi-order relation, and by Theorem 7, G is a comparability graph.

Q.E.D.

Other characterizations of comparability graphs have been discovered by Gallai [1967].

3. Triangulated graphs

A graph G is defined to be a *triangulated graph* if each cycle of length > 3 possesses a chord (i.e. an edge joining two non-consecutive vertices of the cycle). The concept of triangulated graphs is due to Hajnal and Surányi [1958]. A subgraph of a triangulated graph is also a triangulated graph because, otherwise, it would have a cycle of length > 3 without chords, and G would also have a cycle of length > 3 without chords

EXAMPLE 1. A tree is a triangulated graph.

EXAMPLE 2. A cactus with only cycles of length 3 is triangulated because it contains no cycles of length > 3.

EXAMPLE 3. Graph $L(G)$ is defined to be the graph in which each vertex $\bar{e}_i$ represents an edge e_i of G and with two vertices joined together if, and only if, they represent adjacent edges in G. If G is a cactus with only cycles of length 3, then the graph $L(G)$ is triangulated: Otherwise, $L(G)$ contains a cycle $[\bar{e}_1, \bar{e}_2, ..., \bar{e}_k, \bar{e}_1]$ without chords and with $k > 3$, and this cycle corresponds in G to a cycle $(e_1, e_2, e_3, ..., e_k, e_1)$ of length > 3, which contradicts the definition of G.

The structure of triangulated graphs can be clarified by the following theorem, due essentially to Hajnal and Surányi [1958].

Theorem 9. *If G is a triangulated graph, then each minimal articulation set of G is a clique.*

Let A be a minimal articulation set of G. Removing A creates several connected components $C, C', C'', \ldots$. Each vertex $a \in A$ is joined to each of these components. (Otherwise, $A - \{a\}$ would be an articulation set of G, which contradicts the minimality of A.)

Let a_1 and a_2 be two vertices in A. There exists a chain $\mu = [a_1, c_1, c_2, ..., c_p, a_2]$, where

$$c_1, c_2, ..., c_p \in C .$$

Assume that μ is a chain of this type with minimum length. There also exists a chain $\mu' = [a_2, c_1', c_2', ..., c_q', a_1]$, where

$$c_1', c_2', ..., c_q' \in C'.$$

Assume that μ' is a chain of this type with minimum length.

The cycle $\mu + \mu'$ has no chords of the following types:

— $[a_1, c_i]$ $(i \neq 1)$
— $[c_i, c_j]$ $(i \neq j)$ } because μ would not be the shortest chain
— $[a_2, c_i]$ $(i \neq p)$

— $[c_i, c_j']$ because C and C' are two distinct connected components of G_{X-A}
— $[a_2, c_j']$ $(j \neq 1)$
— $[c_i', c_j']$ $(i \neq j)$ } because μ' would not be the shortest chain.
— $[a_1, c_j']$ $(j \neq q)$

Since the graph is triangulated, cycle $\mu + \mu'$, which has a length of at least 4, possesses a chord. This chord must necessarily be $[a_1, a_2]$.

Thus, any two vertices of A are adjacent, and A is a clique.

Q.E.D.

Corollary 1. *For a convex graph G, the following properties are equivalent:*
(1) *G is triangulated,*
(2) *G is a clique, or else every minimal cutset of G is a clique.*

(1) $\Rightarrow$ (2): follows immediately from Theorem 9.
(2) $\Rightarrow$ (1): if a minimal cutset between two non-adjacent vertices is not a clique, then a chordless cycle can be found, as in Theorem 9.

Corollary 2. *If G is a triangulated graph not a clique then there are two non-adjacent vertices which each belong to only one maximal clique ('simplicial vertices').*

We may assume G is a convex graph of order n not a clique. Suppose that the theorem holds for all graphs of order $< n$ (proof by induction). Since $G \neq K_n$, there exists a minimal cutset A. Let C and C' be two distinct connected components of G_{X-A}. By induction, $G_{A \cup C}$ has two non-adjacent simplicial vertices b and c, and we may assume that $c \in C$. Similarly, there is a simplicial vertex $c' \in C'$ of $G_{A \cup C'}$. Thus, c and c' are necessarily simplicial in G, and also non-adjacent.

Q.E.D.

Corollary 3. *G is triangulated if and only if every subgraph has a simplicial vertex.*

Corollary 4. *Every triangulated graph has a clique which meets all maximal stable sets.*

Let x be a simplicial vertex, and let C be a maximal clique containing x. Then $C \cap S \neq \emptyset$ for every maximal stable set S. This is obvious if $x \in S$. If $x \notin S$, then $C \cap S = \emptyset$ implies that $S \cup \{ x \}$ is stable, contradicting the maximality of S.

Corollary 5 (Hajnal–Surànyi's theorem [1958]). *If G is triangulated, then $\alpha(G) = \theta(G)$.*

Suppose that $\alpha = \theta$ for all graphs with fewer vertices than G. Let C be the maximal clique of G which contains the simplicial vertex x. Then $\alpha(G_{X-C}) = \alpha(G) - 1$, since C necessarily meets all maximal stable sets. Also, $\theta(G_{X-C}) = (G) - 1$, for any minimal covering of G by cliques must necessarily contain the only clique to cover x, namely C. Since $\alpha(G_{X-C}) = \theta(G_{X-C})$, by induction, we have $\alpha(G) = \theta(G)$.

Theorem 10 (Rose [1970]). *Let $G = (X, E)$ be a triangulated graph. Then there exists a circuit-free 1-graph $H = (X, \Gamma)$ which can be obtained from G by orienting each edge, such that each set of the form $\{ x \} \cup \Gamma(x)$ is a clique, and every maximal clique is of this form.*

Index the vertices x_i of G so that, for all k, vertex x_k is simplicial in $G_{X - \{ x_1, x_2, ..., x_{k-1} \}}$. Orient each edge $[x_i, x_j]$ from x_i to x_j if $i < j$. Thus, G has no circuits, and $\{ x \} \cup \Gamma(x)$ is a clique. Let C be a maximal clique. Then C is spanned by a path in H by Rédei's theorem (Theorem 6, Ch. 10). Let x be the first vertex on this path, then $\{ x \} \cup \Gamma(x) = C$.

Q.E.D.

Corollary 1 (Berge–Duchet [1983]). *Every triangulated graph has a stable set which meets all maximal cliques.*

The graph H defined in Theorem 10 has no circuits and therefore has a kernel S (Theorem 4, Ch. 14). Any maximal clique $C = \{ x \} \cup \Gamma(x)$ meets S, for if $x \notin S$, then $\Gamma(x) \cap S \neq \emptyset$, and thus $C \cap S \neq \emptyset$.

Corollary 2. (Berge [1960]). *Every triangulated graph has* $\gamma(G) = \omega(G)$.

By Corollary 1 there exists a set S which meets all maximum cliques, thus $\omega(G_{X-S}) = \omega(G) - 1$. By induction, there is a colouring of G_{X-S} with $\omega(G)$ $- 1$ colours, hence a colouring of G with $\omega(G)$ colours.[1]

Q.E.D.

The argument of Corollary 1 can be extended to a large class of graphs due to Chvátal [1983]: we shall call a 1-graph H *perfectly ordered* if it has no circuits, and if there are no four vertices a, b, c, d such that $h_{\{a, b, c, d\}} = \{(a, b), (a, d), (b, c)\}$. A simple graph is *perfectly orderable* if one can obtain a perfectly ordered graph by orienting the edges. Comparability graphs, triangulated graphs, and complements of triangulated graphs are all perfectly orderable. Chvátal showed:

Lemma. *Let* $H = (X, \Gamma)$ *be a circuit-free 1-graph. Then for all* $A \subset X$, H_A *has a kernel meeting all maximal cliques if and only if* H *is perfectly ordered.*

Let S be such a kernel of H perfectly ordered. Denote by 1, 2, ..., n the vertices of H, so that $(x, y) \in U$ implies $x > y$, and let $p(x) = \min\{y/y \in (x) \cup S\}$. It is sufficient to show that *every clique* C *with* $C \cap S = \varnothing$ *contains an* x_0 *with* $p(x_0)$ *adjacent to all vertices in* C.

If this is false for C (yet true for all subgraphs of C), then there is a correspondence between $x \in C$ and σx in C such that $p(\sigma x)$ is not adjacent to x but adjacent to all vertices in $C - \{x\}$. The application σ is an injection from C to C, and thus a permutation of the elements of C (with no fixed point). Let a be the vertex in C with maximum $p(a)$ and let b be the vertex σa. But then with $C = p(b)$ and $d = p(a)$ the subgraph induced by $\{a, b, c, d\}$ is the forbidden configuration, contradicting the fact that H is perfectly ordered.

Q.E.D.

Hence: *every perfectly orderable graph has a stable set meeting all maximal cliques.*

4. Interval graphs

Consider a family $\mathscr{A} = (A_1, A_2, ..., A_n)$ of intervals on a line. The *representative graph* of $\mathscr{A}$ is defined to be a simple graph G in which each vertex a_i corresponds to an interval A_i, and with two vertices joined together if, and

---

[1] There exist in the literature other characterizations of triangulated graphs (Laskar, Gabril, Buneman, Duchet, etc. ...) which can bè found in: M. Golumbic [1980] or in Duchet's article in C. Berge and V. Chvatal [1984].

only if, the two corresponding intervals intersect. Such a graph is also called an *interval graph*. G. Hajós 1957 and N. Wiener were the first to study interval graphs, and so far two topological characterizations have been found; the first is due to Lekkerkerker and Boland [1962], and the other is due to Gilmore and Hoffman [1964]. For bipartite interval graphs, see also Kotzig [1963].

EXAMPLE 1. Each student visits the university library once a day, and at the end of the day he submits the list of the names of the students met in the library while he was there. The problem is to find the order in which the students entered the library. Construct a graph G in which each vertex represents a student, with two vertices joined together if, and only if, the corresponding students were present in the library at the same time. This graph is an interval graph, because it represents the intervals of time during which the students were present in the library. The theorems of this section will give all the possible solutions.

EXAMPLE 2. In genetics, tests can be performed to determine if two chromosomes overlap one another, and the problem is to prove or disprove that a set of chromosomes are linked together in linear order. Construct the graph G whose edges are the pairs of overlapping chromosomes; if this graph is not an interval graph, it follows that the chromosomes cannot be linked in linear order.

EXAMPLE 3. Consider the following problem that occurs in psychology: Given a finite number of points $x_1, x_2, ..., x_n$ on a line and an infinite family Ω of intervals, two points x_i and x_j are said to be *indistinguishable* if there is an interval in family Ω that contains both of them. The indistinguishable pairs determine a graph, and we may ask for the characteristic properties of these graphs.

In fact, such a graph represents a family of intervals $I_1, I_2, ..., I_n$, where interval I_i corresponds to point x_i and is defined by

$$ I_i = [x_i, +\infty] \cap \bigcup \{ \omega_t / \omega_t \in \Omega, \omega_t \ni x_i \} . $$

If $x_i < x_j$, point x_i and point x_j are indistinguishable if, and only if, $x_j \in I_i$, which is equivalent to $I_i \cap I_j \neq \varnothing$.

Theorem 11. *Every interval graph G is triangulated.*

Suppose that there is a cycle $[a_1, a_2, ..., a_p, a_1]$ without chords. Let A_i be the interval corresponding to vertex a_i; since interval A_k does not overlap with interval A_{k-2}, the initial endpoints of the A_i constitute a monotone sequence; and therefore, A_p cannot overlap with A_1, which contradicts that $[a_p, a_1]$ is an edge of G.

Q.E.D.

Remark. The converse is not true: Graph G in Fig. 16.4 is triangulated, but we shall show that G is not an interval graph.

Clearly, the intervals A_1, A_2, A_3 are pairwise disjoint and may be placed in this order on the line. But, then, interval B_3 that intersects intervals A_1 and

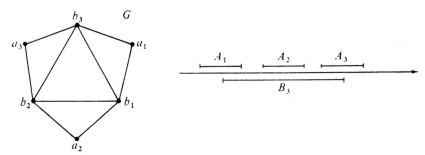

Fig. 16.4

A_3 must also intersect interval A_2, which contradicts that vertices a_2 and b_3 are non-adjacent.

Corollary. *Every interval graph G is α-perfect and γ-perfect.*

Since G is triangulated, G is α-perfect (Corollary 2 to Theorem 9), and γ-perfect (Corollary 1 to Theorem 9).

Q.E.D.

APPLICATION 1 (Gallai). *If $\mathscr{A}$ is a finite family of intervals on a line, and if k is the maximum possible number of pairwise disjoint intervals in $\mathscr{A}$, then there exist k points on the line such that each interval contains at least one of these points.*

Let graph G represent these intervals. Each clique in G corresponds to a family of intervals having one point in common, by the Helly theorem.[1] The result follows.

APPLICATION 2. *If $\mathscr{A}$ is a family of intervals on a line, and if k is the maximum number of intervals that together have a non-empty intersection, then the*

---

[1] Helly's theorem for intervals can be stated as follows: If a family of intervals does not contain two disjoint intervals, then all the intervals have a common point. A simple proof is given in Chapter 17, Section 3, Example 1.

intervals can be coloured with k colours such that no two intervals with the same colour intersect.

The first application can be related to Example 1. The minimum number of photographs of the library that are needed so that each student is photographed at least once equals the maximum number of students who were pairwise not present together in the library.

We shall now study characterizations of interval graphs.

Lemma 1. *If G is an interval graph, then its complementary graph $\bar{G}$ is a comparability graph.*

If graph G represents a family of intervals $\mathscr{A}$, two vertices x and y are linked together in $\bar{G}$ if, and only if, they represent disjoint intervals of $\mathscr{A}$. Direct edge $[x, y]$ from x to y if interval y is to the right of interval x on the line. This produces a 1-graph (X, U) such that:

$$(x, y) \in U \quad \Rightarrow \quad (y, x) \notin U,$$
$$(x, y) \in U, (y, z) \in U \quad \Rightarrow \quad (x, z) \in U.$$

Thus, $\bar{G}$ is a comparability graph.

$$\text{Q.E.D.}$$

The following lemmas treat a simple graph G with the following properties:

(1) *Every cycle of length 4 has a chord,*
(2) *the elementary graph $\bar{G}$ is a comparability graph.*

Let $G = (X, E)$ be a graph with these properties. Let $\mathscr{C}$ denote the family of the maximal cliques of G. Let $\bar{G} = (X, U)$ denote the complementary graph of G, assuming that the edges of $\bar{G}$ are directed transitively.

Lemma 2. *Let $C_1, C_2 \in \mathscr{C}$; there exists an arc in $\bar{G}$ that joins together set C_1 and set C_2. Furthermore, all the arcs in G that join C_1 and C_2 have the same direction.*

If C_1 and C_2 are two distinct maximal cliques of G, then G contains two non-adjacent vertices $a \in C_1$ and $c \in C_2$. (Otherwise, $C_1 \cup C_2$ would be a clique.) For example, let $(a, c) \in U$.

Let bd be another edge of $\bar{G}$ with $b \in C_1$, $d \in C_2$ and $a \neq b$.

If $c = d$, then edge bd has the same direction as edge ac (from C_1 to C_2) because, otherwise, $\bar{G}$ would not be a comparability graph.

If $c \neq d$, then either ad or bc is an edge of $\bar{G}$ (since, otherwise, the cycle $[a, d, c, b, a]$ of G would contain no chords). Without loss of generality, let this edge be ad. Then, $(a, d) \in U$. Hence,

$$(b, d) \in U.$$

Thus, in both cases, edge bd is directed from C_1 to C_2, as is edge ac.

Q.E.D.

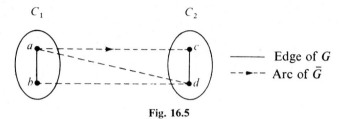

Fig. 16.5

Lemma 3. *Let H be a 1-graph whose vertices represent the cliques of $\mathscr{C}$ and with an arc from C_1 to C_2 if there exist two vertices $a \in C_1$ and $c \in C_2$ in $\bar{G}$ with $(a, c) \in U$. Then, H is a complete, transitive 1-graph.*

By Lemma 2, H is complete and anti-symmetric. It remains to show that if (C_1, C_2) is an arc of H and (C_2, C_3) is an arc of H, then (C_1, C_3) is an arc of H.

We shall suppose that (C_1, C_2) is not an arc of H (and we shall show that this leads to a contradiction). Then, (C_3, C_1) is an arc of H.

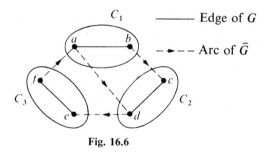

Fig. 16.6

For example, suppose that:

$$a, b \in C_1, \quad c, d \in C_2, \quad e, f \in C_3$$

$$(b, c) \in U, \quad (d, e) \in U, \quad (f, a) \in U.$$

In this case, ad is an edge of $\bar{G}$ because, otherwise, the cliques $\{a, d\}$ and C_3 would contradict Lemma 2. Thus, from Lemma 2, $(a, d) \in U$. Since U is a transitive relation, $(a, e) \in U$. But, then, $(f, a) \in U$ and $(a, e) \in U$ implies that $(f, e) \in U$. This contradicts that both vertices e and f belong to C_3.

Q.E.D.

Theorem 12. (Gilmore, Hoffman [1964]). *A simple graph G is an interval graph if, and only if, the following two conditions hold:*

(1) *every cycle of length 4 has a chord,*
(2) *the complementary graph $\bar{G}$ is a comparability graph.*

Necessity. Condition (1) is necessary because a graph that represents a family of intervals is triangulated (Theorem 11). Condition (2) is necessary from Lemma 1.

Sufficiency. Let $G = (X, E)$ be a simple graph that satisfies conditions (1) and (2). Let $\mathscr{C} = \{ C_1, C_2, ..., C_q \}$ be the family of the maximal cliques of G. As in Lemma 3, form the 1-graph H. Graph H is complete, anti-symmetric and transitive. From Theorem (5, Ch. 10), H contains a unique hamiltonian path μ. Suppose that the cliques of $\mathscr{C}$ are indexed so that $\mu = [C_1, C_2, ..., C_q]$. Consequently, (C_i, C_j) is an arc of H if, and only if, $i < j$.

Note that G represents the sets

$$I_x = \{ i \,/\, C_i \in \mathscr{C}, C_i \ni x \},$$

because two vertices x and y are joined together in G if, and only if,

$$I_x \cap I_y \neq \varnothing.$$

To show that I_x is an interval, it suffices to show that

$$\left. \begin{array}{c} x \in C_p \\ x \in C_q \\ p < r < q \end{array} \right\} \quad \Rightarrow \quad x \in C_r.$$

If $x \notin C_r$, there exists a vertex $y \in C_r$ such that xy is an edge of $\bar{G}$ (see Fig. 16.7).

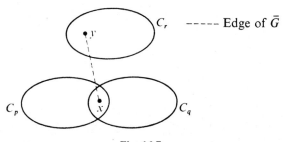

C_r ----- Edge of $\bar{G}$

C_p x C_q

Fig. 16.7

Clearly, $(x, y) \in U$ (since $r > p$) and $(y, x) \in U$ (since $r < q$); this gives the required contradiction.

$$\text{Q.E.D.}$$

5. Cartesian product and cartesian sum of simple graphs

In Chapter 14, three operations were defined for 1-graphs: normal product, Cartesian product, and Cartesian sum.

If $G = (X, E)$ and $H = (Y, F)$ are two simple graphs, the *Cartesian sum* of graphs G and H is defined to be a simple graph $G + H$ whose vertex set is $X \times Y$, and with two vertices xy and $x'y'$ joined together if, and only if,

either $x = x'$, $[y, y'] \in F$,   or $[x, x'] \in E, y = y'$.

Let the Kronecker delta be defined on graph G as follows: $\delta_G(x, x') = 1$ if $x \neq x'$, and $\delta_G(x, x') = 0$, otherwise. The number of edges that join vertices xy and $x'y'$ in graph $G + H$ can be written as

$$m_{G+H}(xy, x' y') = \delta_G(x, x') m_H(y, y') + \delta_H(y, y') m_G(x, x') .$$

The *cartesian product* of graphs G and H is defined to be a simple graph $G \times H$ whose vertex set is $X \times Y$, and with two vertices xy and $x'y'$ joined together if, and only if $[x, x'] \in E$ and $[y, y'] \in F$. The number of edges joining these two vertices in graph $G \times H$ can be written as

$$m_{G \times H}(xy, x' y') = m_G(x, x') m_H(y, y') .$$

The *normal product* (or simply, *product*) of graphs G and H is defined to be a simple graph $G.H$ whose vertex set is $X \times Y$, and with two vertices xy and $x'y'$ joined together if and only if, either

$$x = x', \qquad [y, y'] \in F ,$$

or

$$[x, x'] \in E , \qquad y = y' ,$$

or

$$[x, x'] \in E , \qquad [y, y'] \in F .$$

The number of edges joining these two vertices can be written as

$m_{G.H}(x, y) =$

$$= m_G(x, x') m_H(y, y') + \delta_G(x, x') m_H(y, y') + m_G(x, x') \delta_H(y, y') .$$

Note that these operations are commutative. If the definitions are extended to more than two graphs, it can be shown that the operations are associative and distributive (C. Picard [1970]).

These definitions were first introduced to study the chromatic number and the stability number. The relationships between these operations and the main fundamental numbers are described in this section.

Proposition 1. *Let G and H be two graphs. Then*

$$\omega(G + H) = \max\{\omega(G), \omega(H)\}.$$

Let $C_1, C_2, ..., C_p$ be the maximal cliques of G, and let $D_1, D_2, ..., D_q$ be the maximal cliques of H; then the maximal cliques of $G + H$ are the sets $C_i \times \{y_j\}$ and the sets $\{x_i\} \times D_j$. Hence,

$$\omega(G + H) = \max\{|\{x_i\} \times D_j|, |C_i \times \{y_j\}|\} = \max\{\omega(G), \omega(H)\}.$$

Q.E.D.

Proposition 2 (Vizing [1963], Aberth [1964]). *Let G and H be two graphs. Then*

$$\gamma(G + H) = \max\{\gamma(G), \gamma(H)\}.$$

1. Let $r = \max\{\gamma(G), \gamma(H)\}$; colour G and H with r colours $0, 1, 2, r - 1$. Let $g(x) = k$ if vertex x is coloured k. For each $xy \in X \times Y$, let

$$g(xy) \equiv g(x) + g(y) \quad (\bmod. r).$$

Consequently, $g(xy)$ defines a colouring of $G + H$ because if xy and $x'y'$ are adjacent and have the same colour, then either $x = x'$ and vertices y and y' are adjacent in H, or $y = y'$ and vertices x and x' are adjacent in G. In the first case,

$$g(x) + g(y) \equiv g(x) + g(y') \quad (\bmod. r)$$

and $g(y) = g(y')$, which is impossible because $[y, y'] \in F$. A similar result follows for the second case. Hence, $G + H$ is r-colourable.

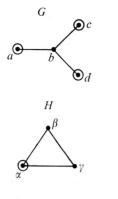

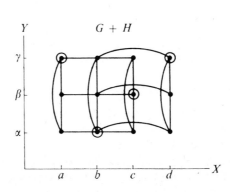

Fig. 16.8

2. Graph $G + H$ cannot be $(r - 1)$-colourable, because, graph $G + H$ contains a subgraph isomorphic to G and a subgraph isomorphic to H. Hence, $\gamma(G + H) = r$.

Q.E.D.

Proposition 3. *Let G and H be two graphs. Then*

$$\alpha(G + H) \geqslant \alpha(G)\,\alpha(H) \,.$$

If S is a maximum stable set of G, and if T is a maximum stable set of H, then the set $S \times T$ is a maximum stable set of $G \times H$. Hence,

$$\alpha(G + H) \geqslant |\, S \times T \,! = \alpha(G)\,\alpha(H)$$

Q.E.D.

In Fig. 16.8, the vertices of a maximum stable set are circled, and it is easily seen that

$$\alpha(G + H) = 4 > \alpha(G)\,\alpha(H) = 3 \,.$$

Proposition 4. *If graphs G and H respectively have orders $n(G)$ and $n(H)$, then*

$$\alpha(G + H) \leqslant \min \left\{ \alpha(G)\,n(H), \alpha(H)\,n(G) \right\} \,.$$

Let S_0 be a maximum stable set of $G + H$. Its intersection with set $X \times \{ y_j \}$ cannot have more than $\alpha(G)$ vertices; hence, $|\, S_0 \,| \leqslant |\, Y \,|\, \alpha(G)$. Similarly, $|\, S_0 \,| \leqslant |\, X \,|\, \alpha(H)$.
The formula follows.

Q.E.D.

Proposition 5. *If graphs G and H respectively have order $n(G)$ and $n(H)$, then*

$$\theta(G + H) \leqslant \min \left\{ n(G)\,\theta(H), n(H)\,\theta(G) \right\} \,.$$

Let $\mathscr{C} = (C_1, C_2, ..., C_q)$ be a minimum partition of G. The sets $C_i \times \{ y_j \}$ are cliques of $G + H$ and cover $X \times Y$. Thus,

$$\theta(G + H) \leqslant q \times |\, Y \,| = n(H)\,\theta(G) \,.$$

Q.E.D.

Remark. Vizing [1965] proved a similar inequality for the dominance numbers, i.e.:

$$\beta^*(G + H) \leqslant \min \left\{ \beta^*(G)\,n(H), \beta^*(H)\,n(G) \right\} \,.$$

He also conjectured:

Conjecture. $\beta^*(G + H) \geqslant \beta^*(G)\,\beta^*(H)$.

Theorem 13. *If only q colours are available, then the maximum number of vertices in G that can possibly be coloured with these q colours so that no two adjacent vertices have the same colour is equal to* $\alpha(G + K_q)$.

Let $K_q = (Y, F)$ be a q-clique and let $G = (X, E)$ be a simple graph. We shall show that a stable set of $G + K_q$ determines a set $A \subset X$ of vertices in G that can be coloured with q colours. (This result is illustrated in Fig. 16.8 for $q = 3$.)

Since the sets $\{x_i\} \times Y$ are all cliques in graph $G + K_q$, a stable set S_0 of $G + K_q$ has at most one vertex in each of them. Put

$$A = \{x / x \in X ; \; S_0 \cap (\{x\} \times Y) \neq \varnothing\}.$$

For $x \in A$, put $g(x) = j$ if, and only if,

$$S_0 \cap (\{x\} \times Y) = \{xy_j\}.$$

Function $g(x)$ is a q-colouring of G_A because two adjacent vertices x and x' in G_A cannot have the same colour (since S_0 is stable).

Conversely, if $g(x)$ is a q-colouring of a subgraph G_A, then the set of vertices $xy_{g(x)}$ with $x \in A$ is a stable set of $G + K_q$. Thus, there is a one-to-one correspondence between the stable sets of $G + K_q$ and the partial q-colouring of G

Q.E.D.

Corollary. *A graph G of order n is q-colourable if, and only if,* $\alpha(G + K_q) = n$. The proof follows immediately.

For example, graph G in Fig. 16.8 is 3-colourable, since

$$\alpha(G + K_3) = 4 = n(G).$$

We shall now present a similar result for the cartesian product $G \times H$:

Proposition 6. *Let G and H be two graphs. Then*

$$\gamma(G \times H) \leqslant \min\{\gamma(G), \gamma(H)\}.$$

Let $q = \min\{\gamma(G), \gamma(H)\}$, and suppose that $\gamma(G) = q$. Let $g(x)$ be a q-colouring of graph G. Put

$$g(xy) = g(x) \qquad (xy \in X \times Y).$$

Thus, $g(xy)$ is a q-colouring of $G \times H$ because if xy and $x'y'$ are two adjacent vertices, then $g(xy) \neq g(x'y')$ (since x and x' are adjacent in G). Thus, $\gamma(G \times H) \leqslant q$.

Q.E.D.

The following similar results are available for the normal product $G \cdot H$.

Proposition 7. *Let G and H be two graphs. Then*

$$\gamma(G.H) \geqslant \max\{\gamma(G), \gamma(H)\}.$$

Let $q = \max\{\gamma(G), \gamma(H)\}$, $q = \gamma(G)$; then q colours are necessary to colour $G.H$ since $G.H$ contains a subgraph isomorphic to G.

Q.E.D.

Proposition 8. *Let G and H be two graphs. Then*

$$\omega(G.H) = \omega(G)\,\omega(H).$$

Let C be a maximum clique of G, and let D be a maximum clique of H. The set $C \times D$ is a clique of $G.H$ because if x, $x' \in C$ and y, $y' \in D$, and if $xx' \neq yy'$, one of the three following cases occurs:

$x = x'$ and $[y, y'] \in F$, and then, xy and $x'y'$ are neighbours in $G.H$, or
$y = y'$ and $[x, x'] \in E$, and then, xy and $x'y'$ are neighbours in $G.H$, or
$[x, x'] \in E$ and $[y, y'] \in F$, and then, xy and $x'y'$ are neighbours in $G.H$.

Hence,

$$\omega(G.H) \geqslant |C \times D| = \omega(G)\,\omega(H).$$

Conversely, if C_0 is a maximum clique of $G \cdot H$, let the projection of C_0 on X be C, and let the projection of C_0 on Y be D. Since C and D are cliques in G and H, the set $C \times D$ is a clique in $G \cdot H$. Hence, $C_0 = C \times D$ (since C_0 is a maximal clique). Thus,

$$\omega(G.H) = |C_0| = |C| \times |D| \leqslant \omega(G)\,\omega(H).$$

The required equality follows.

Q.E.D.

Proposition 9. *Let G and H be two graphs. Then*
$$\alpha(G.H) \geqslant \alpha(G)\,\alpha(H).$$

If S and T are maximum stable sets respectively in G and H, then
$$|S| = \alpha(G), \qquad |T| = \alpha(H).$$

The cartesian product $S \times T$ is a stable set in $G.H$, and, consequently, we have

$$\alpha(G.H) \geqslant |S \times T| = |S| \times |T| = \alpha(G)\,\alpha(H)\,.$$

Q.E.D.

An upper bound for $\alpha(G \cdot H)$ will be given in Chapter 19, §2.

Proposition 10. $\theta(G.H) \leqslant \theta(G)\,\theta(H)\,.$

Let $(C_1, C_2, ..., C_p)$ be a minimum partition into cliques of graph G. Let $(D_1, D_2, ..., D_q)$ be a minimum partition into cliques of graph H

In graph $G.H$, set $C_i \times D_j$ is a clique, for $i = 1, 2, ..., p$, and $j = 1, 2, ..., q$. The cliques $C_i \times D_j$ partition graph $G.H$. Therefore,

$$\theta(G.H) \leqslant pq = \theta(G)\,\theta(H)\,.$$

Q.E.D.

APPLICATION (Shannon [1956]). Consider a transmitter that can emit five signals, a, b, c, d, e, and a receiver that can interpret each of these signals in two different ways. Signal a can be interpreted as either p or q, signal b can be interpreted as either q or r, etc., as shown in Fig. 16.9. What is the maximum number of signals that can be used for a code so that there is no possible confusion on reception? This problem reduces to finding a maximum stable set S of a graph G shown in Fig. 16.10, in which two vertices are adjacent if and only if they represent two signals that the receiver can confuse. For example, we can take $S = \{a, c\}$, since graph G in Fig. 16.10 has stability number $\alpha(G) = 2$.

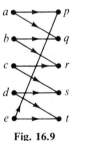

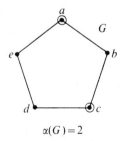

Fig. 16.9   **Fig. 16.10**

$$G\,.\,G = G^2$$

Instead of single letter words, we could use a code of two letter words, provided that no two letter words of this code can lead to confusion on reception. Thus, the letters a and c which cannot be confused can form the

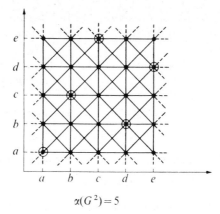

$$\alpha(G^2) = 5$$

Fig. 16.11

code: *aa*, *ac*, *ca*, *cc* which has a vocabulary of $(\alpha(G))^2 = 4$ words. But an even richer code is: *aa*, *bc*, *ce*, *db*, *ed*. It is easily seen that no two of these words can be confused by the receiver. This gives a vocabulary of 5 words.

Note that the words xy and $x'y'$ can be confused if, and only if, these two words are adjacent vertices in the normal product $G . G = G^2$. A code consisting of 2-letter words has a maximum vocabulary of $\alpha(G^2)$ words. More generally, the maximum possible vocabulary for a code of k letter words is the stability number of the product.

$$G^k = \underbrace{G.G.G \dots G}_{k} .$$

With this as motivation, the *capacity of graph G* (or, "zero-error capacity") is defined to be the number

$$c(G) = \sup \sqrt[k]{\alpha(G^k)} .$$

The capacity of the graph G in Fig. 16.10 is known to be between 2 and 3; however, its exact capacity remains unknown. Furthermore, Ljubich [1964] has shown that $^k\sqrt{\alpha(G)^k}$ tends to $c(G)$ when k tends to infinity.

6. The perfect graph theorem

We first present two Lemmas:

Lemma 1. *Let G be an α-perfect graph, and let H be the graph obtained by adding a vertex x' and joining it to all neighbours of some vertex x in G. Then H is also α-perfect.*

It is sufficient to show that $\alpha(H) = \theta(H)$. Consider a partition $\mathscr{C}$ of the vertices of G in $\theta(G) = \alpha(G)$ cliques, and let C_x be the clique containing x in this partition.

If there is in G a maximum stable set containing x, then

$$\alpha(H) = \alpha(G) + 1$$

However, $\mathscr{C} \cup \{ x' \}$ is a partition of H in $\alpha(G) + 1$ cliques, thus

$$\theta(H) = \alpha(H).$$

If there is in G no maximum stable set containing x, then

$$\alpha(H) = \alpha(G).$$

But $C' = C_x - \{ x \}$ meets all maximum stable sets of G, hence

$$\alpha(G_{X - C'}) = \alpha(G) - 1.$$

Therefore, since G is α-perfect,

$$\theta(G_{X - C'}) = \alpha(G) - 1 = \alpha(H) - 1.$$

Thus, H can be partitioned in $\alpha(H)$ cliques by taking the clique $C' \cup \{ x \}$ and the $\alpha(H) - 1$ cliques which partition $X - C'$. Thus $\alpha(H) = \theta(H)$.

Q.E.D.

Lemma 2. *Let G be a graph all of whose subgraphs are α-perfect and satisfy:*

(1) $$\omega(G_A)\alpha(G_A) \geq | A | \qquad (A \subset X).$$

Let H be a graph obtained from G by replacing each vertex x_i by a set $X_i = \{ y_i^1, y_i^2, \ldots \}$ and by joining y_i^s and y_i^t if and only if x_i and x_j are adjacent in G. Then H also satisfies (1).

Consider any such H that does not satisfy (1) and with the smallest number of vertices. We shall show that the existence of such a graph $H = (Y, E)$ leads to a contradiction.

Obviously, $\max | X_i | \neq 1$. Suppose that

$$| X_i | = h \geq 2.$$

Then

$$\omega(H_{Y - X_1}) \leq \omega(H),$$

$$\alpha(H_{Y - X_1}) \leq \alpha(H).$$

Let $y_1 \in X_1$. By minimality of H, the subgraph $H_{Y - y_1}$ satisfies (1), thus

$$|Y| - 1 = |Y - y_1| \leqslant \omega(H_{Y - y_1})\alpha(H_{Y - y_1}) \leqslant \omega(H)\alpha(H) \leqslant |Y| - 1.$$

Thus equality holds, and hence

$$\omega(H_{Y - y_1}) = \omega(H) = p,$$

$$\alpha(H_{Y - y_1}) = \alpha(H) = q,$$

$$|Y| - 1 = pq.$$

Since $H_{Y - X_1}$ can be obtained from $G_{X - X_1}$ by vertex duplication, it follows by Lemma 1 that

$$\theta(H_{Y - X_1}) = \alpha(H_{Y - X_1}) \leqslant q.$$

Hence, $Y - X_1$ can be covered by q cliques $C_1, C_2, ..., C_q$ of H; we may assume these are pairwise disjoint and that $|C_1| \geqslant |C_2| \geqslant \cdots \geqslant |C_q|$.
As well,

$$|C_i| \leqslant \omega(H) = p,$$

$$\sum_{i=1}^{q} |C_i| = |Y| - h = pq - (h - 1).$$

Thus $|C_i| < p$ for at most $h - 1$ values of i, from which

$$|C_1| = |C_2| = \cdots = |C_{q - h + 1}| = p.$$

Let H' be the subgraph of H induced by $C_1 \cup C_2 \cup \cdots \cup C_{q - h + 1} \cup \{y_1\}$.
Then

$$n(H') = p(q - h + 1) + 1 < pq + 1 = |Y|.$$

Thus, by minimality of H, the graph H' (which is constructed like H) satisfies (1). Therefore

$$p\alpha(H') = \omega(H')\alpha(H') \geqslant n(H') = p(q - h + 1) + 1.$$

Thus $\alpha(H') > q - h + 1$. Let S' be a stable set of H' with $q - h + 2$ vertices. Since $C_1, C_2, ..., C_{q - h + 1}, \{y_1\}$ is a partition of H' in $q - h + 2$ cliques, $y_1 \in S'$. Thus $S = S' \cup X_1$ is a stable set of H, and consequently

$$q = \alpha(H) \geqslant |S| = |S'| + |X_1| = q + 2.$$

Contradiction.

Q.E.D.

Theorem 14. (Lovász [1972]). *For a graph G, the following conditions are equivalent:*

(1) $\omega(G_A)a(G_A) \geqslant \omega(G_A)a(G_A) \geqslant |A|$ $(A \subset X)$
(2) $\gamma(G_A) = $ $\gamma(G_A) = \omega(G_A)$ $(A \subset X)$
(3) $a(G_A) = $ $a(G_A) = \theta(G_A)$ $(A \subset X).$

$(1) \Rightarrow (2).$ By induction. Suppose that the theorem holds for all graphs with order less than n. By induction, every graph G_A for $A \neq X$ is γ-perfect and thus a-perfect.

Let $p = \omega(G)$ and let $\mathcal{S}$ be the family of stable sets of G.

We first show that there exists a stable set S in G such that $\omega(G_{X-S}) < \omega(G)$. If not, then to every set S in $\mathcal{S}$ there is a corresponding clique C_s contained in $X - S$ such that $|C_s| = p$.

Consider the subgraph of G induced by replacing each vertex x_i by a set X_i having as cardinality the number of cliques C_s in G containing x_i.

By Lemma 2, the resulting graph H satisfies

$$\omega(H)a(H) \geqslant n(H).$$

Now

$$n(H) = \sum_i |X_i| = \sum_{S \in \mathcal{S}} |C_S| = p|\mathcal{S}|,$$

$$\omega(H) \leqslant \omega(G) = p,$$

$$a(H) = \max_{T \in \mathcal{S}} \bigcup_{x_i \in T} X_i = \max_{T \in \mathcal{S}} \sum_{S \in \mathcal{S}} |T \cap C_S|$$

$$= \max_{\substack{T \in \mathcal{S} \\ S \neq T}} \sum_{S \in \mathcal{S}} |T \cap C_S| \leqslant |\mathcal{S}| - 1.$$

Thus

$$\omega(H)a(H) \leqslant p(|\mathcal{S}| - 1) < n(H).$$

This is a contradiction.

Hence, there exists a stable set S in G such that

$$\omega(G_{X-S}) \leqslant \omega(G) - 1.$$

The vertices of G can thus be coloured by using one colour for S and $\gamma(G_{X-S}) = \omega(G_{X-S})$ other colours. Thus $\gamma(G) \leqslant 1 + [\omega(G) - 1] = \omega(G)$, i.e. $\gamma(G) = \omega(G)$.

$(2) \Rightarrow (1).$ If G satisfies (2), then for every $A \subset X$ there is a colouring $(A_1, A_2, ..., A_q)$ of G_A with $q = \gamma(G_A) = \omega(G_A)$ colours, hence

$$| A | = \sum_{i=1}^{q} | A_i | \leqslant q\alpha(G_A) = \omega(G_A)\alpha(G_A).$$

(1) $\Rightarrow$ (3). If G satisfies (1), then $\bar{G}$ satisfies (1), and thus (2). Therefore

$$\alpha(G_A) = \omega(\bar{G}_A) = \gamma(\bar{G}_A) = \theta(G_A).$$

(3) $\Rightarrow$ (1). If G satisfies (3), then $\bar{G}$ satisfies (2), and thus (1). Therefore

$$\omega(G_A)\alpha(G_A) = \alpha(\bar{G}_A)\omega(\bar{G}_A) \geqslant | A |.$$

Q.E.D.

This important result shows that 'α-perfect' is equivalent to 'γ-perfect'; by also showing that every minimally imperfect graph G satisfies $n(G) = \alpha(G)\omega(G) + 1$, the result has paved the way for numerous further attempts to characterize minimally imperfect graphs.

Anna Ljubiw [1982] showed that the recognition of imperfect graphs is in $N-P$, regardless of the outcome of the perfect graph conjecture. Padberg [1974] proved that *every minimally imperfect graph G of order n has exactly n stable sets $S_1, S_2, ..., S_n$ and n maximum cliques $C_1, C_2, ..., C_n$, with each vertex belonging to exactly α of the S_i and ω of the C_j. Also $S_i \cap C_j = \emptyset$ if and only if $i = j$.*

For every $\alpha > 1$ and $\omega > 1$, it is possible to construct a graph $C_{\alpha\omega+1}^{\omega-1}$ by taking $n = \alpha\omega + 1$ vertices $x_1, x_2, ..., x_n$ and joining x_i and x_j if and only if $| i - j | < \omega$ (modulo n). Chvatal [1977] thus showed that the perfect graph conjecture is equivalent to: *every minimally imperfect graph G with $\alpha(G) = \alpha$ and $\omega(G) = \omega$ has a partial graph isomorphic to $C_{\alpha\omega+1}^{\omega-1}$.*

Unfortunately, Padberg's conditions can be satisfied by graphs other than the $C_{\alpha\omega+1}^{\omega-1}$. Chvatal, Graham, Perold and Whitesides [1979] have shown how to construct other infinite classes of graphs satisfying Padberg's conditions. Two of these graphs were discovered independently by Bland, Huang and Trotter [1979].

Another related result is a theorem due to Meyniel [1976], who proved: *every graph in which each odd cycle of length > 3 has at least two chords is perfect.* This generalizes two different theorems, one due to Gallai [1962] and the other to Olaru and Sachs [1970]. Ravindra [1982] further showed that such a graph has a stable set meeting all maximal cliques. Burlet and Fonlupt [1984] showed how to recognize Meyniel graphs in polynomial time.

Two graphs $G_1 = (X_1, E_1)$ and $G_2 = (X_2, E_2)$ are said to have the same P_1-*structure* if there is a bijection $f: X_1 \rightarrow X_2$ such that a subset S of X_1 induces a P_4 (the chordless path with four vertices) in G_1 if and only if $f(S)$ induces a

P_4 in G_2. Chvatal [1982] conjectured: *if a graph G has the P_4-structure of a perfect graph, then G is perfect.* It has been shown that if a graph has the P_4-structure of a bipartite graph (Chvátal and Hoang [1983]), a triangulated graph (Hayward [1983]), or a line graph of a bipartite graph (Hoang [1983]), then G is perfect.

In fact, the conjecture has been recently proved (in full generality) by Reed [1983]. This theorem is now referred to as the 'Semi-strong perfect graph theorem', as it both implies the perfect graph theorem, and is implied by the perfect graph conjecture (also known as the strong perfect conjecture).

Readers who are interested in a further study of perfect graphs can consult: C. Berge and V. Chvátal [1984] and M. Golumbic [1980].

References

CHAPTER 2 : **Cyclomatic Number**

G. Demoucron, Y. Malgrange, R. Pertuiset, Graphes planaires. *Revue Française de Rech. Opérat.*, **30**, 1964, 33–47.

T. Gallai, Über reguläre Kettengruppen. *Acta Math. Ac. Sc. Hungar.*, **10**, 1959, 227–240.

A. Ghouila-Houri, Flots et tensions dans un graphe, Thesis, Université de Paris, 1964.

K. Kuratowski, Sur le problème des courbes gauches en topologie. *Fund. Math.*, **15**, 1930, 271–283.

A. Lempel, S. Even, I. Cederbaum, "An Algorithm for Planarity Testing of Graphs". *Théorie des Graphes. Rome I. C. C.* (P. Rosenstiehl, ed.), Dunod Paris, 1967, 215–232.

S. MacLane, A Combinatorial Condition for Planar Graphs. *Fund. Math.* **28**, 1937, 22–32.

G. J. Minty, Monotone Networks. *Proc. Roy. Soc.*, A **257**, 1960, 194–208.

R. Pellet, *Initiation à la théorie des graphes, vocabulaire descriptif.* Entreprise Moderne d'Edition, Paris, 1966.

H. Sachs, *Einführung in die Theorie der endlichen Graphen*, Teubner, Leipzig, 1970.

E. Steinitz, Polyeder und Raumeinteilungen. *Enzykl. Math. Wiss.*, **3**, 1922, 1–139.

W. T. Tutte, A Class of Abelian Groups. *Canad. J. Math.*, **8**, 1956, 13–28.

W. T. Tutte, How to Draw a Graph. *Proc. London Math. Soc.*, **13**, 1963, 743–767.

H. Whitney, Planar Graphs. *Fund. Math.*, **21**, 1933, 73–84.

A. A. Zykov, *Theory of Finite Graphs* (Russian), Nauka, Novosibirsk, 1969.

CHAPTER 3 : **Trees and Arborescences**

T. van Aardenne-Ehrenfest, N. G. de Bruijn, Circuits and Trees in Oriented Linear Graphs. *Simon Stevin*, **28**, 1951, 203–217.

C. Berge, *Principles of Combinatorics*, Academic Press, New York, London, 1971.

R. Bott, J. P. Mayberry, "Matrices and Trees". *Economic Activity Analysis*, Wiley, New York, 1954, 391–407.

A. Cayley, A Theorem on Trees. *Quart. J. Math.* **23**, 1889, 376–378.

P. Camion, Modules unimodulaires. *J. Combinat. Theory*, **4**, 1968, 301–362.

G. Chaty, Graphes fortement connexes c-minimaux. *C. R. Ac. Sc. Paris*, **266**, 1968, 907–909.

G. Chaty, Graphes fortement connexes c-minimaux et graphes sans circuit co-minimaux. *J. Combinat. Theory*, **10**, B, 1971, 237–244.

G. Choquet, Etude de certains réseaux de route. *C. R. Ac. Sc. Paris*, **206**, 1938, 310–311.

L. E. Clarke, On Cayley's Formula for Counting Trees. *Proc. Cambridge Phil. Soc.*, **59**, 1963, 509–517.

J. P. Crestin, Un algorithme d'orientation des arêtes d'un graphe connexe fini. Miméographe du Séminaire sur les Problèmes Combinatoires, I. H. P., 1969.

P. J. Kelly, A Congruence Theorem for Trees. *Pacific J. of Math.* **7**, 1957, 961–968.

J. B. Kruskal, On the Shortest Spanning Subtree of a Graph. *Proc. Am. Math. Soc.*, **71**, 1956, 48–50.

B. Leclerc, P. Rosenstiehl, "Arbres". *Structures algébriques et combinatoires*, chapter I, Unesco 1970.

V. V. Menon, On the Existence of Trees with Given Degrees. *Sankhya*, **26**, 1964, 63–68.

J. W. Moon, "Enumerating Labelled Trees". *Graph Theory and Theoretical Physics*, (F. Harary) Academic Press, London, New York, 1967, 261–271.

J. W. Moon, On the Second Moment of the Complexity of a Graph. *Mathematika*, **11**, 1964, 95–98.

L. Nesbeský, Algebraic Properties of Trees. *Acta Univ. Carolinae Philologica Monograph*, **25**, 1969.

P. V. O'Neil, The Number of Trees in a Certain Network. *Notices Amer. Math. Soc.*, **10**, 1963, 569.

C. F. Picard, *Graphes et questionnaires.* Gauthier-Villars, Paris, 1972.

C. de Polignac, Théorie des ramifications. *Bull. Soc. Math. France*, **8**, 1880, 120.

C. Ramanujacharyulu, Trees and Tree-equivalent Graphs. *Canad. J. Math.* **17**, 1965, 731–733.

A. Rényi, Some Remarks on the Theory of Trees. *Magy. Tudom. Akad. Mat. Kut. Intéz. Kölz.*, **4**, 1959, 73–85.

P. Rosenstiehl, "L'arbre minimum d'un graphe". *Théorie des Graphes, Rome I. C. C.* (P. Rosenstiehl, ed.), Dunod, Paris, 1967, 357–368.

P. Rosenstiehl, Graph Problems Solved by Finite Automate Network, *Abstract, Calgary International Conf.* 1969.

B. Roy, Cheminement et connexité dans les graphes. Thesis, Université de Paris, 1962.

H. I. Scoins, The Number of Trees with Nodes of Alternate Parity. *Proc. Camb. Phil. Soc.*, **58**, 1962, 12–16.

J. Sedláček, Regular Graphs and their Spanning Trees. Res. paper 97, Calgary, 1970.

H. N. V. Temperley, On the Mutual Cancellation of Cluster Integrals. *Proc. Phys. Soc.*, **83**, 1964, 3–16.

W. T. Tutte, The Dissection of Equilateral Triangles into Equilateral Triangles, *Proc. Cambridge Phil. Soc.*, **44**, 1948, 203–217.

H. M. Trent, A Note on the Enumeration and Listing of All Possible Trees in a Connected Linear Graph. *Proc. Nat. Ac. Sciences*, **40**, 1954, 1004–1007.

L. Weinberg, Number of Trees in a Graph. *Proc. IRE*, **46**, 1958, 1954–1955.

CHAPTER 4 : Path Problems

S. B. AKERS, On the Construction of (d, k)-graphs. *IEEE Transactions on Electronic Computers*, **19**, 1965, 488.

M. BALINSKI, On the Structure of Convex Polyhedra. *Pac. J. of Math.*, **11**, 1961, 431–434.

M. BECKMANN, C. B. McGUIRE, C. B. WINSTEN, *Studies in the Economics of Transportation*. Yale University Press. New Haven, 1956.

R. BELLMAN, The Theory of Dynamic Programming. *Bull. A. M. S.*, **60**, 1954, 503.

B. BOLLOBÁS, "Graphs with Given Diameter". *Theory of Graphs (Tihany)*, Acad. Press 1968, 29–36.

M. BORILLO, Quelques remarques sur la détermination de l'algorithme optimal pour la recherche du plus court chemin. *Revue Française de Rech. Opérat.*, **33**, 1964, 385–388.

J. BOSÁK, A. KOTZIG, Š. ZNÁM, "On some Metric Problems in Graph Theory", in *Beiträge zur Graphentheorie* (Sachs, Voss, Walther, ed.), Leipzig, 1968, 33–39.

J. BOSÁK, A. KOTZIG, Š. ZNÁM, Strongly Geodetic Graphs. *J. of Combinat. Theory*, **5**, 1968, 170–176.

D. BRATTON, Efficient Communication Networks. *Cowles Comm. Disc. Paper*, 2119, 1955.

G. B. DANTZIG, On the Shortest Route Through a Network. *Manag. Sc.*, **6**, 1960, 187–189.

J. C. DERNIAME, C. PAIR, *Problemes de cheminement dans les graphes*, Monographie (8), Dunod, 1971.

E. W. DIJKSTRA, A Note on Two Problems in Connection with Graphs. *Numerische Mathematik*, **1**, 1959, 269–271.

S. E. DREYFUS, An Appraisal of Some Shortest-path algorithms. *Operations Res.*, **17**, 1969, 395–412.

P. ERDÖS, H. SACHS, Reguläre Graphen gegebener Taillenweite mit minimaler Knotenzahl, *Wiss. Z. Univ. Halle. Math.*, **12**, 1963, 251–258.

A. GHOUILA-HOURI, Diamètre maximal d'un graphe fortement connexe. *C. R. Ac. Sc. Paris* **250**, 1960, 4 254–4 256.

M. K. GOLDBERG, The Radius of a Graph (Russian). *Uspehi Mat. Nauk* **20**, 1965, n° 5.

M. K. GOLDBERG, The Diameter of a Strongly Connected Graph (Russian). *Doklady*, **170**, 4, 1966.

B. GRÜNBAUM, *Convex Polytopes*. Wiley, New York, 1967.

C. HEUCHENNE, Sur une certaine correspondance entre graphes. *Bull. Soc. Roy. Sc. Liège*, **33**, 1964, 743–753.

A. HOFFMAN, R. SINGLETON, On Moore Graphs with Diameters 2 and 3. I. B. M. *J. of Res.*, **4**, 1960, **5**, 497–504.

C. JORDAN, Sur les assemblages de lignes. *J. Reine Angew. Math.* **70**, 1869, 185–190.

A. KAUFMAN and Y. MALGRANGE, Recherche des chemins et circuits hamiltoniens d'un graphe. *Revue Française de Rech. Opérat.*, **26**, 1963, 61–73.

V. KLEE, Diameters of Polyhedral Graphs. *Canad. J. Math.*, **16**, 1964, 602–614.

E. LUCAS, *Récréations mathématiques*, 1, Blanchard, Paris, 1921.

K. MAGHOUT, Thesis, University of Paris, 1962.

U. S. R. MURTY, On Some Extremal Graphs. *Acta Math. Acad. Sc. Hungaricae*, **19**, 1968, 69–74.

U. S. R. MURTY, Extremal Non-separable Graphs of Diameter Two. *Univ. of Waterloo Mimeograph*, 1968.

U. S. R. MURTY, K. VIJAYAN, On Accessibility in Graphs. *Sankhya*, **A 26**, 1965, 270–302.

C. PAIR, "Sur des algorithmes pour les problèmes de cheminements dans les graphes finis". *Théorie des Graphes, Rome I. C. C.* (P. Rosenstiehl, ed.), Dunod, Paris, 1967, 271–300.

R. RADNER, A. TRITTER, Communication in Networks. *Cowles Comm. paper*, 2098, 1954.

P. ROSENSTIEHL, Existence d'automates finis capables de s'accorder bien qu'arbitrairement connectés et nombreux. *I. C. C. Bulletin*, 1966, **5**, 245–261.

B. ROY, Contribution de la Théorie des Graphes à l'étude de certains problèmes linéaires. *C. R. Acad. Sc. Paris*, **248**, 1959, 2437.

B. ROY, Cheminement et connexité dans les graphes. Thesis, University of Paris, 1962.

B. ROY, *Algèbre moderne et Théorie des Graphes*. Dunod, Paris, 1969.

H. SACHS, Über selbstkomplementäre Graphen, *Publ. Math. Debrecen*, **9**, 1962, 270–288.

G. TARRY, Le problème des labyrinthes. *Nouvelles Annales de Math.*, **14**, 1895, 187–189.

V. G. VIZING, The Number of Edges in a Graph with Given Radius (Russian). *Doklady*, **173**, 1967, n° 6.

Š. ZNÁM, Decomposition of the Complete Directed Graph into Factors with Given Diameters. *Abstract Calgary Conference*, 1969.

CHAPTER 5: **Flow Problems**

C. BERGE, Sur la déficience d'un réseau infini. *C. R. Acad. Sciences*, **245**, 1957, 1206–1207.

C. BERGE, Sur l'équivalence du problème du transport généralisé et du problème des réseaux. *C. R. Ac. Sciences*, **251**, 1960, 324–325.

C. BERGE, Les problèmes de flot et de tension. *Cahiers du Centre d'Etudes de Rech. Opérat., Bruxelles*, **3**, 1961, 69–83.

C. BERGE and A. GHOUILA-HOURI, *Programmes, Jeux et Réseaux de Transport* Dunod, Paris, 1962.
(English translation: Methuen, London; Wiley, New York, 1965.— German translation: Teubner, Leipzig, 1967.—Spanish translation: Compania ed. Continental, México, 1965.)

R. FAURE, *Eléments de Recherche Opérationnelle*. Gauthier-Villars, Paris, 1968.

L. R. FORD and D. R. FULKERSON, Maximal Flow Through a Network. *Canadian J. of Math.*, **8**, 1956, 399.

L. R. FORD and D. R. FULKERSON, Dynamic Network Flow. *Rand Corp. Paper*, **P967**, 1956.

L. R. FORD and D. R. FULKERSON, A Network Flow Feasibility Theorem and Combinatorial Applications. *Canad. J. Math.*, **11**, 1959, 440–450.

L. R. FORD and D. R. FULKERSON, *Flows in Networks*. Princeton Press, Princeton, 1962 (French translation by J. C. Arinal, Gauthier-Villars, Paris, 1967).

D. Gale, A Theorem on Flows in Networks. *Pacific J. Math.*, **7**, 1957, 1073.

G. de Ghellinck, Aspects de la notion de dualité en Théorie des Graphes. *Cahiers du Centre d'Etudes de Rech. Opérat.*, *Bruxelles*, **3**, 1961, 94–102.

A. Ghouila-Houri, Sur l'existence d'un flot ou d'une tension prenant ses valeurs dans un groupe abélien. *C. R. Ac. Sciences*, **250**, 1960, 3931–3932.

A. Ghouila-Houri, "Recherche du flot maximum dans certains réseaux lorsqu'on impose une condition de bouclage". *Proc. Second Int. Conf. Op. Res.*. London, 1960, 156–157.

A. Ghouila-Houri, *Flots et tensions dans un graphe.* Ann. Ec. Normale **81**, 1964, 267–339.

A. J. Hoffman (*in* C. Berge, *Théorie des Graphes*, Paris, 1958, p. 80).

J. C. Herz, Cours. Mimeo, Lille, 1967.

C. Ramanujacharyulu, Transport Networks with Pairs of Homologous Arcs. *Cahiers Centre Et. Rech. Operat. Bruxelles*, **5**, 1963, 203–210.

C. Ramanujacharyulu, Trees and Tree Equivalent Graphs. *Canad. J. Math.*, **170**, 1965, 731–733.

B. Roy, Thesis, Métra, 1962.

B. Roy, *Algèbre moderne et Théorie des Graphes*, I, Dunod, Paris, 1969; II, Paris, 1970.

P. Slepian, *Mathematical Foundations of Network Analysis.* Springer Tracts, No. 16, Springer Verlag, Berlin, Heidelberg, New York, 1968.

L. Weinberg, A Survey of Linear Graphs; Fundamentals and Applications to Network Theory. *Matrix Tensor Quarterly*, **14**, 1963–1964, 103–115.

CHAPTER 6: **Degrees and Demi-degrees**

C. Berge, *Espaces topologiques*, p. 193. Dunod, Paris, 1966 (p. 184, English translation, Oliver and Boyd, 1963).

L. W. Beineke, F. Harary, Local Restrictions for Various Classes of Directed Graphs. *J. London Math. Soc.*, **40**, 1965, 87–95.

Jack Edmonds, Existence of *k*-edge Connected Ordinary Graphs with Prescribed Degrees. *J. Res. Nat. Bur. Standards*, B **68**, 1964, 73–74.

P. Erdös, T. Gallai, Graphen mit Punkten vorgeschriebenen Grades. *Mat. Lapok*, **11**, 1960, 264–274.

D. R. Fulkerson, Zero-one Matrices with Zero Trace. *Pacific J. Math.*, **10**, 1960, 831–836.

D. R. Fulkerson, A. J. Hoffman, M. H. McAndrew, Some Properties of Graphs with Multiple Edges. *Canad. J. Math.*, **17**, 1965, 166–177.

D. Gale, A Theorem on Flows in Networks. *Pacific J. Math.*, **7**, 1957, 1073–1082.

G. H. Hardy, J. E. Littlewood, G. Polya, *Inequalities.* Cambridge, 1952.

H. G. Landau, On Dominance Relations. *Bull. Math. Biophys.* **15**, 1953, 143–148.

S.-Y.R. Li, Graphic sequences with Unique Realization. *J. Combin. Theory Ser.* **B 19**, 1975, 42–68.

V. V. Menon, On the Existence of a Tree with Given Degrees. *Sankhya*, **26**, 1964, 63–68.

J. W. Moon, An Extension of Landau's Theorem on Tournaments. *Pac. J. Math.*, **13**, 1963, 1343–1345.

R. F. MUIRHEAD, Some Methods Applicable to Identities and Inequalities of Symmetric Algebraic Functions of *n* Letters. *Proc. Edinburgh Math. Soc.*, **21**, 1903, 144–150.

A. RAMACHANDRA RAO, Some Extremal Problems and Characterizations in the Theory of Graphs. Thesis, Indian Stat. Inst., Calcutta, 1969.

S. B. RAO, A. RAMACHANDRA RAO, Existence of 3-connected Graphs with Prescribed Degrees. *Pac. J. Math.*, **33**, 1970, 203–207.

H. J. RYSER, Combinatorial Properties of Matrices of Zeros and Ones. *Canad. J. Math.*, **9**, 1957, 371–377.

H. J. RYSER, Matrices of Zeros and Ones. *Bull. A. M. S.*, **66**, 1960, 442–464.

J.K. SENIOR, Degree Constraints, *Amer. J. Math.*, 1951, 663–682.

CHAPTER 7 : Matchings

J. AKIYAMA, M. KANO, Factors and Factorization of Graphs, a Survey (to appear).

M. BALINSKI, On Maximum Matching, Minimum Coverings and Their Connections. *J. Combinat. Theory* (to appear).

C. BERGE, Two Theorems in Graph Theory. *Proc. Nat. Ac. Sciences*, U. S. A., **43**, 1957, 842.

C. BERGE, Sur le couplage maximum d'un graphe. *C. R. Acad. Sciences*, **247**, 1958, 258–259.

C. BERGE, "Alternating chains", in: *Graph Theory and Computing* (R. C. Read) Acad. Press, 1972.

C. BERGE, "A Theorem Related to the Chvátal Conjecture". *Proc. 5th British Comb. Conf.* (C. J. A. Nash-Williams, J. Sheehan, eds.) Utilitas Math., Winnipeg, 1976, 35–39.

C. BERGE, M. LAS VERGNAS, On the Existence of Subgraphs with Degree Constraints. *Proc. Koninklijke Nederland. Akad. Wetenschappen, Amsterdam,* **A 81**, 1978, 165–176.

A. L. DULMAGE and N. S. MENDELSOHN, Some Graphical Properties of Matrices with Non-negative Entries. Mimeo, Univ. Alberta, 1961.

Jack EDMONDS, Covers and Packings in a Family of Sets. *Bull. A. M. S.*, **68**, 1962, 494–499.

Jack EDMONDS, Paths, Trees and Flowers. *Canad. J. Math.*, **17**, 1965, 449–467.

P. ERDÖS, T. GALLAI, On Maximal Paths and Circuits of Graphs, *Acta Math. Acad. Sc. Hungar.*, **10**, 1959, 337–356.

P. ERDÖS, L. POSÁ, On the Maximal Number of Disjoint Circuits in a Graph. *Publ. Math. Debrecen*, **9**, 1962, 3–12.

P. ERDÖS, T. GALLAI, On the Minimal Number of Vertices Representing the Edges of a Graph. *Publ. Math. Mag. Tud. Ak.*, **6**, 1961, 181–203.

M. E. FISHER, Statistical Mechanics of Dimers on a Plane Lattice. *Phys. R.*, **124**, 1961, 278–286.

J. FOLKMAN, D. R. FULKERSON, "Edge Colouring in Bipartite Graphs". *Combinatorial Mathematics and their Applications* (Bose, Dowling ed.), Univ. of N.C. Press, Chapel Hill, 1969.

R. L. GRAHAM, L. H. HARPER, Some Results on Matching in Bipartite Graphs, *SIAM., J.* **17**, 1969, 1017–1022.

R. P. GUPTA, *Studies in the Theory of Graphs*. Tata Institute of Fund. Res., Bombay, 1967, Thesis, Calcutta, 1967.

R. P. GUPTA, "A Decomposition Theorem for Bipartite Graphs". *Théorie des Graphes Rome I. C. C.*, (P. Rosenstiehl, ed.), Dunod, Paris, 1967.

M. HALL, Distinct Representatives of Subsets. *Bull. Am. Math. Soc.*, **54**, 1948, 922–926.

P. HALL, On Representations of Subsets. *J. London Math. Soc.*, **10**, 1934, 26–30.

P. R. HALMOS and H. E. VAUGHAN, The Marriage Problem. *Am. J. of Math.*, **72**, 1950, 214–215.

A. J. HOFFMAN, H. W. KUHN, Systems of Distinct Representatives and Linear Programming. *Am. Math. Monthly*, **63**, 1956, 455–466.

P. W. KASTELEYN, The Statistics of Dimers on a Lattice. I The Number of Dimer Arrangements on a Quadratic Lattice. *Physica*, **27**, 1961, 1209–1225.

P. W. KASTELEYN, "Graph Theory and Crystal Physics". *Graph Theory and Theoretical Physics* (F. Harary, ed.), Academic Press, London, New York, 1967.

D. KÖNIG, Graphen und Matrizen. *Mat. Fiz. Lapok*, **38**, 1931, 116–119.

D. KÖNIG, *Theorie der endlichen und unendlichen Graphen*, Leipzig, 1935.

D. KÖNIG and S. VALKO, Über mehrdeutige Abbildungen von Mengen. *Math. Annalen*, **95**, 1926, 135.

H. W. KUHN, The Hungarian Method for the Assignment Problem. *Naval Research Quarterly*, **2**, 1955, 83.

M. LAS VERGNAS, A Note on Matchings in Graphs. *Cahiers du C.E.R.O.*, **17**, 1975, 257–260.

L. LOVÁSZ, On Coverings of Graphs. *An. Akad. Hung.*, **4**, 1968, 231–236.

H. B. MANN, H. J. RYSER, Systems of Distinct Representatives. *Am. Math. Monthly*, **60**, 1953, 397.

P. MEDGYESSY, Über nicht primitive reguläre Graphen dritten Grades. *Publ. Math.*, **1**, 1950, 183–185.

J. F. NAGLE, "A New Method for Combinatorial Problems on Graphs". *Théorie des Graphes Rome I. C. C.* (P. Rosenstiehl, ed.), Dunod, Paris, 1967, 257–262.

R. Z. NORMAN, M. O. RABIN, An Algorithm for a Minimum Cover of a Graph. *Proc. Amer. Math. Soc.*, **10**, 1959, 315–319.

O. ORE, Graphs and Matching Theorems. *Duke Math. J.*, **22**, 1955, 625–639.

J. M. PLA, Une application de la théorie des graphes en physique mathématique Mimeo, IHP Séminaire sur les Problèmes Combinatoires, 4 March 1970.

R. RADO, Factorization of Even Graphs. *Quarterly J. of Math.*, **20**, 1949, 95–104.

A. RÉNYI, New Methods and Results in Combinatorial Analysis (Hungarian). *Magyar Tud. Akad. Mat. Fiz.*, **16**, 1966, 159–177.

B. RUY, *Algèbre Moderne et Théorie des Graphes I.* Dunod, Paris, 1969; *II.* Dunod, Paris, 1970.

C. WITZGALL, C. T. ZAHN, Modification of Edmonds' Maximum Matching Algorithm. *J. of Res. N. B. S.*, **69 B**, 1965, 91–98.

CHAPTER 8 : *c*-Matchings

B. ALSPACH, A 1-Factorization of the Line-Graphs of Complete Graphs. *J. Graph Theory*, **6**, 1982, 441–445.

M. L. BALINSKI, On Perfect Matchings. *S.I.A.M. Rev.*, **12**, 1970, 570–572.

C. BERGE, Two Theorems in Graph Theory. *Proc. Nat. Ac. Sciences* USA, **43**, 1957, 842–844.

C. BERGE, Sur le couplage maximum d'un graphe. *C. R. Acad. Sciences, Paris*, **247**, 1958, 258–259.

C. BERGE, *Lectures on Graph Theory*. Tata Institute of Fund. Res., Bombay, 1967.

G. CORNUEJOLS, W. PULLEYBLANK, A Matching Problem with Side Conditions. *Discrete Math.*, **29**, 1980, 135–159.

A. ERRERA, Du coloriage des cartes. *Mathesis*, **36**, 1922, 56–60.

T. GALLAI, On Factorization of Graphs. *Acta Math. Hung.*, **1**, 1950, 133–153.

S. KUNDU, Generalization of the k-Factor Theorem. *Discrete Math.*, **9**, 1974, 173–179.

M. LAS VERGNAS, An Extension of Tutte's l-Factor Theorem. *Discrete Math.*, **23**, 1978, 241–255.

L. LOVÁSZ, Three Short Proofs in Graph Theory. *J. Comb. Theory B*, **19**, 1975, 111–113.

L. LOVÁSZ, M.D. PLUMMER, 'On Bicritical Graphs'. *Infinite and Finite Sets*, North-Holland, Amsterdam, 1975, 1051–1079.

F. G. MAUNSELL, A Note on Tutte's Paper. *J. London Math. Soc.*, **27**, 1952, 127.

P. J. McCARTY, Matchings in Graphs. *Bull. Australian Math. Soc.*, **9**, 1973, 141–143.

D. NADDEF, W. R. PULLEYBLANK, Matchings in Regular Graphs. *Discrete Math.*, **34**, 1981, 283–291.

L. NEBESKY, Some Sufficient Conditions for the Existence of a l-Factor. *J. Graph Theory*, **2**, 1978, 251–255.

J. PETERSEN, Die Theorie der regulären Graphen. *Acta Math.*, **15**, 1891, 193–220.

J. PLESNIK, Remark on Matchings in Regular Graphs. *Acta Fac. Rerum Natur. Univ. Comenian. Math.*, **34**, 1979, 63–67.

D. P. SUMMER, l-Factors and Anti-Factor Sets. *J. London Math. Soc.*, **13**, 1976, 351–359.

C. THOMASSEN, A Remark on the Factor Theorem of Lovász and Tutte. *J. Graph Theory*, **5**, 1981, 441–442.

W. T. TUTTE, The Factorization of Linear Graphs. *J. London Math. Soc.*, **22**, 1947, 107–111.

W. T. TUTTE, The Factorization of Locally Finite Graphs. *Canad. J. Math.*, **2**, 1950, 44–49.

W. T. TUTTE, The Factors of Graphs. *Canad. J. Math.*, **4**, 1952, 314–328.

W. T. TUTTE, A Short Proof of the Factor Theorem for Finite Graphs. *Canad. J. Math.*, **6**, 1954, 347–352.

CHAPTER 9: Connectivity

F. BÄBLER, Über die Zerlegung regulärer Streckenkomplexe ungerader Ordnung. *Comm. Math. Helvetici*, **10**, 1938, 275–287.

C. BERGE, Sur le nombre minimum d'arcs à inverser pour rendre un graphe fortement connexe. *Cahiers du C.E.R.I.*, **25**, 1983, 183–186.

F. T. BOESCH, The Strongest Monotone Degree Condition for n-Connectedness of a Graph. *J. Comb. Theory B*, **16**, 1974, 162–165.

J. A. BONDY, Properties of Graphs with Constraints on Degrees, *Studia Sc. Math. Hung.*, **4**, 1969, 473–475.

G. CHARTRAND, F. HARARY, "Graphs with Prescribed Connectivities". *Theory of Graphs* (P. Erdös, G. Katona, ed.), Akad. Kiadó, Budapest, 1968, 61–63.

G. CHARTRAND, S. F. KAPOOR, H. V. KRONK, A Sufficient Condition for *n*-connectedness of Graphs. *Mathematika*, **15**, 1968, 51–52.

M. CHEIN, Graphes régulièrement décomposables. *Revue Fr. Informatique Rech. Opérat.*, **2**, 1968, 27–42.

G. A. DIRAC, In abstrakten Graphen vorhandene vollständige 4-Graphen und ihre Unterteilungen. *Math. Nachr.*, **22**, 1960, 61–85.

L. EGYED, Über die wohlgerichteren unendlichen Graphen. *Math. Phys. Lapok*, **48**, 1941, 505–509.

R. C. ENTRIGER, P. J. SLATER, A Theorem on Critically 3-Connected Graphs. *Nanta Math.*, **11**, 1978, 141–145.

T. GALLAI, Elementare Relationen bezüglich der Glieder. *Magyar Tud. Ak. Mat.*, **9**, 1964, 235–236.

E. J. GRINBERG, J. J. DAMBIT, (Russian), *Lato. Math. E*, **2**, 1966, 65–70.

R. HALIN, A Theorem on *n*-connected Graphs. *J. Combinat. Theory*, **7**, 1969, 150–154.

Y. O. HAMIDOUNE, Quelques problèmes de connexité dans les graphes orientés. *J. Comb. Theory*, **13**, 1981, 1–10.

Y. O. HAMIDOUNE, On a Conjecture of Entriger and Slater. *Discrete Math.*, **41**, 1982, 323–326.

F. HARARY, The Maximum Connectivity of a Graph. *Proc. Nat. Ac. Sc. USA*, **48**, 1962, 1142–1146.

F. HARARY, *Graph Theory*. Addison-Wesley, 1969.

M. LAS VERGNAS, Une propriété forte de connexité en Théorie des Graphes. *C. R. Ac. Sc. Paris*, A 266, 1968, 561–563.

M. LAS VERGNAS, Sur le nombre de circuits dans un graphe fortement connexe. *Cahiers du C.E.R.O.*, **17**, 1975, 261–265.

W. MADER, "Connectivity and Edge-Connectivity in Finite Graphs". *Survey in Combinatorics*, Proc. 7th British Comb. Conf. (B. Bollobas, ed.) Cambridge, 1979.

W. MADER, "On *k*-Critically *n*-Connected Graphs". *Progress in Graph Theory* (J.A. Bondy, U.S.R. Murty, eds.) Academic Press, Toronto, 1984, 389–398.

S. B. MAURER, P. J. SLATER, On *k*-Critical *n*-Connected Graphs. *Discrete Math.*, **20**, 1977, 255–262.

K. MENGER, Zur allgemeinen Kurventheorie. *Fund. Math.*, **10**, 1926, 96.

D. M. MESNER, M. E. WATKINS, Some Theorems about *n*-vertex Connected Graphs. *J. Math. Mech.*, **16**, 1966, 321–326.

C. St. J. A. NASH-WILLIAMS, "Well Balanced Orientations of Finite Graphs and Unobstrusive Odd Vertex Pairings". *Recent Progress in Combinatorics* (Tutte, ed.), Academic Press, 1969, 133–149.

C. St. J. A. NASH-WILLIAMS, On Orientations, Connectivity and Odd Vertex Pairings in Finite Graphs. *Canad. J. Math.*, **12**, 1960, 555–567.

M. D. PLUMMER, On Minimal Blocks. *Trans. Am. Math. Soc.* **134**, 1968, 85–94.

A. RAMACHANDRA RAO, An Extremal Problem in Graph Theory. *Israel J. Math.*, **6**, 1968, 261–266.

H. E. Robbins, A Theorem on Graphs with an Application to a Problem of Traffic Control. *Am. Math. Monthly*, **46**, 1939, 281–283.

W. T. Tutte, A Theory of 3-connected Graphs. *Indag. Math.*, **23**, 1961, 441–455.

W. T. Tutte, *Connectivity in Graphs*, University of Toronto Press, Toronto-London, 1966.

D. L. Wang, D. J. Kleitman, On the Existence of *n*-Connected Graphs with Prescribed Degrees. *Networks*, **3**, 1973, 225–239.

M. E. Watkins, Connectivity of Transitive Graphs. *J. Comb. Theory*, **8**, 1970, 23–29.

L. Weinberg, A Simple and Efficient Algorithm for Determining Isomorphism of Planar Triply Connected Graphs. *IEEE Trans.* CT-13, 1966, 142–148.

H. Whitney, Congruent Graphs and the Connectivity of Graphs. *Am. J. Math.*, **54**, 1932, 150–168.

CHAPTER 10: Hamiltonian Cycles

B. Alspach, Cycles of Each Length in Regular Tournaments. *Canad. Math. Bull.*, **10**, 1967, 283–286.

J. C. Arditti, R. Cori, Hamilton Circuits in the Comparability Graph of a Tree. *Combinatorial Theory and its Applications, Balatonfüred* (Erdös, Rényi, Turán-Sus ed.), North-Holland Publ. Co., Amsterdam-London, 1970, 41–52.

M. Balinski, On the Graph Structure of Convex Polyhedra. *Pac. J. Math.*, **11**, 1961, 431–434.

D. Barnette, E. Juković, Hamiltonian circuits on 3-polytopes, *J. Comb. Theory*, **2**, 1970, 54–59.

C. Berge, "Contribution de la Théorie des Graphes à l'étude des relations d'ordre". *Ordres totaux finis* (M. Barbut, ed.), Gauthier-Villars, Paris, 1971, 143–162. I. C. C. Res. Report 67/2, 1967.

J. C. Bermond, Graphes orientés fortement *k*-connexes et graphes *k-arc*-hamiltoniens. *C. R. Acad. Sc. Paris*, Série A, **271**, 1970, 141–144.

J. C. Bermond, Hamiltonian Decompositions of Graphs. *Advances in Graph Theory* (B. Bollobas, ed.) Ann. Discrete Math., **3**, 1978, 21–28.

J. C. Bermond, V. Faber, Decomposition of K_n^* Into *k*-Circuits, *J. Comb. Theory* **B**, **21**, 1976, 146–155.

J. C. Bermond, C. Thomassen, Cycles in Digraphs — A Survey. *J. Graph Theory*, **5**, 1981, 43, 145–157.

J. A. Bondy, Properties of Graphs with Constraints on Degrees. *Studia Sc. Math. Hung.*, **4**, 1969, 473–475.

J. A. Bondy, "Cycles in Graphs". *Combinatorial Structures and their Applications*, Gordon and Breach, New York, 1970, 15–18.

J. A. Bondy, 'Hamilton Cycles in Graphs and Digraphs', *Proc. 9th South Eastern Conf. in Combinatorics*, Boca Raton, Congress. Numer., **21**, Utilitas Math. Winnipeg, 1978, 3–28.

J. A. Bondy, A Remark on Two Sufficient Conditions for Hamilton Cycles. *Discrete Math.*, **22**, 1978, 191–193.

J. A. Bondy, V. Chvátal, A Method in Graph Theory. *Discrete Math.*, **15**, 1976, 111–136.

J. A. BONDY, C. THOMASSEN, A Short Proof of Meyniel's Theorem. *Discrete Math.*, **19**, 1977, 195–197.

J. BOSÁK, "Hamiltonian Lines in Cubic Graphs". *Théorie des Graphes ICC* (P. Rosenstiehl, ed.), Dunod, Paris, 1967, 35–46.

T. A. BROWN, Simple Paths on Convex Polyhedra, *Pacif. J. Math.* **11**, 1961, 1211–1214.

R. BRUALDI, Private communication, 1980.

P. CAMION, Quelques propriétés des chemins et des circuits hamiltoniens dans la Théorie des Graphes. *Cahiers du Centre d'Etudes de Rech. Opérat.*, *Bruxelles*, **2**, 1960, 10–15.

P. CAMION, Chemins et circuits hamiltoniens des graphes complets. *C. R. Acad. Sciences, Paris*, **249**, 1959, 2151–2152.

R. CANTONI, On Consequences of the Existence of a Hamiltonian Cycle (Italian). *Ist. Lombardo sc. lett. rend. c. sc. mat. nat.*, **14**, 1950, 371–387.

R. CANTONI, On Cubic Hamilton Graphs (Italian), id., **98**, 1964, 319–326.

G. CHARTRAND, S. F. KAPOOR, The Cube of Every Connected Graph is Hamiltonian. *J. Res. N. B. S.*, **73 B**, 1969, 47–48.

G. CHARTRAND, S. F. KAPOOR, H. V. KRONK, A Generalization of Hamiltonian-Connected Graphs. *J. Math. Pures et Appliquées.* **48**, 1969, 109–116.

M. CHEIN, Graphes Régulièrement Décomposables. *Revue Fr. Rech. Opérat.*, **2**, 1968, 27–42.

Y. J. CHU, C. H. LIN, Z. G. YU, Hamiltonian Graphs. Academia Sinica, 1982.

V. CHVÁTAL, P. ERDÖS. A Note on Hamiltonian circuits, *Discrete Math.* **2**, 1972, 111–113.

V. CHVÁTAL, On Hamilton's Ideals. *J. Comb. Theory* **12 B**, 1972, 163–168.

K. ČULIK, A Remark to the Construction of Cubic Graphs Containing Hamiltonian Lines. *Časop. pěstov. mat.*, **89**, 1964, 385–389 (Russian).

G. A. DIRAC, Some Theorems on Abstract Graphs. *Proc. London Math. Soc.*, **2**, 1952, 69–81.

P. ERDÖS, Remarks on a Paper of Pósa. *Publ. Math. Inst. Hung. Ac. Sc.*, **7**, 1962, 227–228.

P. ERDÖS, T. GALLAI, On Maximal Paths and Circuits of Graphs. *Acta Math. Ac. Sc. Hung.*, **10**, 1959, 337–356.

P. ERDÖS, A. M. HOBBS, Hamiltonian Cycles in Regular Graphs of Moderate Degree. *J. Comb. Theory B*, **23**, 1977, 139–142.

H. FLEISCHNER, In the Square of a Graph, Hamiltonicity.... *Monatsch. Math.*, **82**, 1976, 125–149.

H. FLEISCHNER, A. M. HOBBS, A Necessary Condition for the Square of a Graph to be Hamiltonian. *J. Comb. Theory B*, **19**, 1975, 97–118.

A. GHOUILA-HOURI, Une condition suffisante d'existence d'un circuit hamiltonien. *C. R. Acad. Sc.*, **251**, 1960, 495–497.

B. GRÜNBAUM, *Convex Polytopes* (Chap. 17), Wiley, New York, 1967.

B. GRÜNBAUM, Polytopes, Graphs and Complexes, *Bull. A.M.S.*, **76**, 1970, 1131–1201.

B. GRÜNBAUM, T. S. MOTZKIN, Longest Paths in Polyhedral Graphs. *J. London Math. Soc.*, **37**, 1962, 152–160.

R. GRÜNBAUM, "Polytopes and Graphs". *Studies in Graph Theory II*, Studies Math., **12**, Math. Ass. America, 1965, 201–224.

A. GYÁRFÁS, J. GÉRENCZER, Private communication, 1968.

J. C. HERZ, J. J. DUBY, F. VIGUÉ, "Recherche systématique des graphes hypoha-miltoniens". *Théorie des Graphes ICC* (P. Rosenstiehl, ed.), Dunod-Gordon and Breach, 1967, 153–159.

A. M. HOBBS, J. MICHEM, The Entire Graph of a Bridgeless Connected Plane Graph is Hamiltonian. *Discrete Math.*, **16**, 1976, 41–50 and 231–240.

J. HUNTER, Non-Hamiltonian Graphs and their Duals, *Ph.D. Thesis, Rensselaer Polytechn. Inst.* 1962.

B. JACKSON, Hamilton Cycles in Regular 2-Connected Graphs. *J. Comb. Theory B*, **29**, 1980, 27–46.

B. JACKSON, Long Paths and Cycles in Oriented Graphs. *J. Graph Theory*, **5**, 1981, 145–157.

O. S. JACKSON, Cycles and Paths in Tournaments. Thesis, University of Aarhus, 1972.

E. JUCOVIČ, A Note on Paths in Quadrangular Polyhedral Graphs, *Č. Pěst. Mat.*, **93**, 1968, 69–72 (Slovakian).

J. J. KARAGANIS, On the Cube of a Graph. *Canad. Math. Bull.*, **11**, 1968, 295–296.

A. KOTZIG, The Construction of Cubic Hamilton Graphs. *Časop. pestov mat.*, **87**, 1962, 477–488 (Russian).

H. V. KRONK, "Variations on a Theorem of Pósa". *The Many Facets of Graph Theory* (G. Chartrand, S. F. Kapoor, ed.), Springer Verlag, 1969, 193–197.

H. V. KRONK, A Note on *k*-path Hamiltonian Graphs. *J. Combinatorial Theory*, **7**, 1969, 104–106.

S. KUNDU, Generalization of the *k*-Factor Theorem, *Discrete Math.*, **9**, 1974, 173–179.

M. LAS VERGNAS, Sur l'existence des cycles hamiltoniens dans un graphe *C. R. Acad. Sc. Paris*, **270**, 1960, A-1361–1364.

M. LAS VERGNAS, Sur une propriété des arbres maximaux dans un graphe, *C.R. Acad. Sc. Paris*, **272**, 1971, 1297–1300.

M. LAS VERGNAS, Thesis, University of Paris, 1972.

J. LEDERBERG, Hamilton Circuits of Convex Trivalent Polyhedra, *Am. Math. Monthly*, **74**, 1967, 522–527.

M. LEWIN, On Maximal Circuits in Directed Graphs. *J. Comb. Theory B*, **18**, 1975, 175–179.

K. MAGHOUT, Thesis, Université de Paris, 1962.

H. MEYNIEL, Une condition suffisante d'existence d'un circuit hamiltonien dans une graph orienté. *J. Comb. Theory B*, **14**, 1973, 137–147.

H. MEYNIEL, Une condition sufisante d'existence d'un circuit hamiltonian dans un graphe orienté. *J. Comb. Theory B*, **14**, 1973, 137–147.

J. W. MOON, *Topics on Tournaments*. Holt, New York, 1968.

J. W. MOON, L. MOSER, On Hamiltonian Bipartite Graphs. *Israel J. of Math.*, **1**, 1963, 163–165.

C. St. J. A. NASH-WILLIAMS, On Hamiltonian Circuits in Finite Graphs. *Proc. Am. Math. Soc.*, **17**, 1966, 466–467.

C. St. J. A. NASH-WILLIAMS, "Hamiltonian Circuits in Graphs and Digraphs". *The Many Facets of Graph Theory* (G. Chartrand, S. F. Kapoor, ed.), Springer Verlag, 1969, 237–243.

C. St. J. A. NASH-WILLIAMS. Edge-disjoint Hamiltonian Circuits in Graphs with Vertices of Large Valency. *Studies in Pure Mathematics* (L. Mirsky, ed.), Acad. Press, 1971.

O. Ore, Arc Coverings of Graphs. *Ann. Mat. Pura Appl.*, **55**, 1961, 315–321.

O. Ore, Hamilton-connected Graphs. *J. de Math. Pures et Appl.*, **42**, 1963, 21–27.

O. Ore, *The Four Color Problem.* Academic Press, New York, 1967.

M. Overbeck-Larish, Hamiltonian Paths in Oriented Graphs. *J. Comb. Theory B*, **21**, 1976, 76–80.

M. Overbeck-Larish, A Theorem on Pancyclic Oriented Graphs. *J. Comb. Theory B*, **23**, 1977, 168–173.

L. Pósa, A Theorem Concerning Hamilton Lines. *Magyar Tud. Akad. Mat. Kutato Int. Közl.*, **7**, 1962, 225–226.

H. Raynaud, Sur les chemins hamiltoniens dans les graphes orientés. Mimeo, Université du Mans, 1970.

L. Rédei, Ein kombinatorischer Satz. *Acta Litt. Szeged*, **7**, 1934, 39–43.

H. Sachs, "Construction of Non-hamiltonian Planar Regular Graphs of Degrees 3, 4, 5". *Théorie des Graphes ICC* (P. Rosenstiehl ed.), Dunod, Paris, 1967, 373–382.

H. Sachs, "Ein von Kozyrev und Grinberg angegebener nicht-hamiltonscher kubischer planarer Graph". *Beiträge zur Graphentheorie*, B. G. Teubner, Leipzig, 1968, 127–130.

M. Sekanina, Graphs. *Publ. Fac. Sc. Brno*, **412**, 1960, 137–141.

C. A. B. Smith, Private communications, 1946 (*in* C. Berge, *Théorie des graphes*, Paris, 1958, p. 185).

F. Supnick, L. V. Quintas, Extreme Hamiltonian Circuits, Resolution of the Convex Add-case. *Proc. Am. Math. Soc.*, **15**, 1964, 454–456.

C. Thomassen, An Ore-Type Condition Implying a Digraph to Be Pancyclic. *Discrete Math.*, **19**, 1977, 85–92.

T. Tillson, A Hamiltonian Decomposition of K^*_{2m}. *J. Comb. Theory B*, **29**, 1980, 68–74.

J. Turner, Point-Symmetric Graphs With a Prime Number of Point. *L.C.T.*, **3**, 1967, 136–145.

W. T. Tutte, On Hamilton Circuits. *J. London Math. Soc.*, **21**, 1946, 98–101.

W. T. Tutte, A Theorem on Planar Graph. *Trans. Amer. Math. Soc.*, **82**, 1956, 99–116.

W. T. Tutte, A Non-Hamiltonian Planar Graph. *Acta Math. Acad. Sc. Hung.*, **11**, 1960, 371–375.

W. T. Tutte "Hamiltonian Circuits", in *Graph Theory and Computing* (R. C. Read), Academic Press, New York, 1972, 295–301.

H. Walther, Ein kubischer planarer zyklisch fünffach zusammenhängender Graph, der keinen Hamiltonkreis besitzt. *Wiss. Zeit. Hochschule für Elektrotechnik. Ilmenau*, **11**, 1965, 163–166.

C.-Q. Zhang, Arcs-Disjoint Circuits in Digraphs. *Discrete Math.*, **41**, 1982, 79–96.

A. A. Zykov, *Theory of Finite Graphs*, Nauka, Novosibirsk, 1969.

CHAPTER 11 : Covering Edges with Chains

T. Van Aardenne-Ehrenfest, N. G. de Bruijn, Circuits and Trees in Oriented Linear Graphs. *Simon Stevin*, **28**, 1951, 203–217.

A. S. Amitsur, J. Levitzki, Minimal Identities for Algebras. *Proc. Am. Math. Soc.*, **1**, 1950, 449–463.

B. AUERBACH, R. LASKAR, On Decomposition of r-Partite Graphs Into Edge Disjoint Hamilton Circuits. *Discrete Math.*, **14**, 1976, 265–270.

F. BÄBLER, Über die Zerlegung regulärer Streckenkomplexe ungerader Ordnung. I, *Comment. Math. Helvet.*, **10**, 1938, 275–287; II, *Comment. Math. Helvet.*, **26**, 1952, 117–118; III, *Comment. Math. Helvet.*, **28**, 1954, 155–161.

H. B. BELCK, Reguläre Faktoren von Graphen. *J. Reine Angew. Math.*, **188**, 1950, 228–259.

A. BOUCHET, Private Communication, 1976.

N. G. DE BRUIJN, A Combinatorial Problem. *Proc. Nederl. Akad. Wetensch.*, **49**, 1946, 758–764.

G. CHARTRAND, Graphs and Their Associated Line-graphs, Thesis, Michigan State University, 1964.

A. DONALD, An Upper Bound for the Path Number of a Graph. *J. Graph Theory*, **4**, 1980, 189–202.

P. ERDÖS, A. W. GOODMAN, L. PÓSA, The Representation of a Graph by Set Intersections, *Canad. J. Math.*, **18**, 1966, 106–112.

L. EULER, *Commentationes Arithmeticae Collectae*, St. Petersburg, 1766, 337–338.

FLYE Sainte-Marie, Note sur un problème relatif à la marche du cavalier sur l'échiquier. *Bull. Soc. Math. de France*, **5**, 1876, 144–147.

R. P. GUPTA, A Contribution to the Theory of Finite Oriented Graphs. *Sankhya*, A **27**, 1965, 401–404.

A. KOTZIG, Eulerian Lines and the Decompositions of the Regular Graphs of Even Degree. *Mat. fyz. Časop.*, **6**, 1956, 133–136 (Slovakian).

A. KOTZIG, Beiträge zur Theorie der endlichen gerichteten Graphen. *Wiss. Martin-Luther Univ. Halle-Wittenberg*, **10**, 1961, 118–125.

A. KOTZIG, "Über Zyklen in Turnieren". *Beiträge zur Graphentheorie* (Sachs, Voss, Walther), Teubner, Leipzig, 1968, 85–89.

L. LOVÁSZ "On Covering of Graphs". *Theory of Graphs*, Tihany (Erdös-Katona, ed.). Academic Press, New York, 1968, 231–236.

C. St. J. A. NASH-WILLIAMS, Decomposition of Finite Graphs into Open Chains. *Canad. J. Math.*, **13**, 1961, 157–166.

C. St. J. A. NASH-WILLIAMS, Euler Lines in Infinite Directed Graphs. *Canad. J. Math.*, **18**, 1966, 692–714.

C. St. J. A. NASH-WILLIAMS, An Unsolved Problem Concerning Decomposition of Graphs into Triangles. Mimeo, University of Waterloo, 1970.

J. PETERSEN, Die Theorie der regulären Graphen. *Acta Math.*, **15**, 1891, 193–220.

A. SAINTE-LAGÜE, Les réseaux unicursaux et bicursaux. *C. R. Acad. Sciences*, Paris, **182**, 1926, 747–748.

M. P. SCHÜTZENBERGER, Note in: C. BERGE, *Théorie des Graphes et ses applications*, chapter 17, page 164 (French edition, Dunod, Paris, 1958); page 170 (English edition, Methuen and Wiley, London-New York, 1962); page 213 (Spanish edition, Com. éd. Cont., Mexico 1962); page 181 (Rumanian edition, éd. Tehnica, Bucarest 1969); page 184 (Russian edition, éd. Litt. Etr., Moscou 1962).

R. G. SWAN, An Application of Graph Theory to Algebra. *Proc. Am. Math. Soc.*, **14**, 1963, 367–373.

CHAPTER 12: **Chromatic Index**

C. BERGE, Graph Theory. *Am. Math. Monthly*, **71**, 1964, 471–481.

C. BERGE, Lectures on Graph Theory. Mimeo, Tata Institute of Fundamental Research, Bombay, 1967.

J. FIAMČIK, E. JUCOVIČ, Colouring the Edges of a Multigraph. *Arch. Math.* **21**, 1970, 446–448.

J. FOLKMAN, D. R. FULKERSON, "Edge Colorings in Bipartite Graphs". *Combinatorial mathematics and their applications* (B. C. Bose, T. A. Dowling, ed.), University of North Carolina Press, Chapel Hill, 1969.

J. C. FOURNIER, Methode et théorème général de coloration des arêtes d'un multigraphe. *J. Math. Pures Appl.*, **56**, 1977, 437–453.

J.C. FOURNIER, Un théorème général de coloration. *Problèmes Combinatoires en Théorie des Graphes, Coll. Internat. C.N.R.*, **260**, Orsay 1976, Editions du C.N.R.S., 1978, 153–155.

R. P. GUPTA, Studies in the Theory of Graphs. Thesis, Tata Inst. Bombay, 1967.

R. P. GUPTA, The Chromatic Index and the Degree of a Graph. *Notices Am. Math. Soc.*, **13**, No. 6, 1966, Abstract GGT-429.

A. J. W. HILTON, D. DE WERRA, Sufficient Conditions for Balanced and for Equitable Edge-Colorings of Graphs. Memo ORWP 82/3, Ecole Polytechnique de Lausanne.

R. LASKAR, W. HARE, Chromatic Number for Certain Graphs. *J. Londor. Math. Soc.*, **4**, 1972, 489–492.

E. LUCAS, *Récréations Mathématiques*. A. Blanchard, Paris, 1892/1924.

O. ORE, *The Four-Color Problem*. Academic Press, New York, 1968.

C. E. SHANNON, A Theorem on Coloring the Lines of a Network. *J. Math. Phys.*, **28**, 1949, 148–151.

L. VIGNERON, Coloration des réseaux cubiques. *C. R. Ac. Sc. Paris*, **223**, 1946, 705; **223**, 1946, 770; **249**, 1959, 2462.

V. G. VIZING, On an Estimate of the Chromatic Class of a p-graph (Russian). *Diskret. Analiz.* **3**, 1964, 25–30.

V. G. VIZING, Critical Graphs with a·Given Chromatic Class (Russian). *Diskret. Anal.*, **5**, 1965, 9–17.

V. G. VIZING, On Chromatic Class (Russian). *Cybernetika*, **3**, 1965, 29–39. **3**, 1965, 29–39.

D. DE WERRA, A Few Remarks on Chromatic Scheduling. *Combinatorial Programming* (B. Roy, ed.), Reidel, Boston, 1975, 337–342.

D. DE WERRA, Colorings. *Math. Programm.*, **15**, 1978, 236–238.

D. DE WERRA, Regular and Canonical Colorings. *Discrete Math.*, **27**, 1979, 309–316.

D. DE WERRA, Obstructions for Regular Colorings. *J. Comb. Theory B*, **32**, 1982, 326–335.

CHAPTER 13: **Stability Number**

B. ANDRÁSFAI, "On Critical Graphs". *Théorie des Graphes Rome I. C. C.* (P. Rosenstiehl, ed.), Paris, Dunod, 1967, 9–19.

L. W. BEINEKE, F. HARARY, M. D. PLUMMER, On the Critical Lines of a Graph. *Pacific J. of Math.*, **21**, 1967, 205–212.

C. BERGE, Two Theorems on Graph Theory. *Proc. Nat. Acad. Sc. USA*, **43**, 1957, 842–844.

C. BERGE, Graph Theory, *Am. Math. Monthly*, **71**, 1964, 471–481.

C. BERGE, Problèmes de coloration en Théorie des Graphes. *Publ. Inst. Stat. Université de Paris*, **9**, 1960, 123–160.

C. BERGE, "Une propriété des graphes k-stables-critiques". *Combinatorial structures and their applications*, 7–11, Gordon and Breach, New York, 1970.

C. BERGE, Regularisable Graphs. *Advances in Graph Theory* (B. Bollobas ed.), *Ann. Discrete Math.*, **3**, North-Holland, Amsterdam, 1978, 11–21.

C. BERGE, Some Common Properties for Regularisable Graphs, Edge-Critical Graphs and B-Graphs. *Graph Theory and Algorithms* (N. Saito, T. Nishiezeki, eds.), Lecture Notes in Computer Sc., **108**, Springer, New York, 1981, 108–124.

C. BERGE, Diperfect Graphs. *Combinatorica*, **2**, 1982, 213–222.

C. BERGE, Path-Partitions in Directed Graphs. *Combinatorial Mathematics* (C. Berge, D. Bresson, P. Camion, J.F. Maurras, F. Sterboul, eds.), Ann. Discrete Math., **17**, North-Holland, Amsterdam, 1983, 59–64.

A. BRUEN, R. DIXON, The n-Queens Problem. *Discrete Math.*, **12**, 1975, 393–395.

P. ERDÖS, T. GALLAI, On the Minimal Number of Vertices Representing the Edges of a Graph. *Publ. Math. Inst. Hung. Ac. Sc. (Mag. Tud. Akad.)*, **6**, 1961, 181–203.

P. ERDÖS, A. HAJNAL, J. MOON, Mathematical Notes. *Am. Math. Monthly*, **71**, 1964, 1107–1110.

T. GALLAI, Kritische Graphen I. *Publ. Math. Inst. Hung. Acad. Sc.*, **8**, 1963, 165–192.

T. GALLAI, A. N. MILGRAM, Verallgemeinerung eines Graphentheoretischen Satzes von Rédei. *Acta Sc. Math.*, **21**, 1960, 181–186.

A. GEORGE, On Line-Critical Graphs, Thesis, Vanderbilt Univ., 1971.

A. GHOUILA-HOURI, Sur la conjecture de Berge. Mimeo, Séminaire Probl. Combinat. IHP, 1960.

S. W. GOLOMB, *Polyominoes*, Scribner's, New York, 1965.

A. HAJNAL, A Theorem on k-saturated Graphs. *Canad. Math. J.*, **17**, 1965, 720–724.

A. HAJNAL, J. SURÁNYI, Über die Auflösung von Graphen in vollständige Teilgraphen. *Ann. Univ. Sc. Budapestinensis*, **1**, 1958, 113–121.

A. HAJNAL, E. SZEMERÉDI, Proof of a Conjecture of P. Erdös, *Combinat. Theory and its applications, Balatonfüred* (Erdös, Rényi, V. T. Sós, ed.), North Holland, Amsterdam, 1970, 601–623.

F. HARARY, M. D. PLUMMER, On Indecomposable Graphs. *Canad. J. of Math.*, **19**, 1967, 800–809.

E. J. HOFFMAN, J. C. LOESSI, R. C. MOORE, Constructions for the Solution of the m Queens Problem. *Math. Mag.*, **42**, 1969, 66–72.

T. KLOVE, The Modular n-Queen Problem. *Discrete Math.*, **19**, 1977, 289–291.

T. KLOVE, The Modular n-Queen Problem II. *Discrete Math.*, **36**, 1981, 33–81.

M. LAS VERGNAS, Sur les arborescences dans un graphe orienté. *Discrete Math.*, **15**, 1976, 27–40.

N. LINIAL, Covering Digraphs by Paths, *Discrete Math.*, **23**, 1978, 257–272.

J. C. MEYER. Ensembles stables maximaux dans les hypergraphes, *C.R. Acad. Sc. Paris* **274**, 1972, 144–147.

J. W. MOON, On Independent Complete Subgraphs in a Graph. *Canad. J. of Math.*, **20**, 1968, 95–102.

M. D. PLUMMER, On a Family of Line-critical Graphs, *Monatsh. Math.*, **71**, 1967, 40–48.

M. D. PLUMMER, Some Covering Concepts in Graphs, *J. Comb. Theory*, **8**, 1970, 91–98.

P. TURÁN, An Extremal Problem in Graph Theory (Hungarian). *Mat. Fiz. Lapok*, **48**, 1941, 436–452.

P. TURÁN, On the Theory of Graphs. *Colloquium Math.*, **3**, 1954, 19–30.

W. WESSEL, Kanten-kritische Graphen mit der Zusammenhangszahl 2, *Manuscripta Math.*, **2**, 1970, 309–334.

K. ZARANKIEWICZ, Sur les relations symétriques dans l'ensemble fini. *Colloquium Math.*, **1**, 1947, 10–15.

K. ZARANKIEWICZ, On a Problem of Turán Concerning Graphs. *Fund Math.*, **41**, 1954, 137–145.

A. A. ZYKOV, On Some Properties of Linear Complexes. *Math. Sb.*, **24**, 1949, 163–188 (*A.M.S. Transl.*, **79**, 1952).

CHAPTER 14 : Kernels and Grundy Functions

O. ABERTH, On the Sum of Graphs. *Revue Fr. Rech. Opérationnelle*, No. 33, 1964, 353–358.

E. W. ADAMS and D. C. BENSON, Nim Type Games. *Carnegie Inst. of Technology, Techn. Report*, **13**, 1956.

C. BERGE, La fonction de Grundy d'un graphe infini. *C. R. Acad. Sciences, Paris*, **242**, 1956, 1404–1405.

C. BERGE, *Théorie générale des jeux à n personnes*. Mémorial des Sciences Math., **138**, Paris, 1957 (Gauthier-Villars, ed.); Russian translation (V. F. Kolchina): éditions Physico-Mathématiques, Moscou, 1961.

C. BERGE, "Topological Games with Perfect Information". *Contributions to the Theory of Games* (3), *Annals of Math. Studies*, **39**, 1957, 165.

C. BERGE and M. P. SCHÜTZENBERGER, Jeux de Nim et solutions. *C. R. Acad. Sciences, Paris*, **242**, 1956, 1672.

C. BERGE, A. RAMACHANDRA RAO, A Combinatorial Problem in Logic. *Discrete Math.*, **17**, 1977, 23–26.

C. L. BOUTON, Nim, a Game with a Complete Mathematical Theory, *Ann. of Math.*, **3**, 1902, 35–39.

Y. CHAO, On a Problem of Claude Berge. *Proc. Am. Math. Soc.*, **14**, 1963, 80–82.

V. CHVÁTAL, L. LOVÁSZ, Every Directed Graph Has a Semi-Kernel. *Hypergraph Seminar* (Berge, Ray-Chaudhuri, eds.) Lecture Notes in Math., **411**, 1972, 175.

P. DUCHET, Graphes noyau-parfaits. *Ann. Discrete Math.*, **9**, 1980, 93–101.

P. DUCHET, H. MEYNIEL, A Note on Kernel-Critical Graphs. *Discrete Math.*, **22**, 1981, 103–105.

P. DUCHET, H. MEYNIEL, Une généralisation du théorème de Richardson sur l'existence de noyaux dans les graphes orientés. *Discrete Math.*, **43**, 1983, 21–27.

H. Galeana-Sanchez, V. Neumann-Lara, On Kernels and Semi-Kernels of Digraphs. *Discrete Math.*, **48**, 1984, 67–76.

P. M. Grundy, Mathematics and Games. *Eureka*, **2**, 1939, 6–8.

P. M. Grundy, C. A. B. Smith, Disjunctive Games with the Last Player Loosing. *Proc. Cambridge Phil. Soc.*, **52**, 1956, 527–533.

R. K. Guy, C. A. B. Smith, The *G*-values of Various Games. *Proc. Cambridge Phil. Soc.*, **52**, 1956, 514–526.

J. Hanak, On Generalized Bergean Pay-off Functions at Complete Games. *Res. Memorandum* No. 63, May 1969, *University of J. E. P.*, *Brno*.

J. C. Holladay, "Cartesian Product of Termination Games". *Contributions to the Theory of Games* (3). *Ann. of Math. St.*, **39**, 1957, 189–199.

J. C. Kenyon, Nim-like Games and the Sprague-Grundy Theory, Thesis, Univ. Calgary, Alberta, April 1967.

F. Maffray, Sur l'existence de noyaux dans les graphes parfaits, Thesis, Paris 6, 1984.

E. H. Moore, A Generalization of the Game called Nim. *Ann. of Math.*, **11**, 1909, 93.

M. H. McAndrew, On the Product of Directed Graphs. *Proc. Am. Math. Soc.*, **14**, 1963, 600–606.

J. von Neumann, O. Morgenstern, *Theory of Games and Economic Behavior*. Princeton University Press, Princeton, 1944.

C. F. Picard, "Latticoid Product, Sum and Product of Graphs". *Combinatorial structures and their applications* (Calgary, June 1969), Gordon and Breach, New York, 1970, 321–322.

C. F. Picard, Distributivité d'opérations de graphes. Mimeo, Conf. Séminaire Problèmes Combinatoires, I. H. P., 1970.

A. Pultr, "Tensor Products on the Category of Graphs". *Combinatorial structures and their applications* (Calgary, June 1969), Gordon and Breach, New York, 1970, 327–329.

M. Richardson, Solutions of Irreflexive Relations. *Annals of Math.*, **58**, 1953, 573–580.

M. Richardson, Extensions Theorems for Solutions of Irreflexive Relations. *Proc. Nat. Acad. of Sciences U. S. A.*, **39**, 1953, 649–651.

B. Roy, *Algèbre moderne et théorie des Graphes*. Volume 2, chapter VI, Dunod, Paris, 1970.

S. Rudeanu, Notes sur l'existence et l'unicité du noyau d'un graphe. *Revue Française Rech. Opérat.*, **33**, 1964, 20–26 and **41**, 1966, 301–310.

B. Sands, N. Sauer, R. Woodrow, On Monochromatic Paths in Edge Coloured Digraphs. *J. Comb. Theory B*, **33**, 1982, 271–275.

C. A. B. Smith, "Graphs and Composite Games". *A Seminar on Graph Theory* (F. Harary-L. Beineke), Holt Rinehart, Winston, New York, 1967, 86–111.

R. Sprague, Bemerkungen über eine spezielle Abelsche Gruppe. *Math. Z.*, **51**, 1947, 82–93.

L. P. Varvak, Existence of a Kernel in a Product (Russian), *Ukr. Mat. Zh.*, **17**, No. 3, 1965, 112–114.

L. P. Varvak, Games on a Sum of Graphs (Russian). *Kiber.*, **4**, No. 1, 1968, 63–66.

V. G. Vizing, A Bound on the External Stability Number of a Graph. *Doklady* A. N. **164**, 1965, 729–731.

V. G. Vizing, The Cartesian Product of Graphs. *Vych. Sis.*, **9**, 1963, 30–43.

P. M. Wechsel, The Kronecker Product of Graphs. *Proc. Am. Math. Soc.*, **13**, 1962, 47–52.

C. P. Welter, The Theory of a Class of Games on a Sequence of Squares, in Terms of the Advancing Operation in a Special Group. *Proc. Kon. Nederl. Akad. N.* (série A), **57**, 1954, 194–198.

N. Y. Withoff, A Modification of the Game of Nim, *Nieuw Arch. voor Wiskunde*, **7**, 1907, 199–202.

CHAPTER 15 : **Chromatic Number**

K. Appel, W. Haken, Every Planar Map is 4-Colourable. *Illinois J. Math.*, **21**, 1979, 429–490 and 491–567.

C. Berge, Les problèmes de colorations en Théorie des Graphes. *Publ. Inst. Stat. Univ. Paris*, **9**, 1960, 123–160.

C. Berge, *k*-Optimal Partitions in Directed Graphs. *European J. Combin.*, **3**, 1982, 97–101.

C. Berge, Diperfect Graphs. *Combinatorica*, **2**, 1982, 213–222.

G. Berman, W. T. Tutte, The Golden Root of a Chromatic Polynomial. *J. Combinatorial Theory*, **6**, 1969, 301–302.

G. D. Birkhoff, A Determinant Formula for Coloring a Map. *Ann. Math.*, **14**, 1912, 42–46.

J. A. Bondy, Disconnected Orientations and a Conjecture of Las Vergnas. *J. London Math. Soc.*, **2**, 1976, 277–282.

G. D. Birkhoff, D. C. Lewis, Chromatic Polynomials, *Trans. Amer. Math. Soc.*, **60**, 1946, 355–451.

R. L. Brooks, On Colouring the Nodes of a Network. *Proc. Cambridge Phil. Soc.*, **37**, 1941, 194–197.

V. Chvátal, A Note on Coefficients of Chromatic Polynomials, *J. Combinat. Theory*, **9**, 1970, 95–96.

V. Chvátal, J. Komlós, Some Combinatorial Theorems on Monotonicity, *Canad. Math. Bull.*, **14** (2), 1971:

H. H. Crapo, The Tutte Polynomials. *Aequationes Math.*, **3**, 1969, 211–219.

M. Dhurandhar, Improvement on Brooks' Chromatic Bound for a Class of Graphs. *Discrete Math.*, **42**, 1982, 51–56.

G. A. Dirac, A Property of 4-chromatic Graphs and Some Remarks on Critical Graphs. *J. London Math. Soc.*, **27**, 1952, 85–92.

G. A. Dirac, Map-colour Theorems. *Can. J. Math.*, **4**, 1952, 480–490.

G. A. Dirac, "Valence-variety and Chromatic Number". *Wiss. Z. Martin-Luther Univ. Halle-Wittenberg*, **13**, 1964, 59–63.

P. Erdös, G. Szekeres, A Combinatorial Problem in Geometry, *Compositio Math.*, **2**, 1935, 463–470.

A. Errera, Exposé historique du problème des quatre couleurs. *Periodica di Mat.*, **7**, 1927, 20–41.

A. Errera, Une contribution au problème des quatre couleurs. *Bull. Soc. Math. France*, **53**, 1925, 42–55.

H. J. Finck, Über die chromatischen Zahlen eines Graphen und seines Komplements. I, II, 1966. *Mitteilung aus dem Institut für Mathematik, Ilmenau.*

J. C. FOURNIER, Effet de la contraction d'une arête sur le genre d'un graphe. *C. R. Acad. Sc. Paris*, **270**, 1970, 743-744.

P. FRANKLIN, Note on the Four Color Problem. *J. Math. Phys.*, **16**, 1938, 172-184.

T. GALLAI, Kritische Graphen, I and II. *Publ. Math. Inst. Hungarian Acad. Sci.*, A **8**, 1963, 165-192; **9**, 1964, 373-395.

T. GALLAI, "On Directed Paths and Circuits". *Theory of Graphs, Tihany* (P. Erdös, G. Katona, ed.), Academic Press, New York, 1968, 115-118.

H. GRÖTZSCH, Ein Dreifarbensatz für dreikreisfreie Netze auf der Kugel. *Wiss. Z. Martin Luther Univ., Halle-Wittenberg, Math. Naturwiss. Reihe*, **8**, 1958, 109-119.

B. GRÜNBAUM, Grötzsche's Theorem on 3-colorings. *Michigan Math. J.*, **10**, 1963, 303-310.

R. P. GUPTA, Studies in the Theory of Graphs. Thesis, Tata Inst. of Fund. Res. (mimeo.), Bombay, 1967.

R. P. GUPTA, Bounds on the Chromatic and Achromatic Numbers of Complementary Graphs. Mimeo, series No. 577, Univ. of North Carolina at Chapel Hill, 63, April 1968.

A. GYÁRFÁS, Private communication, 1980.

H. HADWIGER, Über eine Klassifikation der Streckenkomplexe. *Vierte Naturforsch. Ges. Zurich*, **88**, 1943, 133-142.

G. HAJÓS, Über eine Konstruktion nicht *n*-färbbarer Graphen. *Wiss. Zeitschr. Martin Luther Univ. Halle-Wittenberg*, A **10**, 1961, 116-117.

D. W. HALL, J. W. SIRY and B. R. VANDERSLICE, The Chromatic Polynomial of the Truncated Icosahedron. *Proc. Amer. Math. Soc.*, **16**, 1965, 620-628.

S. T. HEDETNIEMI, Homomorphism of Graphs and Automata. *Tech. Rep., The Univ. of Michigan, Ann Arbor*, 1966.

C. HEUCHENNE, Sur le critère de chromaticité de Hajós-Ore. *Bull. Soc. Roy. Liège*, 1-2, 1968, 10-13.

M. R. IYER, V. V. MENON, On Coloring the *n* × *n* Chessboard. *Am. Math. Monthly*, **73**, 1966, 721-725.

E. L. JOHNSON, A Proof of the Four-Coloring of the Edges of a Regular Three-Degree Graph. *O. R. C. 63-28 (R. R.) Mim. report* 1963. *Operations Research Center, Univ. of Calif.*

G. KREWERAS, Peut-on former un réseau avec des parties finies d'un ensemble dénombrable. *C. R. Acad. Sc. Paris*, **222**, 1946, 1025-1026.

M. LAS VERGNAS, H. MEYNIEL, Kempe Classes and the Hadwiger's Conjecture. *J. Comb. Theory B*, **31**, 1981, 95-104.

M. LAS VERGNAS, The Tutte Polynomial. *Progress in Graph Theory* (J. A. Bondy, U.S.R. Murty, eds.), Academic Press, Toronto, 1984, 381-387.

N. LINIAL, Extending the Greene-Kleitman Theorem to Directed Graphs. *J. Comb. Theory A*, **30**, 1981, 331-334.

L. S. MELNIKOV, V. G. VIZING, New Proofs of Brooks Theorem. *J. Combinatorial Theory*, **7**, 1969, 289-290.

J. MITCHEM, A Short Proof of Catlin's Extension of Brook's Theorem. *Discrete Math.*, **21**, 1978, 213-214.

E. A. NORDHAUS, J. W. GADDUM, On Complementary Graphs. *Ann. Math. Monthly*, **63**, 1956, 175-177.

O. ORE, *The Four Color Problem*. Academic Press, New York, 1968.

F. P. Ramsey, On a Problem of Formal Logic. *Proc. London Math. Soc.*, **30**, 1930, 264–286.

D. K. Ray-Chaudhuri and R. Wilson, Solution of Kirkman's Schoolgirl Problem, *Proc. Symposium Pure Math.*, **19**, A.M.S., Providence, 1971, 187–203.

R. C. Read, An Introduction to Chromatic Polynomials. *J. Combinatorial Theory*, **4**, 1968, 52–71.

G. Ringel, *Färbungsprobleme auf Flächen und Graphen.* Berlin, 1959.

G. Ringel, J. W. T. Youngs, Solution of the Heawood Map-coloring Problem, *J. Comb. Theory*, **7**, 1969, 342–363.

G. C. Rota, On the Foundations of Combinatorial Theory, I, *W. V. G.*, **2**, 1964, 340–368.

B. Roy, Nombre chromatique et plus longs chemins d'un graphe. *Revue AFIRO*, **1**, 1967, 127–132.

M. Schäuble, "Bemerkungen zu einem Kantenfärbungsproblem". *Beiträge zur Graphentheorie*, Leipzig, 1968, 137–142.

W. T. Tutte, On Chromatic Polynomials and the Golden Ration. *J. Combinatorial Theory* (to appear).

W. T. Tutte, A Contribution to the Theory of Chromatic Polynomials. *Canad. J. Math.*, **6**, 1954, 80–91.

K. Wagner, Bemerkungen zu Hadwiger's Vermutung. *Math. Ann.*, **141**, 1960, 433–451.

K. Wagner, Beweis einer Abschwächung der Hadwiger Vermutung. *Math. Ann.*, **153**, 1964, 139–141.

H. Whitney, A Logical Expansion in Mathematics. *Bull. Amer. Math. Soc.*, **38**, 1932, 572–579.

H. Whitney, The Coloring of Graphs. *Ann. of Math.*, **33**, 1932, 688–718.

C. E. Winn, A Case of Coloration in the Four Color Problem. *Am. J. Math.*, **49**, 1937, 515–528.

A. A. Zykov, On Some Properties of Linear Complexes (*Amer. Math. Soc. Transl.* No. 79, 1952). *Mat. Sb.*, **66**, 1949, 163–188.

CHAPTER 16 : **Perfect Graphs**

O. Aberth, On the Sum of Graphs. *Revue Fr. Rech. Opérationnelle*, **33**, 1964, 353–358.

M. H. McAndrew, On the Product of Directed Graphs. *Proc. Am. Math. Soc.*, **14**, 1963, 600–606.

C. Berge, Les problèmes de coloration en théorie des graphes. *Publ. Inst. Stat. Univ. Paris*, **9**, 1960, 123–160.

C. Berge, "Färbung von Graphen, deren sämtliche bzw. deren ungerade Kreise starr sind". *Wissenschaftliche Zeitung, Martin Luther Univ., Halle Wittenberg*, 1961, **114**.

C. Berge, Sur une conjecture relative au problème des codes optimaux, comm. 13ème assemblée générale de l'URSI, Tokyo, 1962.

C. Berge, "Perfect graphs", I. *Six Papers on Graph Theory*, Indian Statistical Institute, Calcutta, 1963.

C. Berge, "Une application de la théorie des Graphes à un problème de codage". *Automata Theory* (E. R. Caianello ed.) Academic Press, London, 1966, 25–34.

C. Berge, "Some Classes of Perfect Graphs". Chap. 5. *Graph Theory and Theoretical Physics* (F. Harary, ed.), Academic Press, New York, 1967.

C. Berge, "The Rank of a Family of Sets and Some Applications to Graph Theory". *Recent Progress in Combinatorics* (Tutte, ed.), Academic Press, New York, 1969, 49–57.

C. Berge, M. Las Vergnas, "Sur un théorème du type König pour hypergraphes". *Annals New York Acad. Sc.*, **175**, 1970, 32–40.

C. Berge, P. Duchet, Strongly Perfect Graphs. *Topics on Perfect Graphs* (C. Berge, V. Chvátal, eds.), *Ann. Discrete Math.*, **21**, North-Holland, Amsterdam, 1984, 45–56.

R. G. Bland, H. C. Huang, L. E. Trotter, Graphical Properties Related to Maximal Imperfection, *Discrete Math.*, **27**, 1979, 11–22.

M. Burlet, J. Fonlupt, Polynomial Algorithm to Recognize a Meyniel Graphs. *Topics on Perfect Graphs* (C. Berge, V. Chvátal, eds.) Ann. Discrete Math., **21**, North-Holland, Amsterdam, 1984, 223–250.

P. Camion, Matrices totalement unimodulaires et problèmes combinatoires. Thesis, Université Libre de Bruxelles, 1963.

Y. Chao, On a Problem of Claude Berge. *Proc. Am. Math. Soc.*, **14**, 1963, 80–82.

V. Chvátal, R. L. Graham, A. F. Perold, S. Whitesides, Combinatorial designs related to the perfect graph conjecture. *Discrete Math.*, **26**, 1979, 83–92.

V. Chvátal, C. Hoang, On the P_4-Structure of Perfect Graphs, J. Techn. Report, S.O.C.S. 83.10, McGill Univ. 1983.

J. C. Fournier, M. Las Vergnas, Une classe d'hypergraphes bichromatiques, *Discrete Math.*, **2**, 1972, 407–410.

D. R. Fulkerson, Notes on Combinatorial Mathematics; Anti-Blocking Polyhedra. Mimeo, *RM 620/1-PR*, February 1970.

T. Gallai, Graphen mit triangulierbaren ungeraden Vielecken. *Magyar. Tud. Akad. Mat Kutató Int. Közl.*, **7**, 1962, 3–36.

T. Gallai, Transitiv orientierbare Graphen. *Acta Math. Acad. Sc. Hungar.*, **18**, 1967, 25–66.

F. Gavril, Algorithms on clique separable graphs, *Discrete Math.*, **19**, 1977, 159–165.

A. Ghouila-Houri, Sur la conjoncture de Berge. Mimeo, Séminaire Probl. Combinat. IHP, 1960.

A. Ghouila-Houri, Caractérisation des Graphes non orientés dont on peut orienter les arêtes de manière à obtenir le graphe d'une relation d'ordre. *C. R. Acad. Sc. Paris*, **254**, 1962, 1370–1371.

A. Ghouila-Houri, Caractérisation des matrices totalement unimodulaires. *C. R. Acad. Sci. Paris*, **254**, 1962, 1192–1193.

R. Giles, L. E. Trotter, A. Tucker, The Strong Perfect Graph Theorem for a Class of Partitionable Graphs. *Topics on Perfect Graphs* (C. Berge, V. Chvatal, eds.), Ann. Discrete Math., **21**, North-Holland, Amsterdam, 1984, 161–167.

P. C. Gilmore, A. J. Hoffman, A Characterization of Comparability Graphs and of Interval Graphs. *Canad. J. of Math.*, **16**, 1964, 539–548.

M. GOLUMBIC, *Algorithmic Theory and Perfect Graphs*, Academic Press, New York, 1980.

A. HAJNAL, J. SURÁNYI, Über die Auflösung von Graphen in vollständige Teilgraphen. *Ann. Univ. Sc. Budapestinensis*, 1, 1958, 113–121.

G. HAJÓS, Über eine Art von Graphen. *Int. Math. Nachrichten.* 11, 1957.

R. B. HAYWARD, On the P_4-Structure of Perfect Graphs. Techn. Report S.O.C.S. 83.22, McGill Univ. 1983.

C. HOANG, On the P_4-Structure of Perfect Graphs. Techn. Report S.O.C.S. 83.20, McGill Univ. 1983.

W. L. HSU, How to Color Claw-Free Perfect Graphs. *Studies on Graphs and Discrete Programming* (P. Hammer, ed.), *Annals of Discrete Math.*, 11, 1982, 189–197.

H. JACOBS, On the Berge's Conjecture Concerning Perfect Graphs. *Combinatorial Structures and their Applications*, Gordon and Breach, New York, 1970, 377–384.

A. KOTZIG, Pair Hajós graphs (Slovakian). *Casopis pest. mat.*, 88, 1963, 236–241.

C. G. LEKKERKERKER, J. Ch. BOLAND, Representation of a Finite Graph by a Set of Intervals on the Real Line. *Fund. Math.*, 51, 1962, 45–64.

LJUBICH, Note on a Problem of Claude Berge (Russian). *Sibir. Math. Zh. (Journal Math. de Sibérie)*, 5, 1964, 961–963.

L. LOVÁSZ, Normal Hypergraphs and the Perfect Graph Conjecture. *Discrete Math.*, 2, 1972, 253–267.

H. MEYNIEL, On the Perfect Graph Conjecture. *Discrete Math.*, 16, 1976, 339–342.

D. J. MILLER, Categorical Product of Graphs. *Canad. J. Math.*, 20, 1968, 1511–1521.

E. OLARU, Zur Theorie der perfekten Graphen. *J. Comb. Theory B*, 23, 1977, 94–105.

M. W. PADBERG, Perfect Zero-Ones Matrices. *Math. Program.*, 6, 1974, 180–196.

M. W. PADBERG, A Characterization of Perfect Matrices. *Topics on Perfect Graphs* (C. Berge, V. Chvátál, eds.), *Ann. Discrete Math.*, 21, North-Holland, Amsterdam, 1984, 169–178.

K. R. PARTHASARTHY, G. RAVINDRA, The Strong Perfect Graph Conjecture is True for $K_{1,3}$-free graphs. *J. Comb. Theory B*, 21, 1976, 212–223.

C. F. PICARD, "Latticoid Product, Sum and Product of Graphs". *Combinatorial structures and their applications*, Gordon and Breach, New York, 1970, 321–322.

C. F. PICARD, Distributivité d'opérations sur les graphes. *C. R. Acad. Sc. Paris*, 270, 1970, A, 1219–1221.

A. PULTR, "Tensor Products on the Category of Graphs". *Combinatorial structures and their applications*, Gordon and Breach, New York, 1970, 327–329.

G. RAVINDRA, Meyniel's Graphs are Strongly Perfect. *Topics on Perfect Graphs* (C. Berge, V. Chvatal, eds.), Ann. Discrete Math., 21, North-Holland, Amsterdam, 1984, 145–148.

H. SACHS, "On the Berge Conjecture Concerning Perfect Graph". *Combinatorial structures and their applications*, Gordon and Breach, New York, 1970, 377–384.

C. E. SHANNON, The Zero-error Capacity of a Noisy Channel. *Comp. Information Theory, IRE Trans.*, 3, 1956, 3–15.

L. SURÁNYI, The Covering of Graphs by Cliques. *Studia Sc. Math. Hung.*, **3**, 1968, 345–349.

L. E. TROTTER, Linear Perfect Graphs. *Math. Program.*, **12**, 1977, 255–259.

A. C. TUCKER, Critical Perfect Graphs and Perfect 3-Chromatic Graphs. *J. Comb. Theory B*, **23**, 1977, 143–149.

A. C. TUCKER, Berge's Strong Perfect Graph Conjecture. *Ann. New York Acad. Scie.*, **319**, 1979, 530–535.

V. G. VIZING, The Cartesian Product of Graphs. *Vyč. Sis.*, **9**, 1963, 30–43.

P. M. WECHSEL, The Kronecker Product of Graphs. *Proc. Am. Math. Soc.*, **13**, 1962, 47–52.

E. S. WOLK, A Note on the Comparability Graph of a Tree. *Proc. Am. Math. Soc.*, **16**, 1965, 17–20.